CROP DISEASES AND THEIR MANAGEMENT

Integrated Approaches

CROP DISEASES AND THEIR MANAGEMENT

Integrated Approaches

Edited by

Gireesh Chand, PhD
Santosh Kumar, PhD

Department of Plant Pathology,
Bihar Agricultural University,
Sabour, Bhagalpur–813210, Bihar, India

Apple Academic Press Inc. | Apple Academic Press Inc.
3333 Mistwell Crescent | 9 Spinnaker Way
Oakville, ON L6L 0A2 | Waretown, NJ 08758
Canada | USA

First issued in paperback 2021

Exclusive worldwide distribution by CRC Press, a member of Taylor & Francis Group
No claim to original U.S. Government works

ISBN 13: 978-1-77463-581-0 (pbk)
ISBN 13: 978-1-77188-270-5 (hbk)

Library and Archives Canada Cataloguing in Publication

Crop diseases and their management : integrated approaches / edited by Gireesh Chand, PhD, Santosh Kumar, PhD, Department of Plant Pathology, Bihar Agricultural University, Sabour, Bhagalpur–813210, Bihar, India.

Includes bibliographical references and index.
Issued in print and electronic formats.
ISBN 978-1-77188-270-5 (hardcover).--ISBN 978-1-77188-261-3 (pdf)
1. Plant diseases. 2. Phytopathogenic microorganisms--Control. I. Chand, Gireesh, author, editor II. Kumar, Santosh, 1974-, author, editor

SB731.C76 2016   632'.3   C2016-901750-8   C2016-901751-6

Library of Congress Cataloging-in-Publication Data

Names: Chand, Gireesh, editor. | Kumar, Santosh, 1978- editor.
Title: Crop diseases and their management : integrated approaches / editors: Gireesh Chand, Santosh Kumar.
Description: Oakville, ON ; Waretown, NJ : Apple Academic Press, [2016] | Includes bibliographical references and index.
Identifiers: LCCN 2016011312 | ISBN 9781771882705 (hardcover : alk. paper)
Subjects: LCSH: Plant diseases. | Phytopathogenic microorganisms--Control.
Classification: LCC SB601 .C76 2016 | DDC 632'.3--dc23

Apple Academic Press also publishes its books in a variety of electronic formats. Some content that appears in print may not be available in electronic format. For information about Apple Academic Press products, visit our website at **www.appleacademicpress.com** and the CRC Press website at **www.crcpress.com**

CONTENTS

LIST OF CONTRIBUTORS

V. Ramesh Babu
Division of Crop Protection, ICAR-Central Research Institute for Jute and Allied Fibers, Barrackpore, Kolkata–700120, West Bengal, India

Gireech Chand
Department of Plant Pathology, Bihar Agricultural University, Sabour, Bhagalpur–813210, India

K. K. Chandra
Department of Forestry, Wildlife and Environmental Sciences, Guru Ghasidas Vishwavidyalaya, Bilaspur–495009, Chhattisgarh, India

Shivnath Das
Betelvine Research Centre, Islampur, Nalanda–801303, India, E-mail: prabhathau@gmail.com

R. K. De
Division of Crop Protection, ICAR-Central Research Institute for Jute and Allied Fibers, Barrackpore, Kolkata–700120, West Bengal, India

Upma Dutta
Sher-e-Kashmir University of Agriculture Science and Technology, Jammu (J&K), India

Sandhya Kant
Former Research Scholar, Department of Botany, D.A.V. College, Kanpur–208001, India

C. P. Khare
Division of Plant Pathology Indira Gandhi Agricultural University, Raipur, Chhattisgarh, India

A. Kotesthane
Division of Plant pathology Indira Gandhi Agricultural University, Raipur, Chhattisgarh, India

Amarendra Kumar
Department of Plant Pathology, Bihar Agricultural University, Sabour, Bhagalpur-813210, Bihar, India

Anil Kumar
Department of Plant Breeding and Genetics, Bihar Agricultural University, Sabour, 813210, Bihar, India

Manoj Kumar
Department of Genetics and Plant Breeding, Bihar Agricultural University, Sabour, Bhagalpur 813210, Bihar, India

Prabhat Kumar
Betelvine Research Centre, Islampur, Nalanda–801303, India, E-mail: prabhathau@gmail.com

Rakesh Kumar
Sher-e-Kashmir University of Agriculture Science and Technology, Jammu (J&K), India

S. Kumar
Department of Plant Pathology, Bihar Agricultural University, Sabour–813210, Bhagalpur, Bihar, India

Sanjay Kumar
Department of Plant Breeding and Genetics, Bihar Agricultural University, Sabour, 813210, Bihar, India

Santosh Kumar
Jute Research Station, Katihar, Bihar Agricultural University, Sabour, Bihar, India;
E-mail: santosh35433@gmail.com

Sunil Kumar
AICRP on Soybean, School of Agricultural Sciences and Rural Development, Nagaland University, Medziphema – 797106, Nagaland, India

Sunita Kumari
Krishi Vigyan Kendra, Kishanganj, Bihar Agricultural University, Sabour, Bihar, India

P. N. Meena
Division of Crop Protection, ICAR-Central Research Institute for Jute and Allied Fibers, Barrackpore, Kolkata–700120, West Bengal, India

Y. R. Meena
Jute Research Station (Katihar), Bihar Agriculture University, Sabour, Bihar, India

R. K. Mishra
The Energy and Resources Institute (TERI), India Habitat Centre, Lodi Road, New Delhi, India

Udit Narain
Department of Plant Pathology, C. S. Azad University of Agriculture and Technology, Kanpur–208002, India

Sushma Nema
Division of Plant Pathology, Jawaharlal Krishi Vishwa Vidyalaya, Jabalpur, Madhya Pradesh, India

M. D. Ojha
College of Horticulture, Bihar Agriculture University, Sabour, Bhagalpur, Bihar, India;
E-mail: drmdojha@gmail.com

Kanika Pagoch
Sher-e-Kashmir University of Agriculture Science and Technology, Jammu (J&K), India

Ajit Kumar Pandey
Betelvine Research Centre, Islampur, Nalanda–801303, India, E-mail: prabhathau@gmail.com

V. S. Pandey
National Seed Corporation, Beej Bhawan, Pusa Complex, New Delhi, India

M. Pathak
College of Horticulture and Forestry, Central Agricultural University, Pasighat – 791102, Arunachal Pradesh, India, E-mail: rcshakywar@gmail.com

Bishun D. Prasad
Department of Plant Breeding and Genetics, B.A.C., Sabour, Bihar, India

Sangita Sahni
Department of Plant Pathology, T.C.A., Dholi, Muzaffarpur, Bihar, India

R. C. Shakywar
College of Horticulture and Forestry, Central Agricultural University, Pasighat – 791102, Arunachal Pradesh, India, E-mail: rcshakywar@gmail.com

N. D. Sharma
Division of Plant Pathology Indira Gandhi Agricultural University, Raipur, Chhattisgarh, India

P. K. Sharma
National Bureau of Agriculturally Important Microorganisms (ICAR), Mau, Uttar Pradesh, India

Ashok Kumar Singh
Sher-e-Kashmir University of Agriculture Science and Technology, Jammu (J&K), India

Birender Singh
Department of Plant Breeding and Genetics, Bihar Agricultural University, Sabour, 813210, Bihar, India

Dinesh Singh
Program Coordinator, Krishi Vigyan Kendra, P.G. College, Ghazipur, Uttar Pradesh, India

K. M. Singh
College of Horticulture and Forestry, Central Agricultural University, Pasighat – 791102, Arunachal Pradesh, India, E-mail: rcshakywar@gmail.com

Mahesh Singh
Department of Plant Pathology, Narendra Dev University of Agriculture and Technology, Kumarganj, Faizabad–224229, Uttar Pradesh, India

Mamta Singh
Subject Matter Specialist-Plant Breeding and Genetics, Krishi Vigyan Kendra, Sagar, Madhya Pradesh, India

R. P. Singh
Subject Matter Specialist-Plant Protection, Krishi Vigyan Kendra, P.G. College, Ghazipur, Uttar Pradesh, India

V. K. Singh
Jute Research Station (Katihar), Bihar Agriculture University, Sabour, Bihar, India

J. N. Srivastava
Bihar Agricultural University, Sabour, Bhagalpur, Bihar, India

V. S. Thrimurthi
Division of Plant pathology Indira Gandhi Agricultural University, Raipur, Chhattisgarh, India

P. K. Tiwari
Division of Plant pathology Indira Gandhi Agricultural University, Raipur, Chhattisgarh, India

A. N. Tripathi
Division of Crop Protection, Central Research Institute for Jute and Allied Fibers, Barrackpore, Kolkata – 700120, West Bengal, India

R. B. Verma
Bihar Agricultural University, Sabour–813210, Bhagalpur, Bihar, India

V. K. Yadav
College of Agriculture, Kundeshwar, Tikamgarh, Madhya Pradesh, India

LIST OF ABBREVIATIONS

AFP	acquisition-feeding period
AM	arbuscular mycorrhizal
AMF	arbuscular mycorrhizal fungi
BGM	botrytis gray mold
BND	bud necrosis disease
BNV	bud necrosis virus
BSDM	brown stripe downy mildew
CABI	Centre for Agriculture and Biosciences International
CBDV	Colocasia bobone disease virus
CLCuBV	Cotton leaf curl Bangalore virus
CLS	Cercospora leaf spot
EDB	ethylene dibromide
ELISA	enzyme linked immunosorbant assay
EPS	extracellular polysaccharide
GCR	groundnut chlorotic rosette
GGR	groundnut green rosette
HPV	high plains virus
ICAR	The Indian Council of Agricultural Research
IDM	integrated disease management
IFP	inoculation-feeding period
IYSV	iris yellow spot virus
JLMV	jute leaf mosaic virus
JYLM	jute leaf yellow mosaic
MCDV	maize chlorotic dwarf virus
MDMV	maize dwarf mosaic virus
MSV	maize streak virus
MYVMV	mesta yellow vein mosaic virus
NSKE	neem seed kernel extract
OYDV	onion yellow dwarf virus
PC	phenotype conversions

PCNB	pentachloronitrobenzene
PDA	potato dextrose agar
PDI	percent disease index
PGPR	plant growth promoting rhizobacteria
PGR	plant growth regulator
PM	powdery mildew
PNRSV	prunus necrotic ring spot virus
PSMV	pigeon pea sterility mosaic virus
RDM	Rajasthan downy mildew
RGNF	Rajiv Gandhi National Fellowship
RT-PCR	reverse transcriptase polymerase chain reaction
SCN	soybean cyst nematode
SDM	sorghum downy mildew
SLB	southern leaf blight
SMV	Sunn-hemp mosaic virus
SSPs	stand support poles
TSWV	tomato spotted wilt virus
TZC	tetrazolium chloride
UGC	University Grant Commission
ULCV	urd bean leaf crinkle virus
VAMF	vesicular arbuscular mycorrhizal fungi
WHO	World Health Organization

PREFACE

The Indian Council of Agricultural Research (ICAR) in view of the suggestions of the Fourth Deans Committee has recently modified the course curriculum for both undergraduate and postgraduate students of Agricultural Sciences. The Council has prescribed three courses each of three credits, namely, *Diseases of Field Crops*, *Diseases of Fruit and Flowering Crops*, and *Diseases of Plantation, Spice and Medicinal Plants*; and *Diseases of Vegetable Crops* for postgraduate students as optional courses. These three courses are usually opted for by students of various disciplines such as Entomology, Nematology, Horticulture, Vegetable Sciences, and Agronomy as optional with the objective to receive updated information on various important plant diseases that concern the country. These courses are gaining popularity with enhanced emphasis on Field and Horticulture in the current five-year plan. The students find these courses helpful in various competitive examinations and later during service periods as Plant Protection Officer, District Horticulture Officer and other Government jobs in the Directorate of Agriculture, Horticulture and Plant Protection. But average students from various disciplines other than Plant Pathology usually do not have the sound technical base required to follow the lectures in these courses. Moreover, they do not need to go through the details of the diseases like physiology of pathogenesis, structural composition of the pathogen, etc. For them knowledge on diagnostic symptoms, mode of perpetuation of the pathogen and dissemination, favorable conditions for diseases development and latest management strategy seem to be adequate. Precisely these are the recommendations of the Deans Committee too. However, so far there is no single volume wherein the students may have all the reading materials prescribed in these three courses. The libraries of most of the State Agricultural Universities can hardly cater to the need of the students in this regard. Easy access to Internet facilities for students is still a distant dream in such universities. Besides students coming from rural agricultural colleges are not comfortable with the English language. They have only one semester to complete the course along with four or five other courses. While we have been teaching such students for last 15 years, we have prepared class notes that suit them. We were encouraged by the results and response of the students. We hope students with similar background in other colleges and universities may find this guide book equally helpful to prepare for these three courses. Moreover, plant protection experts, vegetable specialists, horticultural officers, and extension workers may utilize this guide book as a valuable resource.

Gireesh Chand, PhD
Santosh Kumar, PhD

ABOUT THE EDITORS

Gireesh Chand, PhD, is currently Associate Professor-cum Senior Scientist in the Department of Plant Pathology at Bihar Agricultural University, India. He was previously Assistant Professor at N.D. University of Agriculture and Technology, Faizabad, India. His research specialization is in the study of molecular plant pathology of crop diseases and their management. He has a decade-long career dedicated to teaching undergraduate and postgraduate classes, research, and extension activities and has guided several MSc and PhD theses. He is the author or co-author of six books on plant pathology and has published more than hundred research papers, book chapters, and popular articles. He has presented fifty research papers at national and international seminars and symposia. With research specialization in the study of molecular plant pathology of crop diseases and their management, he has led more than 10 research projects.

He received his PhD in plant pathology from C.S. Azad University of Agriculture and Technology in Kanpur. India. He has received a number of awards and honors, including the P.R. Verma Award (2000), a PhD Research Fellowship (2002), SPPS Fellow Award (2009), ISHA Best Student-Guide Award (2010), Prof. M.J. Narsingham Award (2011), Young Scientist Award (2013), Best Paper Presentation Award (2014), Excellence in Teaching Award (2014), Dr. M. M. Alam Medal (2015), and the SSDAT Fellow Award (2015). He visited Beijing, China, through the International Financial Support Scheme (2013) by DST, New Delhi.

Santosh Kumar, PhD, is presently working as Assistant Professor-cum-Junior Scientist in the Department of Plant Pathology, Bihar Agricultural University, Sabour, Bhagalpur, India. He received his PhD in plant pathology from G.B. Pant University of Agriculture and Technology, Pantnagar, India. Dr. Kumar was awarded Senior Research Fellowship, IARI, New Delhi, Rajiv Gandhi National Fellowship (RGNF) through the University Grant Commission (UGC), New Delhi, during the PhD program. He has taught many undergraduate and postgraduate courses and is actively involved in teaching activity. He received a Young Scientist Award (2015) from BRIAT, Allahabad. He has published many research papers, review papers, book chapters, and popular articles in international and national journals. He has also published one book and two extension bulletins on mushroom production. He has attended and presented at several national and international symposia and has received several best poster awards. His research interests include management of diseases of pulses and rice, biocontrol, and mushrooms.

PART I

CEREAL AND PULSE CROPS

CHAPTER 1

DISEASES AND MANAGEMENT OF MAIZE

R. P. SINGH,[1] MAMTA SINGH,[2] and DINESH SINGH[3]

[1]*Subject Matter Specialist-Plant Protection, Krishi Vigyan Kendra, P.G. College, Ghazipur, Uttar Pradesh, India*

[2]*Subject Matter Specialist-Plant Breeding and Genetics, Krishi Vigyan Kendra, Sagar, Madhya Pradesh, India*

[3]*Program Coordinator, Krishi Vigyan Kendra, P.G. College, Ghazipur, Uttar Pradesh, India*

CONTENTS

1.1   INTRODUCTION

Maize or corn (*Zea mays* L.) is an important cereal food crop in the world with highest production and productivity as compared to rice and wheat. It is the most versatile emerging crop having wider adaptability under varied agro-climatic conditions. It is quick growing, high yielding and provides palatable and nutritious forage, which can be fed at any stage of growth without any risk to animals. On an average, it contains 9–10% crude protein, 60–64% neutral detergent fiber, 38–41% acid detergent fiber, 28–30% cellulose and 23–25% hemicelluloses on dry matter basis when harvested at milk to early-dough stage. The average productivity and potential yield of the crop in India is lower than the developed countries (Yadav, 2012, 2013). There are many biotic constraints (Table 1.1) of maize production including fungi, bacteria, viruses, phytoplasma and nematodes (Thind and Payak, 1978; Sharma et al., 1982; Owolade et al., 2000; CIMMYT, 2004; Negeri et al., 2011; Vincelli, 2008; Ali et al., 2012; Todd et al., 2010; Norton, 2011; Sweet and Wright, 2008; Bhatia and Munkvold, 2002).

1.2   FUNGAL PATHOGENS

1.2.1   SEED ROTS AND SEEDLING BLIGHT

1.2.1.1   Causative Agent and Disease Development

The most common and destructive pathogens are associated with seed rots and seedling blight includes *Pythium, Fusarium, Acremonium, Sclerotium, Rhizoctonia, Diplodia*, etc. (McGee, 1988; Vincelli, 2008). These fungi may lie dormant on maize crop residue or in soil and are carried on seed. Soil temperature of about 50°F is favorable for seedling

TABLE 1.1 Summery of Some of the Disease Infecting Maize

Name of the disease/ parasites	Pathogens/causative agents	Symptoms	Control
A. Fungal diseases			
Grey leaf spot	*Cercospora zeae-maydis* Tehon and Daniels	At early stage appear on leaves as small, pinpoint lesions surrounded by yellow halos, but as lesions mature, they elongate into narrow, rectangular, brown to gray spots. Lesions expand parallel to leaf veins and may become 1.5–2 inches long.	Use clean seed and resistant varieties. Follow crop rotation and destroy infected stubbles. Spry suitable fungicides.
Eye spot	*Kabatiella zeae* Narita and Hiratsuka	Small circular lesions with tan to gray centers that are surrounded by a red and yellow halo; if severe epidemic, lesions may grow together and can lead to death of large areas of tissue.	Clean cultivation, crop rotation, resistant hybrids, foliar fungicides.
Anthracnose	*Colletotrichum graminicola* (teleomorph=*Glomerella graminicola*) Cesati and Wills	Oval to elongate spots on leaves and irregular in shape, rusty brown lesions that have a yellowish halo; dark, hair like structures (setae) can often be seen on the leaf using a hand lens.	Resistant hybrids; clean tillage; rotation; balanced soil fertility.
Polysora rust	*Puccinia polysora* Underw	Light golden brown circular to oval pustules appear on leaf and densely spread. Development of pustules on lower surface is more as compared to upper surface.	Resistant cultivars, foliar fungicides.
Crazy top	*Sclerophthora macrospora* Saccardo	Excessive tillering (up to 10 tillers per plant), stunting, rolling of leaves. Tassels may develop as tiny ears and twisted.	Avoid planting in low, wet areas and follow good drainage.

TABLE 1.1 Continued

Name of the disease/ parasites	Pathogens/causative agents	Symptoms	Control
Gibberella stalk and ear rot	*Gibberella zeae* (Schwein) Petch (anamorph=*Fusarium graminearum*) Schwabe	Plants wilting and leaves changing color from light to dull green; lower stalk turns straw yellow; internal stalk tissue breaks down; interior of stalk has a red discoloration, if fungal infection affects the ears, it produces a red mold at the tips of the ear.	Resistant hybrids; clean tillage; rotation; balanced soil fertility. Control insects, especially stem and ear borers.
Pythium root rot	*Pythium gaminicola* Subramanian	Yellow and stunted on above ground parts. Roots have obvious discolored lesions, root cortex will come away when pulled gently, exposing the white stele; can also cause damping-off of seedlings.	Improve soil drainage, crop rotation, systemic fungicides use as seed treatment prior to planting.
Northern corn leaf spot	*Bipolaris zeicola* (Stout) Shoemaker *(Syn. Drechlera zeicola, Cochliobolus carbonum* R. R. Nelson)	Lesions are variable in size and shape; may be circular, oval, or linear and range from flecks up to 3/4 in. long; tan to brown in color.	Crop rotation, use resistant cultivars and foliar fungicides.
Southern rust	*Puccinia polysora* Underw	Southern rust occurs primarily on the upper leaf surface. Pustules are smaller, orange-brown and in more densely clustered form. In severe cases cause defoliation and premature senescence.	Earlier planting date and/or shorter-season hybrids may reduce risk of yield loss. Use resistant hybrids and foliar fungicides.
Physoderma brown spot	*Physoderma maydis* (Miyabe) Miyabe	Lesions initially appear small, round to oblong and yellow and occur primarily on leaves and leaf sheaths, later coalesce into larger reddish-brown blotches. Chocolate or purple oval blotches also occur on the midrib of infected leaves.	Manage residue through both crop rotation and tillage. Avoid highly-susceptible hybrids.

TABLE 1.1 Continued

Name of the disease/ parasites	Pathogens/causative agents	Symptoms	Control
Diplodia leaf streak or spot	*Stencarpella macrospora* (Earle) Sutton	Lesions are large (up to 10 cm long), gray-green and elliptical with a water-soaked appearance. Older lesions produce black spots.	Corn residue management. Crop rotation and suitable fungicides.
Fusarium kernel or ear rot	*Fusarium moniliforme* f. sp. *Subglutinans* Wollenweb & Reinking (teleomorph: *Gibberella fujikuroi* f. sp. *G. subglutinans*)	Pinkish or discolored caps of individual kernels, sometimes with a pinkish mold growth.	Crop rotation and clean cultivation.
Trichoderma ear rot and root rot	*Trichoderma viride* Pers.; Fr. (Teleomorph: *Hypocrea* sp.)	Dark green fungal growth is found on and between kernels and husks, often covering the entire ear.	Clean cultivation. Crop rotation.
Penicillium ear rot (blue eye, blue mold)	*Penicillium* spp. *P. chrysogenum Thom* *P. expansum* Link *P. oxalicum* Currie and Thom	Powdery green to blue green mold develops on and between kernels. Infection usually begins at the ear tips and primarily occurs on ears with mechanical or insect damage.	Clean cultivation. Crop rotation.

TABLE 1.1 Continued

Name of the disease/ parasites	Pathogens/causative agents	Symptoms	Control
Banded leaf and sheath blight	*Rhizoctonia solani f. sp. Sasakii* Kuhn (teleomorph= *Thanatephorus cucumeris* (Frank) Donk	Lesions appear as concentric bands on lower leaves and sheaths. The affected plant produces large, gray, tan or brown discolored areas alternating with dark brown bands. The developing ear is completely damaged and dried up prematurely.	Use tolerant varieties viz. Pratap Kanchan 2, Pratap Makka 3, Pratap Makka 5 and Shaktiman 1, 3. Foliar spray (30–40 days old crop) of tolcofos-methyl @ 1g/L of water.
B. Bacterial diseases			
Goss's bacterial wilt	*Clavibacter michiganensis* sub sp. *nebraskensis* (Vidaver and Mandel) Davis et al.	Long, gray/tan lesions with wavy margins that follow the leaf veins. Within these lesions, dark green to black, water soaked spots appeared.	Use resistant hybrids. Crop rotation follows with soybean, small grains and alfalfa.
Holcus spot	*Pseudomonas syringae* Van Hall	Initially dark green, water soaked circular spots across near the tips of lower leaves, later turn dark brown with yellow halo.	Crop rotation, clean cultivation.
Stewart's wilt	*Erwinia stewartii* (*Pantoae stewartii*) (Smith) Dye	Bleached tassels, cavities in the stalks near the soil line appear. Leaf blight symptoms include pale-green to yellow, linear, elongated lesions (streaks) with irregular or wavy margins that run parallel to the veins. Leaf blight is often associated with feeding scars caused by flea beetles.	Use resistant hybrids. Insecticides applied as a seed treatment.

TABLE 1.1 Continued

Name of the disease/parasites	Pathogens/causative agents	Symptoms	Control
Bacterial stalk rot	*Erwinia carotovora,* (Jones) Bergey et al. *Erwinia chrysanthemi*	Basal internodes develop soft rot and give a water soaked appearance. A mild sweet fermenting odor accompanies such rotting. Leaves some time show signs of wilting and within a few days lodge or topple down.	Use resistant hybrids. Avoid waterlogging and poor drainage.
C. Viral diseases			
Maize chlorotic dwarf virus	*Maize chlorotic dwarf virus* (MCDV)	Infected plants are stunted to various degrees and leaves may turn partially or completely yellow and/or red, depending on the severity of the disease. MCDV is transmitted by leafhoppers, which acquire the virus from infected Johnson grass.	Good weed management and need based foliar application of insecticides.
High plains virus	*High Plains Virus* (HPV)	Small yellowish flecks appear as lines running parallel to leaf veins. Infected seedlings turn bright yellow and quickly die. Plants may be stunted, older leaves may become red and then necrotic, and ear and kernel size may be reduced. It is spread by the wheat curl mite (*Aceria tosciella*).	Use resistant cultivars, weed management and need based application of acaricides.
Maize streak virus	*Maize Streak Virus* (MSV)	Initially colorless spots appear on the lowest youngest leaves. In severe cases lesions broken and chlorotic stripes appeared. It is transmitted by leafhopper.	Clean cultivation. Regular scouting of the crop for leafhoppers followed by registered insecticide.

TABLE 1.1 Continued

Name of the disease/ parasites	Pathogens/causative agents	Symptoms	Control
D. Phytoplasma disease			
Corn stunt spiroplasma	*Spiroplasma kunkelii*	Margins of whorl leaf turn yellow followed by reddening of older leaves and yellow striping, which runs the length of leaves. Plants are stunted, have multiple tillers and produce numerous small ear shoots. Root systems are reduced. It is transmitted by leafhopper (Hemiptera: Cicadellidae) species, mainly by the corn leafhopper, *Dalbulus maidis*.	Use resistant varieties and need based foliar application of insecticides.
Maize bushy stunt	*Maize Bushy Stunt phytoplasma, syn. Maize Bushy Stunt mycoplasma*	Marginal chlorosis on young leaves and tips gradually turn purple-red as they approach maturity. Bushy and proliferation of tillers appears. The pathogen is transmitted by the cicadellid leafhoppers *Dalbulus maidis, D. elimatus,* and other species of *Dalbulus*.	Use resistant varieties and need based foliar application of insecticides.

blight for germination is very slow and the soil-borne pathogens can grow and invade seeds and seedlings. In addition to the effect of cold and wet soil, seed factors-age, degree of finish or maturity, mechanical damage, and genetic resistance affect the severity of seed rot and seedling blight (Haggag, 2013).

1.2.1.2   Symptoms

Seed rots occur before germination. Seeds are soft and brown and may be overgrown with fungi. Seedling blights may be either pre-emergence, in which the seed germinate but the seedling is killed before it emerges from the soil, or post-emergence, in which the seedling emerges through the soil surface before developing symptoms (Vincelli, 2008).

1.2.1.3   Management

The disease can effectively control by seed treatment by captan, thiram, metalaxyl/mefenoxan. Clean cultivation, removal and burning of left out plant debris and crop rotation are practical in keep away in the disease development (Vincelli, 2008; Haggag, 2013).

1.2.2   *TURCICUM LEAF BLIGHT (NORTHERN LEAF BLIGHT)*

1.2.2.1   Causative Agent and Disease Development

Turcicum leaf blight, initiated by the fungus *Exserohilum turcicum* (Pass) K.J. Leonard and E.G. Suggs [(Teleomorph: *Septoshaeria turcica* (Lutterell) K.J. Leonardo and E.G. Suggs]. The fungus overwinters as mycelium and conidia in and on leaf debris. During warm, moist winter in early summer, new conidia are produced on old residue and the new conidia are carried by the wind or rain to lower leaves of young maize plants. Turcicum leaf blight is favored by moderate temperature between 18–27°C and prolonged wetness (Khatri, 1993; Leonard et al., 1985).

1.2.2.2 Symptoms

Appearance with small, water-soaked spot appear on the lower leaves and progress upwards. Lesions elongated becoming elliptical or cigar-shaped and are typically gray green in color approximately 5–10 cm length and 1 cm width. Sometimes the disease spots are larger or a few of them join together to form large different spots. The laminas death would result in early death of the plants. As the lesions mature they become tan with distinct dark zone of fungal sporulation (Chenulu and Hora, 1962; Ullstrup, 1966; Perkins and Pedersen, 1987).

1.2.2.3 Management

Pandurangegowda et al. (2002) reported that the maize composites NAC 6002, NAC 6003 and NAC 6004 were resistant against *E. turcicum* under artificial inoculated field conditions. Both quantitative and qualitative types of resistance operate against this resistance. The quantitative or polygenic resistance controls the number but not the size of the lesions on plants, the qualitative or specific in expressed as chlorotic lesions (Jha and Dhawan, 1970). Harlapur (2005) observed carboxin powder as seed treatment (2 g/kg seed) followed by two sprays of mancozeb (0.25%) resulting in significantly minimum percent disease intensity and maximum grain yield. The treatment with mancozeb 0.25% and combination treatments of carbendazim and mancozeb, for example, saff 0.25% recorded the lowest percent disease index (PDI) reducing the disease by 73.0% and 72.1%, respectively (Reddy et al., 2013).

1.2.3 MAYDIS LEAF BLIGHT (SOUTHERN LEAF BLIGHT)

1.2.3.1 Causative Agent and Disease Development

Maydis leaf blight is caused by *Drechslera maydis* Nisicado & Miyake, Subram A. Jan [teleomorph: *Cochliobolus heterostrophus* (Drechs.) Drechs.; Anamorph: *Bipolaris maydis* (Nisicado) Shoemaker Syn. *Helminthosporium maydis*]. The epidemic was caused by race

T of *C. heterostrophus*, which is highly virulent to maize (Nasir, et al., 2012) with male sterile T type cytoplasm (Dewey et al., 1988). The currently predominant form of *C. heterostrophus* is race O, which can cause yield losses upto 40% (Byrnes et al., 1989; Fisher et al., 1976). Both race T and O of the pathogen occurs in India, however the most prevalent one remains to be race O. Moreover, in absence of male sterility in Indian maize program, race T does not pose any threat to maize cultivation (Hooker, 1972, 1978). Maydis leaf blight is favored by warm temperatures (68–90°F) and high humidity (CIMMYT, 2004). The fungus overwinters in maize debris as spore or mycelium. Spores are spread by wind or splashing water to growing plants. After infection and colonization, sporulation from these primary lesions serves as source of secondary spread and infection as long as weather conditions are favorable for disease development.

1.2.3.2 Symptoms

Leaves show grayish, tan, and parallel straight sided or diamond shaped, 1–4 cm long lesions with buff or brown borders or with prominent color banding or irregular zonation. Symptoms may be confined to leaves or may develop on sheaths, stalks, husks, ears and cobs. The lesions are longitudinally elongated typically limited to a single inter vascular region, often coalescing to form more extensive dead portions. Mature lesions may coalesce, producing a complete burning of large areas of the leaves (CIMMYT, 2004).

1.2.3.3 Management

Host plant resistance is an effective, economic and environmentally safe component approach to keep maydis leaf blight below the threshold level (Ali et al., 2012). Resistance to race T is conditioned by both cytoplasmic factors and nuclear genes, cytoplasmic component being the most important. Almost two inbred lines CM 104 and CM 105 have been demonstrated to possess durable resistance to turcicum and maydis leaf blight (Sharma and Payak, 1990). Application of mancozeb (0.25%) thrice at 10 days interval is effective against southern blight management. (Miller, 1970;

Payak and Sharma, 1985). Foliar application of two fungicides viz., propiconazole (0.1%) or chlorothalonil (0.2%) is highly effective in reducing disease severity and avoiding yield loss (Nasir et al., 2012).

1.2.4   COMMON RUST

1.2.4.1   Causative Agent and Disease Development

Common rust is caused by the fungus Puccinia sorghi, and the disease often develops when susceptible varieties are grown under cool, wet weather conditions. The disease is favored by high humidity with cool evening temperatures (14–18°C), followed by moderate daytime temperatures. If environmental conditions are favorable, new urediniospores can be produced every 7 to 8 days after initial infection.

1.2.4.2   Symptoms

Rusts produce quite distinctive reproductive structures called pustules that erupt through the surface of leaves, stalks, or husks and produce spores called urediniospores. It serves as secondary inoculums throughout the growing season. This rust is recognized by oval to elongate cinnamon-brown pustules (blister-like spots) scattered over both surfaces of the leaves. As corn matures, the pustules become brownish-black. The pustules may appear on any of the aboveground parts of the plant, but they are most abundant on the leaves. The pustules of this rust break through the epidermis early in their development.

1.2.4.3   Management

Most inbred lines are susceptible, but a few have shown resistance to one or more physiologic races of this rust. Resistant hybrids and fungicides can be economical in high value corn such as seed corn if considerable rust is present on the lower leaves prior to silking and the weather forecast is for unseasonably cool, wet weather. Generalized or mature plant resistance has been studies in details (Kim and Brewbaker, 1976,

Sharma and Payak, 1979, Shah and Dillard, 2006). Several hybrids and other materials such as Ganga 1, Deccan 103, Buland, Sheetal, HHM 1, HHM 2, HQPM 1 and Nithyashree possessing resistance have been released for cultivation. Spray of mancozeb @ 2.5 g/L of water at first appearance of pustules and prefer early maturing varieties for cultivation.

1.2.5 COMMON SMUT

1.2.5.1 Causative Agent and Disease Development

The fungus is seed-borne but not seed transmitted (CABI, 2007; McGee, 1988) and caused by *Ustilago maydis* (CIMMYT, 2004). Common smut is favored by humid, temperate environments, by high nitrogen level (CIMMYT, 2004). *Ustilago maydis* infections can originate from telio-spores overwintering in crop debris and soil, or can be introduced with unshelled seed corn or manure from animals fed infected corn stalks. In soil, the teliospores can survive for several years. Spores are spread by wind and splashing water. Infection of the plant can be facilitated by the presence of mechanical wounds (wounds caused by strong wind, hail, insects, cultivation, spraying or de-tasseling).

1.2.5.2 Symptoms

The fungus attacks on ears, stalks, leaves, and tassels. The smut gall is composed of a mass of black, greasy, or powdery spores enclosed by a smooth, greenish-white to silvery-white membrane. Galls on the ears may be up to 5 inches in diameter. Leaf infections result in small pustules, usu-ally on the midrib, that cause some distortion of the leaf. As the spores mature, the outer covering of the gall becomes dry and papery and disinte-grates, releasing the spores passively.

1.2.5.3 Management

Avoiding extremely susceptible hybrids is the most effective means of management. Management through crop rotation and gall destruction has been recommended, but it is unlikely that such methods would be effective or even practical where corn is grown extensively. In small garden plots, the gall removal may help reduce the abundance of spores.

1.2.6 PYTHIUM STALK ROT

1.2.6.1 Causative Agent and Disease Development

Pythium stalk rot of maize is caused by *Pythium aphanidermatum*. The disease is favored by extended periods of hot, humid weather (optimum 90°F or 32°C). The disease is most common in river-bottom fields whose air and soil drainage is poor and humidity is high. The incidence of this disease is significantly influenced by both environmental and host factors (Agrios, 2005, CIMMYT, 2004).

1.2.6.2 Symptoms

The disease is generally first recognized when plant fall over. The rotted part of the stalk is usually only single internodes just above the soil line. Damaged internodes commonly twist before the plant lodge. The diseased area appears in brown color, water soaked, soft and collapsed. The stalk is not broken off completely by the disease and plants alive for several weeks because the vascular bundles remain intact. Pythium stalk rot may also cause top die back (CIMMYT, 2004, Agrios, 2005).

1.2.6.3 Management

Crop rotation prior to planting and sow quality seeds of maize is important to minimize the presence of the Pythium stalk rot of maize (Haggag, 2013). Planting of hybrid Ganga safed-2, high starch, DMH 103 and composite Suwan 1 is recommended for disease prone areas. Good field drainage, planting time between 10 and 20th July in North India, plant population of not more than 50,000/ha reduce the disease. Application of 75% captan @ 12 g/100 L of water and bleaching powder (33% chlorine) @ 10 kg/ha as soil drench help in the control of these stalks. Various biological agents such as *Gliocladium catenulatum* and *Trichoderma* species are also being investigated for control of *Pythium aphanidermatum* (Deadman et al., 2006).

1.2.7 CHARCOAL STALK ROT

1.2.7.1 Causative Agent and Disease Development

Charcoal rot is caused by the soil borne fungus *Macrophomina phaseolina* (Tassi.) Goidanich (McGee, 1988). The fungus also attacks many other hosts, which helps in its perpetuation (Farr et al., 1989; Ali and Dennis, 1992). The fungus overwinters in the soil and infects the host at susceptible crop stage through roots and proceeds towards stems as sclerotia. The disease development is maximum during grain filling stage and is favored by warm temperature (30–40°C) and low soil moisture (CIMMYT, 2004; Sweets and Wright, 2008).

1.2.7.2 Symptoms

The disease first attacks the roots of seedlings and young plants. Lesions are brown and water soaked and later become black. When the plant approaches maturity the disease spreads into crown and lower internodes of the stalk. Infected stalks may be recognized by grayish on the surface of lower internodes. Internal parts of the stalk are shredded and grayish black. Small black fungal bodies (sclerotia) are present in the pith of the affected stalks. Roots are also invaded and show sclerotia in the disorganized tissue. Kernels are also infected and turn completely black (CIMMYT, 2004; Sweets and Wright, 2008).

1.2.7.3 Management

The disease can frequently be minimized by maintaining soil moisture during dry periods after tasseling, where irrigation is available. Dodd (1977) has postulated that resistance to maturity related stalk rots complex inheritance patterns linked to environmental and physiological interaction in plants. A photosynthetic stress translocation balance concept of predisposition to root and stalk rots has been developed by Dodd (1980). The charcoal rot and many other root pathogens could be controlled by the antagonistic fungi and bacteria, the vesicular arbuscular mycorrhizal fungi (VAMF) (Elad et al., 1986; Shankar and Sharma, 2001; Schonbeck, 1987; Mohan, 2000) and plant extracts (Osman et al., 1996; Raja and Kurucheve, 1999).

1.2.8 DOWNY MILDEWS

1.2.8.1 Causative Agent and Disease Development

Several species of the genera *Peronosclerospora, Sclerospora,* and *Sclerophthora* are responsible for downy mildews (CIMMYT, 2004). These include Crazy top downy mildew (*Sclerophthora macrospora*), Brown stripe downy mildew (*Sclerophthora rayssiae* var. *zeae*), Java downy mildew (*Peronosclerospora maydis*), Philippine downy mildew (*Peronosclerospora philippinensis*), and Sorghum downy mildew (*Peronosclerospora sorghi*) (McGee, 1988; De Leon, 1991). The important species causing downy mildew in maize in India are the Sorghum downy mildew (SDM; *Peronosclerospora sorghi*), Brown stripe downy mildew (BSDM; *Sclerophthora rayssiae* var. *zeae*) and Rajasthan downy mildew (RDM; *Peronosclerospora hetropogoni*). The diseases are most prevalent in warm, humid regions (Krishnappa et al., 1995; Hooda et al., 2012).

1.2.8.2 Symptoms

Symptom expression is greatly affected by plant age, pathogen species and environment. Usually, there are chlorotic, stripings or partial symptoms in leaves and leaf sheaths along with dwarfing (Adenle and Cardwell, 2000). Downy mildew becomes conspicuous after development of a downy growth on or under leaf surfaces. This condition is the result of conidia formation, which commonly occurs in the early morning. Some species causing downy mildew also induce tassel malformations, blocking pollen production and ear formation. Leaves may be narrow, thick, and abnormally erect. In early symptoms plants are stunted and may die. (CIMMYT, 2004).

1.2.8.3   Management

Rogue and destroy infected plants as they appear in the field. Avoid maize-sorghum crop rotation in field where disease has occurred. Cultivation of resistant varieties/hybrids viz. Ganga 5, Ganga 11, Satlaj, Deccan 1, Deccan 103, Composite suwan 1, PAU 352, Pratap Makka 3, Gujarat Makka 4, Shalimar KG 1, Shalimar KG 2, PEMH-5, Bio 9636, NECH-X-1280, DMH 1, NAC 6002, COH (M) 4, COH (M) 5 and Nithyashree is recommended for downy mildew endemic areas. Seed should invariably be treated with metalaxyl @ 2.5g/kg seed and need based foliar sprays of systemic fungicide such as metalaxyl @ 2–2.5 g/L of water is recommended at first appearance of disease symptoms.

1.3   BACTERIAL PATHOGEN

1.3.1   BACTERIAL LEAF BLIGHT

1.3.1.1   Causative Agent and Disease Development

Bacterial leaf blight is caused by *Pseudomonas avenae* subsp. *avenae* (Manns). The bacterium caused leaf blight but not stalk rot, shank rot or ear rot. The bacterium does not survive in infected leaves on dead plants for two months after maturity and do not survive for two weeks in infected green leaves buried in soil. *P. avenae* is pathogenic to many other *cvs.* of *gramineous* spp. (oats, barley, wheat, some millets and sorghum). In glass house test, it was not isolated from plants other than maize (Sumner and Schaad, 1977; Giester and Rees, 2004).

1.3.1.2 Symptoms

Water-soaked linear lesions on leaves as they emerge, lesions turn brown and may subsequently turn gray or white; lesions may have a red border after the leaves are mature. Lesions do not tend to extend any further; no new lesions tend to appear after tasseling; if corn variety is susceptible, mature leaves may shred after maturity.

1.3.1.3 Management

Resistant hybrids should be planted in areas where the disease is prevalent.

1.3.2 MAIZE DWARF MOSAIC

1.3.2.1 Causative Agent and Disease Development

Maize dwarf mosaic is caused by several strains of the maize dwarf mosaic virus (MDMV). Maize and sorghum are the main crop hosts of MDMV; however, Johnson grass and other wild grasses are also hosts. Some strains of MDMV overwinter in Johnson grass and are spread from Johnson grass to maize by the aphid vectors. The virus is transmitted by people, animals, or machines moving through the fields, and also by at least 15 species of aphids, including the corn leaf aphid, the green bug, and the green peach aphid (Haggag, 2013).

1.3.2.2 Symptoms

Symptoms first appear on the youngest leaves as an irregular, light and dark-green mottle or mosaic, which may develop into narrow streaks along the veins. As plants mature, the leaves become yellowish-green. Plants with these symptoms are sometimes stunted with excessive tillering, multiple ear shoots, and poor seed set. Early infection may predispose maize to root and stalk rots and premature death. Symptoms can appear in the field within 30 days after seedling emergence.

1.3.2.3 Management

Many commercial maize hybrids are highly tolerant of the disease and no control is needed. Use good weed management that especially reduces infestations of Johnson grass, as well as other grassy weeds. It is critical to plant as early as possible to avoid later buildup of insects and increased disease. Rotate with cotton, soybean or other non-grass crops.

1.3.3 NEMATODES

More than 40 species of nematodes have been reported to feed on maize in worldwide. Several nematodes have been found associated with maize in India are *Heterodera zeae*, *Pratylenchus zeae* Graham; *Tylenchorhychus vulgare* Upadhyay, Swarup and Sethi; *T. masoodii* Siddiqui; *Helicotylenchus dihysteria* (Cobb) Sher; *Hoplolaimus indicus* Sher and *Meloidogyne incognita* (Kofoid and White). The cyst nematode is serious problem in state of Bihar, Madhya Pradesh, Rajasthan and Uttar Pradesh (Koshy and Swarup, 1971).

1.3.3.1 Symptoms

Evidence of injury may vary with species of nematode, its population level, soil type and soil moisture. The major symptoms are stunting, restricted root growth, lesions or galls on roots, stubby roots, chlorosis and wilting.

1.3.3.2 Management

Cultural practices, including crop rotation with non-host or less susceptible plants, and prevention of soil compaction, which restricts downward root growth, are often good nematode management practices. Carbofuran (3G) @ 20 kg/ha and seed treatment with carbosulfan @ 2–3% of seed have been found to be effective in reducing nematode damage under field conditions. Against cyst nematode, sorghum, pearl millet, pulses, vegetables, wheat and oat have been suggested as alternative crops in crop rotation (Sharma, 2009). Resistant corn hybrids may be available for some nematodes.

KEYWORDS

- **bactericide**
- **foliar disease**
- **fungicide**
- **maize**
- **seed disease**
- **virus**

REFERENCES

1. Agrios, G. N. (2005). *Plant Pathology*. Fifth Edition, Academic Press, New York.
2. Ali, F., Muneer, M., Xu, J., Durrishahwar, Rahman, H., Lu, Y., Hassn, W., Ullah, H., Noor, M., Ullah, I., Yan, J. (2012). Accumulation of desirable alleles for southern leaf blight (SLB) in maize (Zea mays L.) under the epiphytotic of *Helminthosporium maydis. Australian J. Crop Sci.*, 6, 1283–1289.
3. Ali, S. M., Dennis, J. (1992). Host range and physiological specialization of *Macrophomina phaseolina* isolated from field pea in South Australia. *Australian J. Exptl. Agr.*, 32, 1121–1125.
4. Andenle, V. O., Cardwell, K. F. (2000). Seed transmission of *Peronosclerospora sorghi*, causal agent of maize downy mildew in Nigeria. *Plant Pathology*, 49, 628–635.
5. Bhatia, A., Munkvold, G. P. (2002). Relationships of environmental and cultural factors with severity of gray leaf spot on maize. *Plant Disease*, 86, 1127–1133.
6. Byrnes, K. J., Pataky, J. K., White, D. G. (1989). Relationships between yield of three maize hybrids and severity of southern leaf blight caused by race of *Bipolaris maydis. Plant Disease*, 73, 834–840.
7. CABI (2007). *Crop Protection Compendium, 2007*. Edition Wallingford, UK. Centre for Agriculture and Biosciences International (CABI).
8. Chenulu, V. V., Hora, T. S. (1962). Studies on losses due to Helminthosporium blight of maize. *Indian Phytopathology*, 15, 235–237.
9. CIMMYT (2004). The CIMMYT maize program, 2004. *Maize Disease: A Guide for Field Identification*. 4th edition. Mexico, D.F.: International Maize and Wheat Improvement Centre (CIMMYT).

10. De Leon, C. (1991). Inheritance of resistance of downy mildews in maize. pp. 218–230. In: K. R. Sarkar et al. (Eds.) Maize Genetics Perspectives. *Indian Soc. Genet. Pl. Breed,* New Delhi.

11. Deadman, M., Hasani, H. Al., Sadi, A. Al. (2006). Solarization and biofumigation reduce *Pythium aphanidermatum* on greenhouse cucumbers. *Canadian J.Pl. Pathol.,* 25,411–417.

12. Dewey, R. E., Siedow, J. N., Timothy, D. H., Levings, C. S. (1988). A-13 Kilodalton maize mitochondrial protein in *E. coli.* confers sensitivity to *Bipolaris maydis* toxin. Science, 239, 293–295.

13. Dodd, J. L. (1977). A photosynthetic stress translocation balance concept of corn stalk rot. *Proc. 32 Annu. Corn,Sorghum Res. Conf.,* 32, 122–136.

14. Dodd, J. L. (1980). The role of plant stresses in development of corn stalk rots. *Plant Disease,* 64, 533–537.

15. Elad, Y., Zvieli, Y., Chet, I. (1986). Biological control of *Macrophomina phaseolina* (Tassi.) Goid. by *Trichoderma harzianum. Crop Protection,* 5, 288–292.

16. Farr, D. F., Bills, G. F., Chamuris, G. P., Rossman, A. Y. (1989). Fungi on plants and plant products in the United States. St. Paul. MN, APS, Press.

17. Fisher, D. E., Hooker, A. L., Lim, S. M., Smith, D. R. (1976). Leaf infection and yield loss caused by 4 Helminthosporium diseases of corn. *Phytopathology,* 66, 942–944.

18. Giester, L. J., Rees, J. M. (2004). Bacterial diseases of corn. Clay/Webster, St. Paul, MN, USA.

19. Haggag, W. M. (2013). Corn diseases and management. *J. Appl. Sci. Res.,* 9, 39–43.

20. Harlapur, S. I. (2005). Epidemiology and management of turcicum leaf blight of maize caused by *Exserohilum turcicum* (Pass.) Leonard and Suggs. PhD Thesis. University of Agricultural Sciences, Dharwad, India.

21. Hooda, K. S., Sekher, J. C., Singh, V., Sreeramasetty, T. A., Sharma, S. S., Parnidharan, V., Bunker, R. N., Kaul, J. (2012). Screening of elite lines for resistance against downy mildews. *Maize Journal,* 1, 110–112.

22. Hooker, A. L. (1972). Southern leaf blight of corn, present status and future prospects. *J. Environ. Qual.,* 7, 244–249.

23. Hooker, A. L. (1978). Genetics of disease resistance in maize in Maize Breeding and Genetics (Eds. Welden, D. B.) John Wiley & Sons, New York, Chap. 21.

24. Jha, T. D., Dhawan, N. L. (1970). Genetic analysis of resistance to *Helminthosporium turcicum* in maize inbred lines. In Plant Disease Problems (Eds. S. P. Raychoudhry et al.), Indian Phytopathological Society, New Delhi.

25. Khatri, N. K. (1993). Influence of temperature and relative humidity on the development of *Helminthosporium turcicum* on maize in Western Georgia. *Indian J. Mycol. Pl. Pathol.,* 23, 35–37.

26. Kim, S. K., Brewbaker, J. L. (1976). Source of general resistance of *P. sorghi* of maize in Hawaii. *Plant Dis. Reptr.,* 60, 551.

27. Koshy, P. K., Swarup, G. (1971). Distribution of *Heterodera avenae, H. zeae, H. cajani,* and *Anguina tritici* in India. *Indian J. Nematology,* 1, 106–111.

28. Krishnappa, M., Naidu, B. S., Seetharam, A. (1995). Inheritance of resistance to downy mildew in maize. *Crop Improv.,* 22, 33–37.

29. Leonard, K. S., Duncan, H., Leath, S. (1985). Epidemic of northern leaf blight of corn in North Carolina. *Plant Disease,* 69, 824.

30. McGee, D. C. (1988). Maize Diseases: A reference source for seed technologists. The American Phytopathological Society (APS, Press), St. Paul, MN, 150 pp.

31. Miller, P. R. (1970). Southern corn leaf blight. *Plant Dis. Reptr.,* 54, 1099–1136.

32. Mohan, V. (2000). Endomycorrhizal interaction with rhizosphere and rhizoplane mycoflora of forest tree species in Indian Arid Zone. Indian Forester, 126, 749–755.

33. Nageri, A. T., Coles, N. D., Holland, J. B., Balint-Kurti, P. J. (2011). Mapping QTL controlling southern leaf blight resistance by joint analysis of three related recombinant inbred line populations. *Crop Science,* 51, 1571–1579.

34. Nasir, A., Singh, V. K., Singh, A. (2012). Management of maydis leaf blight using fungicides and phytoextracts in maize. *Maize Journal,* 1, 106–109.

35. Norton, D. C. (2011). Nematode host attack corn in Iowa. *Review.* PM 1027. Iowa State University.

36. Osman, N. A. I., Awaref, A. H., El-Deeh, A. A., Abdel-Magid, M. S. (1996). Effect of garlic extract and volatiles on root rot and pod rot diseases of peanut (*Arachis hypogea* L.). *Egypt J. Appl. Sci.,* 11,162–175.

37. Owolade, B. E., Fawole, B., Osikanlu, Y. O. K. (2000). Fungi associated with maize seed discoloration and abnormalities in South Western Nigeria. *Tropical Agricultural Research and Extension,* 3, 102–105.

38. Pandurangegowda, K. T., Naik, P., Setty, T. A. S., Hattappa, S., Naik, N. P., Juna, M. (2002). High yielding maize composite NAC 6004 resistant to turcium leaf blight and downy mildew. *Environment and Ecology,* 20, 920–923.

39. Payak, M. M., Sharma, R. C. (1985). Maize diseases and their approach to their management. *Tropical Pest Management,* 31, 302–310.

40. Perkins, J. M., Pedersen, W. L. (1987). Disease development and yield losses associated with northern leaf blight on corn. *Plant Disease,* 71, 940–943.

41. Raja, J., Kurucheve, V. (1999). Fungicidal activity of plant and animal products. *Ann. Agri. Res.,* 20, 113–115.

42. Reddy, T. R., Reddy, P. N., Reddy, R. R., Reddy, S. S. (2013). Management of turcicum leaf blight of maize caused by *Exserohilum turcicum* in maize. *Int. J. Sci. Res. Pubs.,* 3, 1–4.

43. Schnobeck, F. (1987). Mycorrhiza and plant health. A contribution to biological protection of plants. *Angewanddt Botanik, Rev. Pl. Path.,* 66, 4093.

44. Shah, D. A., Dillard, H. R. (2006). Yield loss in sweet corn caused by *Puccinia sorghi*: A meta analysis. *Plant Disease,* 90, 1413–1418.

45. Shankar, P., Sharma, R. C. (2001). Management of charcoal rot of maize with *Trichoderma viride.Indian Phytopathology,* 54, 390–391.

46. Sharma, R. C., Payak, M. M. (1979). Resistance to common rust of maize in India. *Maize Genetics Cooperation Newsletter,* 53, 665–666.

47. Sharma, R. C., Payak, M. M. (1990). Durable resistance to two leaf blights in two maize inbred lines. *Theoretical and Applied Genetics,* 80, 542–544.

48. Sharma, R. C. (2009). Maize disease management in India. pp. 145–156. In: R. K. Upadhyay et al. (Eds.) *Integrated Pest and Disease Management.* A.P.S. Pub. Corporation, New Delhi.

49. Sharma, R. C., Payak, M. M., Shankarlingam, S., Laxminarayan, C. (1982). A comparison of two methods of estimating yield loss in maize caused by common rust. *Indian Phytopathology,* 35, 18–20.

50. Sharma, R. C., Rai, S. N., Batsa, B. K. (2005). Identifying resistance to banded leaf and sheath blight of maize. *Indian Phytopathology*, 58, 121–122.
51. Sumner, D. R., Schaad, N. W. (1977). Epidemiology and control of leaf blight of corn. *Phytopathology*, 67, 1113–1118.
52. Sweets, L. E., Wright, S. (2008). Integrated Pest Management. Corn Diseases. Plant Protection Programs. College of Agriculture, Food and Natural Resources. University of Missouri, Columbia. 1–23.
53. Thind, B. S., Payak, M. M. (1978). Evaluation of maize germplasm and estimation of losses to Erwinia stalk rot. *Plant Disease Reporter*, 62, 319–323.
54. Todd, J. C., Ammar, El. D., Redinbaugh, M. G., Hoy, C., Hogenhout, S. A. (2010). Plant host range and leafhopper transmission of maize fine streak virus. *Phytopathology*, 100, 1138–1145.
55. Ullstrup, A. J. (1966). Corn diseases in the United States and their control. Agriculture Handbook No. 199, United States, Department of Agriculture, p.26.
56. Vincelli, P. (2008). Seed and seedling diseases of corn. Agriculture and Natural Resources. UK, Cooperative Extension Service, University of Kentucky, College of Agriculture, Plant Pathology Fact Sheet, 1–2.
57. Yadav, O. P. (2012–13). Project Director Review. Annual Workshop. All India Coordinated Research Project on Maize, ANGRAU, Hyderabad, A.P., India.

CHAPTER 2

DISEASES OF PIGEON PEA (*Cajanus cajan* L. MILLSP.) AND THEIR MANAGEMENT

AJAY KUMAR[1], MAHESH SINGH[2] and GIREESH CHAND[3]

[1,2]*Department of Plant Pathology, Narendra Dev University of Agriculture and Technology, Kumarganj, Faizabad–224229, Uttar Pradesh, India*

[3]*Department of Plant Pathology, B.A.U., Sabour, Bihar, India*

CONTENTS

2.1 INTRODUCTION

Pigeon pea [*Cajanus cajan* (L.) Millspaugh] is one of the major food legume crops of the tropics and sub-tropics. In India, after chickpea, pigeon pea is the second most important pulse crop. India has largest area under pigeon pea 3.90mha with a total production and productivity of 2.89 mt and 741 kg/ha, respectively (DAC, 2011). Pigeon pea commonly known as red gram or arhar is a very old crop of this country. The chromosome number of pigeon pea is 2n = 22 chromosomes. It is a rich source of protein and supplies a major share of the protein requirement of the vegetarian population of the country.

Pigeon pea can be attacked by more than 100 pathogens. These include fungi, bacteria, viruses, nematodes, and mycoplasma-like organisms. Fortunately, only a few of them cause economic losses and the distribution of the most important diseases is geographically restricted. At present farmers mainly grow pigeon pea landraces and it is possible that they have some degree of tolerance to most of the pathogens. This situation could change once the diverse landraces are replaced by a few improved cultivars. The diseases of considerable economic importance at present are fusarium wilt, sterility mosaic, phytophthora blight and alternaria blight.

2.2 FUSARIUM WILT

In India, the infestation occurs in almost every state in which pigeon pea is cultivated, especially in Rajasthan, Maharashtra, Madhya Pradesh, Uttar Pradesh and Bihar. The wilt is also found in many other countries like Kenya, Malawi and Tanzania. The continuous cultivation of pigeon pea in the same area in each year results up to 50% plant mortality due to the disease. In Bihar and U.P., it loss 5–10% of standing crop is common feature every year (Kannaiyan et al., 1984). The disease incidence is high during flowering and pod formation stages.

2.2.1 SYMPTOMS

Symptoms are more pronounced and the damage is greater when the plants have grown up after the rainy season. The susceptible plants are attacked when young (about 5–6 weeks old). The symptoms are variable. Typically the first symptom is the premature yellowing of the leaves. The next symptom is the wilting or withering of the leaves of the diseased plants. Wilting is characterized by gradual, sometimes sudden, yellowing, withering and drying of leaves followed by drying of the entire plant or some of its branches. Patches of diseased plants are scattered throughout the field indicating locations where pathogen is present and infection started. Examination of main root and the base of stem and infected plant after pulling it out from the soil, the blackened longitudinal streaks are seen on the roots. These black streaks represent the vascular bundles of the infected plants plugged with mycelium and fructifications of the fungus. Partial wilting is also common as only one side gets withered.

2.2.2 CAUSAL ORGANISM

The wilt of pigeon pea is caused by Fusarium *udum* Butler. It is facultative parasite and can survive in the soil in the absence of the host through chlamydospores. The mycelium is restricted to the vascular tissues. It is inter and intracellular, hyaline, branched, geniculate and septate. The profuse growth of mycelium within the xylem vessels completely plugged with the lumen of the vessels and thus, the free flow of water is checked resulting in the disease. Mycelium produces spores of three types within host tissues-macroconidia, micro conidia and chlamydospores. The micro conidia are small, elliptical or curved, thin walled, one or two septate and $5{-}15 \times 2{-}4$ μm. The macro conidia are linear, sickle shaped, pointed at both ends, thin walled, 3–4 septate and measures $15{-}50 \times 3{-}5$ μm. The chlamydospores are usually formed in chains within the tissues of the host plant. They are rounded or oval and thick walled and can remain viable for long time.

The perfect stage, *Gibberella indica*, described by Rai and Upadhyay (1982), is usually found on exposed roots and collar region of the stem up to the height of 35 cm above the ground level. The mature perithecia are superficial, aggregate, subglobose to globose, sessile, and smooth walled, dark violet, and 350–550 μm in diameter. Asci are 8-spored, mostely subcylindrical, $60{-}80 \times 6{-}10$ μm, broader in the middle and, with short stalk, a narrow apex, and a center apical pore. The ascospores are elliptical to ovate, $10{-}17 \times 5{-}7$ μm, hyline, commonly 2-celled, rarely 3–4 celled and constricted at the septa.

2.2.3 DISEASE CYCLE

Fusarium udum is facultative parasite and lives saprophytically in the soil without host for a long time. The fungal pathogen attacks host roots by germ tubes arising from asexual spores and reached in vascular tissues to establish infection. The fungus can be grow and multiply in the vascular tissues, causing partial or complete wilting of the host plant. The host plant is killing within few days or weeks. When the crop is harvested, the plants are cut at the soil level leaving the infected root system and stubble to infest the soil. Being the saprophytic, the fungus continues to grow and multiply in the soil and remain until the next crop season. The fungus can survive at 4.0–9.0 pH and soil temperature of as high as 35 °C. If the crop is growing continuously on the same field, the disease incidence increases each year.

2.2.4   DISEASE MANAGEMENT

The *Fusarium* wilt disease is soil borne in nature; it is difficult to control wilt of pigeon pea like other soil borne fungal diseases. There are several methods to check the severity of the fungal disease. The crop rotation for long duration of 3–5 years help in reducing the virulence of the pathogen even it can live saprophytically in the soil. Mixed cultivation of sorghum or tobacco in the same field followed by pigeon pea reduces the disease severity. Seed treatment with carbendazim may provide protection in the early stage of plant growth. The most effective aspect of disease control is cultivation of disease resistant varieties of pigeon pea like NP 15, ICPL 96058, ICP 4769, ICP 7118, ICP 7035, ICP 7182, ICP 8863, ICP 9168, ICP 10958 and ICP 11299.

Amendment of soil with roots of certain leguminous crops (sweet clover), molasses and oil cakes (groundnut cake) markedly increased the antibiotic (bulbiformin) production by *B. subtilis* and a reduction of 88% in the incidence of wilt. Seedlings gained resistance to Fusarium infection when the seeds were inoculated with *B. subtilis* before sowing. *Trichoderma harzianum* effectively suppressed fusarium wilt of pigeonpea. At ICRISAT, three genotypes, ICPL 96047, ICPL 96061 and ICPL 96046 were found resistant to fusarium wilt, powdery mildew and phyllody.

2.3   STERILITY MOSAIC OF PIGEONPEA

Sterility mosaic was first reported from Pusa (Bihar) by Mitra (1931). It is also called *Green plague of pigeon pea* as it is one of the most damaging disease in most of the arhar growing states in the country. The disease is restricted to Asia and has been reported from Bangladesh, Nepal, Thailand, Myanmar and Sri Lanka. In India, this is a serious problem in Uttar Pradesh, Bihar, Gujarat, Karnataka, Tamil Nadu and Andhra Pradesh. In India, alone annual loss of 205,000 tons of grains (worth Rs. 75 crores) is reported by several workers.

2.3.1   SYMPTOMS

The affected crop plants to not die but remain green even after the crop maturity. The disease affected arhar crop looks from distance as a green forage crop standing in the field. Infected plants appear bushy with yellowish green foliage of reduced size and suppression of flowers and pods. Severe stunting, reduction in leaf size, increased number of secondary and tertiary branches arising from the leaf axils and complete or partial cessation of development of the reproductive structures are also observed. Mainly three types of symptoms are reported:

- Severe mosaic of leaflets: Plants do not produce flowers and pods.
- Ring spot, where there is no sterility; this is characterized by green islands surrounded by a chlorotic halo on leaflets. The rings disappear as the plants mature.
- Mild mosaic with partial sterility.

2.3.2 CAUSAL ORGANISM

Most of the workers assume that sterility mosaic is caused by Pigeon pea Sterility Mosaic Virus (PSMV) simply on the basis of symptomatoloty and the transmission of the pathogen by Eriophyid mites (*Aceria cajani*) and by grafting. There is no information on the morphology or properties in vitro of the pathogen.

2.3.3 BIOLOGY OF MITE VECTORS

Eriophyid mite (*Aceria cajani*) is a worm-like, microscopic animal, which is about 200–250 μm in length. The mite has two pairs of legs and does not possess wings and eyes. Their dispersal in field is mainly by wind currents. The mites have short life cycle of less than two weeks. Mites feed with puncturing and sucking types of mouthparts that consists of slender stylets.

2.3.4 DISEASE CYCLE

Perennial and ratoon pigeon pea plants infected with the disease appear to be the only source of Pigeon pea Sterility Mosaic Virus (PSMV) and its vectors. *A. cajani* is the only vector of PSMV. Besides pigeon pea, these mites have been observed onto common weeds such as *Oxalis circulate*

t

and *Cannabis sativa*. However the role of these two weeds in the survival of the vectors or virus is obscure.

2.3.5   MANAGEMENT

- Seed treatment with 10% aldicarb protected the crop till maturity.
- Uses of insecticides (Phorate, carbofuran and metasystox) as foliar spray were effective for limited period but failed to provide protection as later stage of crop growth.
- Resistant varieties: Host resistance is the best solution of the problem. Some pigeon pea varieties show the resistance to the sterility mosaic like Bahar, DA-11, DA-33, ICPL-306, Hy-3C, Pusa-9, Pusa-885, Asha, Sharad (DA-11), Narendra-Arhar-1, etc.

2.4   PHYTOPHTHORA BLIGHT

This disease is also known as stem blight, stem canker and stem rot. It is recently recognized disease of pigeon pea, phytophthora blight was first suspected at IARI, New Delhi in India in 1966 by Williams, Grewal and Amin. A survey was conducted between 1975 and 1980 indicated Phytophthora blight to be widespread with an average incidence of 2.6%. Its incidence was very high (26.30%) in West Bengal. the disease affects the crop at any stage of its growth when environmental conditions are suitable for the pathogen and disease development.

2.4.1   SYMPTOMS

Symptoms depending upon the age and attacked plant part, different kinds of symptoms are produced by the disease. Pigeon pea seedlings become infected with Phytophthora blight as soon as they emerge. Young seedlings are killed within 3 days, and lesion not clearly noticed. The seedlings show crown rot symptoms, topple over, and dry. When the seedlings are about 1 month old, symptoms first appear as water-soaked lesions on the primary and trifoliolate leaves which become purple to dark brown necrotic within 5 days. The leaflet lesions are circular to irregular in shape and can be large as 1 cm in diameter. The whole foliage can become blighted within a week. Stem symptoms usually appear later on the main stem, branches and petioles as brown to dark brown lesions, distinctly different from the healthy green portions. During favorable weather, upper portion of the mature stems and branches may be attached.

2.4.2   CAUSAL ORGANISM

The disease is caused by *Phytophthora drechleri* f.sp. *cajani*. The fungus produces aerial white mycelium on culture medium. The hyphae are hyaline, cottony, coenocytic, branched, smooth, selender with measuring 3–6 µm in diameter. Irregular swellings with tubular projections are present on the hyphae. Sporangiospores are hypha-like, swollen at tips. The sporangia are ovate to pyriform, rarely spherical and with a minute papilla on some substrates. The sporangia measures 41–78 × 28–45 µm. Zoospores mature within the sporangia and are release individually after dissolution of the apical portion of the sporangium. Zoospores are biflagellate, hyaline, ovoid to reniform, tapering slightly at the anterior end. Oogonia are hyaline and purple to brown in color. Antheridia are simple, hyaline, amphigynous and measure 12.5–19 × 10–17 µm. Oospores are spherical to globose and chlamydospores are also formed.

2.4.3   DISEASE CYCLE

The survival of the pathogen appears in the soil and plant debris as oospores. They may germinate to form sporangia or mycelium under favorable conditions. The disease is most severe in rainy season (July to September) on both seedlings and two months old plants. In the rainy season oospores germinate by sporangia and direct by germ tube and infection occurs in young seedlings.

Secondary inoculum comes from the primary infection as a large number of sporangia are produced on the mycelium. Winds, movement of water and raindrop splashes are cause of the spread of secondary inoculum. The optimum temperature for growth and development of sporangia and zoospore germination is around 25°–30°C. In the absence of potassium (K), high doses of nitrogen (N) increased disease incidence. Addition of K decreased disease incidence regardless of the presence of N or phosphorous (P) in the soil. ICRISAT Center indicated that disease development was faster when day and night temperatures were more or less the same, for example, ranging between 20° and 25°C, the weather was cloudy, and relative humidity was between 70 and 80%.

2.4.4 DISEASE MANAGEMENT

Cultural practices is most effective method in the management of this disease like; early or normal sowing, maintain well drained field to avoid water stagnation, use resistant varieties, select the field with no previous record of blight, provide better drainage, practice ridge planting with wide inter row spacing, summer solarization and summer plowing should be done.

Several germplasm of pigeon pea have been resistant against Phytophthora blight like; Pusa A-3, Pant A-83-14, METH 12, COMP-1, ESR-6, AS-3, ICPL 161 and 366. *Trichoderma viridae*, *T. harzianum*, *Bacillius subtilis* and *Pseudomonas fluorescence* are antagonistic on pathogen. Redomil (metalaxyl) @ 3g/kg of seed as seed treatment is most effective fungicide for the control of this disease. Two foliar sprays of metalaxyl at 15 days interval starting from 15 days after germination.

2.5 ALTERNARIA BLIGHT

This leaf spot disease is reported only from India where *Alternaria alternata* has also been reported to cause a similar leaf spot. It suffers greatly from alternaria blight may cause 40–50% reduction in yield, in most pigeon pea growing states of India (Kushwaha et al., 2010).

2.5.1 SYMPTOMS

These cause blighting of leaves and severe defoliation and drying of infected branches. Brown spots on the leaves with concentric rings.

The lesions appear on all aerial plant parts including pods. Defoliation of leaves and death of tender branches. Initially small necrotic spots appear on the leaves, and these gradually increase in size to characteristic lesions with dark and light brown concentric rings with a wary outline and purple margin. As infection progresses, the lesions enlarge and coalesce. The disease is mostly confined to older leaves in adult plants, but may infect new leaves of young plants, particularly in the post-rainy-season crop.

2.5.2 CAUSAL ORGANISM

The genus *Alternaria* was established in 1817 with *A. alternata* (originally *A. tenuis*) as the type isolate. Because of the absence of an identified sexual stage for the majority of *Alternaria* species, this genus was classified into the division of mitosporic fungi or the phylum Fungi Imperfecti. *Alternaria* is the production of large, multicellular, dark-colored (melanized) conidia with longitudinal as well as transverse septa. These conidia are broadest near the base and gradually taper to an elongated beak, providing a club-like appearance. They are produced in single or branched chains on short, erect conidiophores. *Alternaria* forms conidia that arise as protrusions of the protoplast through pores in the conidiophore cell wall. At the onset of conidial development, the apex of the conidiophore thickens and a ring-shaped electron-transparent structure is deposited at the apical dome.

2.5.3 DISEASE CYCLE

Alternaria species are mainly saprophytic fungi. However, some species have acquired pathogenic capacities collectively causing disease over a broad host range. The fungus sporulates well under warm, humid conditions. Late sown crop or post-rainy season favors disease development. *Alternaria* has no known sexual stage or overwintering spores, but the fungus can survive as mycelium or spores on decaying plant debris for a considerable time, or as a latent infection in seeds.

If seed-borne, the fungus can attack the seedling once the seed has germinated. In other cases, once the spores are produced they are mainly spread by wind on to plant surfaces where infection can occur. Typically, weakened tissues, either due to stresses, senescence or wounding, are more susceptible to *Alternaria* infection than healthy tissues. The observation that saprobic *Alternaria* species can become parasitic when they meet a weakened host illustrates that the distinction between saprophytic and parasitic behavior is not always evident.

2.5.4   DISEASE MANAGEMENT

Use resistant varieties such as ICPL 366 and DA 2 are effective. Avoid fields close to perennial pigeon pea. Select the seed from healthy plant and sow early are the best cultural methods for management of the disease.

Sed treatment with Thiram 2.5 g/kg seed. Spray the crop with Mancozeb 75% WP @ 2 g/L of water. If the infection persists then repeat the spray after 15 days interval. Three or four spray of carbendazim 7–7.5 g/L or Mancozeb (Dithane M 45) 75% WP Zineb @ 2 g/L just at the appearance of the disease.

KEYWORDS

- blight
- fungicide
- mite
- pigeon Pea
- virus
- wilt

REFERENCES

1. Agrios, G. N. (2005). *Plant Pathology*. 5th ed. Elsevier Academic Press, London.
2. Bart, P. H., Thomma, J. (2003). *Alternaria* spp.: from general saprophyte to specific parasite Molecular Plant Pathology. Laboratory of Phytopathology, Wageningen University, Binnenhaven 5, Netherlands.

3. Common Names of Diseases, The American Phytopathological Society Retrieved from "http://en.wikipedia.org/w/index.php?title=List_of_pigeonpea_diseases&oldid =581610763".

4. DAC. (2011). *Fourth advance estimates of production of Food grains for 2010–11*. Agricultural statistics division, Directorate of economics and statistics, Department of agriculture and cooperation, Government of India, New Delhi [http://eands.dacnet. nic.in/advance_estimate/3rdadvance_estimates_2010–11 (English)].

5. Ingole, M. N., Ghawade, R. S., Raut, B. T., Shinde, V. B. (2005). "Management of Pigeonpea wilt caused by *Fusarium udum* Butler," *Crop Protection and Productivity*, Vol 1, No. 2, pp. 67–69.

6. Kannaiyan, J., Nene, Y. L., Reddy, M. V., Ryan, J. G., Raju, T. N. (1984). Prevalence of pigeonpea diseases and associated crop losses in Asia, Africa and the Americas. *Trop. Pest Management* 30:62.

7. Kushwaha, A., Srivastava, A., Nigam, R., Srivastava, N. (2010). Management of alternaria blight of pigeonpea crop through chemicals. *Inter. J. Pl. Prot.* 3(2), 313–315.

8. Mandhare, V. K., Suryawanshi, A. V. (2005). "Application of *Trichoderma* species against pigeonpea wilt," *JNKVV Research Journal*, Vol 32, No. 2, pp. 99–100.

9. Rai, B., Upadhyay, R. S. (1982). *Gibberella indica*, the perfect stage of *Fusarium udum*. *Mycologia*, 74, 343.

9. Sharma, P. D. (2006). *Plant Pathology*. Narosa Publishing House Pvt. Ltd., New Delhi.

10. Singh, R. P. (2008). *Plant Pathology*. 8th ed. Kalyani Publisher, New Delhi.

11. Singh, R. S. (2005). *Plant Diseases*. 8th ed. Oxford & IBH Publishing Co., New Delhi.

CHAPTER 3

DISEASES OF CHICKPEA CROP AND THEIR MANAGEMENT

ANIL KUMAR,[1] SANTOSH KUMAR,[2] SANJAY KUMAR,[1] and BIRENDRA KUMAR[1]

[1]*Department of Plant Breeding and Genetics, Bihar Agricultural University, Sabour, 813210, Bihar, India*

[2]*Department of Plant Pathology, Bihar Agricultural University, Sabour, 813210, Bihar, India*

CONTENTS

3.1   INTRODUCTION

Chickpea (*Cicer arietinum* L.) is the second most important crop after beans, commonly known as gram or Bengal gram is a versatile crop among the grain legumes with a total production of 11.6 (Mt) from 13.2m ha. However, it ranks fifth in the productivity after Fababean, pea, lupin, and lentil and ranks first among the pulses both in acreage and production. India accounts for 70.7% of the world Chickpea production followed by Australia (4.4%), Pakistan (4.3%), Turkey (4.2%), Myanmar (4.0%), Ethiopia (2.8%), Iran (2.5%), USA (0.84%), Canada (0.78%), and Maxico (0.62%) (FAO, 2011). Chickpea is used as an important source of protein in human nutrition and cattle feed. Chickpea is considered to have medicinal effects and it is used for blood purification. Chickpea contains 21% protein, 61.5% carbohydrate, 4.5% fat. Its seed is also rich in protein, starch, fiber, calcium, iron and niacin, malic and oxalic acid, which makes it one of the best nutritionally balanced pulses for human consumption (Jukanti et al., 2012). However, like any other pulses, the chickpea seed also contains antinutrional factors, which can be reduced or eliminated by cooking. Chickea enrich the soil fertility by fixing atmospheric nitrogen 141 kg/ha (Rupela, 1987) in the root nodules and improves the soil structure (Asthana and Chaturvedi, 1999). It is also to improve soil fertility by biological nitrogen fixation which helps reduce the input cost for the existing crop and. Chickpea usually receives few inputs other than labor, insecticides and seed. Chickpea is a crop of both tropical and temperate regions. The two distinct forms of cultivated chickpeas are *Desi* (small seeds, angular ram's head shape, and colored with high percentage of fiber) and *Kabuli* (large-seeds, irregular rounded, owl's-head shape, and beige colored seeds with low percentage of fiber) types. *Kabuli* type is grown in temperate regions while the *Desi* type chickpea grown in the semi-arid tropics. Low yield of chickpea attributed to its susceptibility to several fungal, bacterial and viral diseases. In general, estimates of yield losses by individual insects and diseases range from 5 to 10% in temperate regions and 50 to 100% in tropical regions (Van Emden et al., 1988).

3.2 FUSARIUM WILT

Chickpea production is severely affected by Fusarium wilt caused by *Fusarium oxysporum* (Schlechtend.:Fr.) f. sp. *ciceris* (Padwick) Matuo & K. Sato, in most chickpea growing areas of the world which cause annual chickpea yield losses vary from 10 to 15% (Jalali and Chand, 1992). In India, wilt alone causes on an average 10% loss annually and is prevalent in all chickpea growing states (Singh and Dahiya, 1973). In Bihar, it ranges from 2 to 20%. However, it was observed that early wilting causes 77–94% losses while late wilting causes 24–65% loss (Haware and Nene, 1980). It causes complete loss in grain yield if the disease occurs in the vegetative and reproductive stages of the crop (Haware and Nene, 1980; Haware et al., 1990; Halila and Strange, 1996; Navas et al., 2000).

3.2.1 *SYMPTOMATOLOGY*

The disease manifests as mortality of young seedlings (within 25–30 days after sowing) to wilting or death of adult plants. Seedlings that die due to wilt disease can be confused with other diseases of wilt complex, if not examined carefully. Fusarium wilt infected seedlings collapse and lie flat on the ground retaining their dull green color. Adult plants show typical wilt symptoms of drooping of petioles, rachis and leaflets. All the leaves turn yellow and then light brown. The roots of the wilting plants do not show any external rotting but when split open vertically, dark brown discoloration of internal xylem is seen. Vascular discoloration is observed on longitudinal splitting of stem. Seeds harvested from wilted plants were lighter and duller than those from healthy plants (Haware and Nene, 1980).

3.2.2   CAUSAL ORGANISM

The fungus produces white to light orange aerial mycelium and sporo-dochia on incubated seed. The mycelium is profusely branched, covers the entire seed and is white to light pink in color. Sporodochia are rarely produced, but if present, the aerial mycelium completely covers it. Microconidia are abundant, and are produced on short, unbranched mono-phialides (microconidiophores) in small, dry, false heads. Microconidia are hyaline, single celled, over to cylindrical straight to slightly curved, and measuring 2.5 to 3.5 × 5 to 11 μ in size. Macroconidia are fewer then microconidia and produced on branched macroconidiophores. They are fusoid with pointed ends, hyaline, have 3 to 5 septa and measure 3.5 to 4.5 × 25 to 65 μ. Chlamydospores are usually intercalary and are produced singly in pairs or in chains. They are globose to sub-globose, thick walled and smooth surfaced. Chlamydospores like swellings are often seen also on the hyphae.

3.2.3   DISEASE CYCLE AND EPIDEMIOLOGY

The Fusarium wilt is soil and internally seed borne disease, facultative saprophyte and can survive in soil more than 6 years in the absence of susceptible host (Haware et al., 1986). Haware et al. (1982) showed the fungus to be in the helium of the seed in the form of chlamydospore like structures. The primary infection is through chlamydospores or mycelia. The conidia of the fungus are short lived; however, the chlamydospores can remain viable upto next crop season. Plant species other than chickpea may serve as symptomless carriers of the disease. Gupta (1991) reported

Vigna radiata, V. mungo, Cajanus cajan, Pisum sativum and *Lens culinaris* as symptomless carriers of the disease. The pathogen may also parasitize several weeds such as *Cyperus rotendus, Tribulus terrestris, Convolvulus arvensis* and *Cardiospermum halicacabum* (Nene et al., 1980). The soil type, reaction, moisture and temperature are known to influence disease development. Rachana et al. (2002) reported that black soil support highest wilt incidence (75.5%). Wilt incidence in sandy-loam, red and clay soil was found to be 64.4, 59.9 and 46.6%, respectively. According to Sugha et al. (1994b) soil temperature in the range of 24.8–28.5°C and soil moisture above 25% within the water holding capacity of soil were most conducive for chickpea wilt. Below 17°C, infection remains restricted in the root without any wilt symptoms. The importance of the soil temperature has also been substantiated by the observation that late sowing of the crop reduces the incidence of the disease.

3.2.4 DISEASE MANAGEMENT

Deep plowing during summer and removal of host debris from the field reduces inoculum levels. The soil inoculum can be reduced by addition of 15–20 tons of farmyard manure with *Trichoderma* sp. @ 4–5 kg/ha before sowing (Singh and Dubey, 2007). Seed dressing with benlate T (Benomyl + thiram) eradicated seed borne inoculum (Haware et al., 1978). The disease can be managed by seed treatment with various seed dressing fungicides. Two years field data clearly indicates that seed treatment with Bavistin + thiram (1:1) @ 2.5 g/kg seed before sowing decreased seedling mortality 7% and increased seed germination 11.2% and grain yield 2.8 Q/ha (Pal and Singh, 1993). The filamentous fungi, *Trichoderma* have attracted the attention because of their multiprong action against various plant pathogens (Harman et al., 2004). The species of *Trichoderma* have been evaluated against the wilt pathogen and have exhibited greater potential in managing chickpea wilt under glasshouse and field conditions (Kaur and Mukhopadhayay, 1992). Earlier, Padwick (1941) also observed that a species of *Trichoderma* was highly antagonistic to wilt pathogen of chickpea. Mane (1995) isolated and evaluated 88 different fungal and bacterial agents against *F. oxysporum* f. sp. *ciceris* and four isolates of *Trichoderma* spp., three isolates of fluorescent Pseudomonas, one isolate of *Acrophilophora* sp. and three isolates of *Gliocladium* spp. showed antagonistic activity in vitro. Under field conditions, maximum wilt reduction (28.3%) in cultivar Pusa 256 was observed when *T. viride* was applied as seed coating with talc as carrier with gum.

(Nikam et al., 2007) tested four oilseed cakes and observed that groundnut cake followed by neem seed and castor cake were found to be most effective in checking percent wilt incidence by 61.91, 52.39 and 47.62% respectively as against control. Haware et al., 1978 reported seed-borne inoculum can be eradicated by seed dressing with Benlate (benomyl 30% + thiram 30%) at 0.25% rate. Nikam et al., 2007 revealed that foliar sprays of thiram followed by carbendazim and captan proved to be effective in checking the wilt incidence by 42.46, 38.10 and 33.34%, respectively as against control (100% wilting). Kolte et al. (1998) effectively controlled chickpea wilt with seed treatment by *Rhizobium, T. viride, T. harzianum* and *Azotobactor sp.*

The identification and use of host plant resistance has the great potential in the long-term management of wilt. The genotypes H99–9, Pusa 212, JG 315, JG 322, PCS 1(Sel.ICCV-11), PCS 2 (Sel.KPG 142–1), PCS 5 (Sel.BGD-112), and PCS 6 (Sel. Pusa- 1073) showed resistant (<10% wilt incidence) reaction during 3 years (2001–2004) were deposited to Gene Bank, NBPGR, New Delhi with ACC No. 405202–405209 (Dubey and Singh, 2004). In addition to these a large number of wilt resistant cultivars namely Avrodhi, Haryana Channa 1, BGD 72, BGM 547, GNG 469 (Samrat), GNG 663 (Vardan), RSG 693 (Aadhar), KPG-59 (Uday), K 3256 (Pragati), Phule G-87207 (Vishal), Phule G 9425–9, JG 322, GPF 2, PBG 1, Pusa 372 and Pusa 1053 (Chamtkar) were identified. The resistance to wilt for race 2 in WR 315 is controlled by a single recessive gene (Sharma et al., 2005). Tullu et al. (1998) reported single recessive gene for resistance to race 4 and identified a RAPD marker linked with resistance.

3.3 BOTRYTIS GRAY MOLD (BGM)

Botrytis gray mold (BGM) caused by *B. cinerea* Pers. Ex. Fr., is the most potentially important disease of chickpea. The occurrence of Botrytis gray mold on chickpea was first reported by Shaw and Ajrekar in 1915. Since 1967–68 *Botrytis cinerea* has caused vast devastation in chickpea crop grown in parts of West Bengal, Bihar, Uttar Pradesh, Rajasthan, Haryana, Punjab and Himachal Pradesh (Singh, 1997). The disease was responsible for heavy losses in the Indo-Gangetic plains of India during 1979–1982 (Grewal and Laha, 1982) and caused 70–100% losses in yield at Central State farm Hissar and several parts of Punjab.

3.3.1 CAUSAL ORGANISM

BGM of chickpea is caused by *Botrytis cinerea* Pers. Ex. Fr. The asexual stage of the necrotrophic fungus *B. cinerea* (Moniliaceae, Hyphales) is dominant on chickpea crops. *B. cinerea* grown on potato dextrose agar (PDA) has a white, cottony appearance, which turns light gray with age. The mycelium is septate, brown and 8–10 µm wide. Young hyphae are thin and hyaline. Conidia and conidiophores are not in pycnidia or acervuli. Conidiophores lighter brown than hyphae, with hyaline tip, septate, 8–24 µm wide. Tips of conidiophores or their branches are slightly enlarged and bear small pointed sterigmata. Conidia are hyaline, one-celled oval or globose or short cylindrical and borne in clusters at the tips of conidiophores branches.

3.3.2 CHARACTERISTIC SYMPTOMS

All the aerial parts of chickpea are susceptible to the disease with growing tips and flowers being the most vulnerable (Haware, 1998; Bakr et al., 1997).

Initial symptoms appear on stem, leaves, inflorescence and pods as gray or dark brown lesions covered with erect hairy sporophores. Stem lesions are 10–30 mm long, which later girdles the stem completely. Tender branches break off at the points where gray mold causes rotting. Affected leaves and flowers turn into a rotting mass. In the field, the disease first appears in isolated patches when the crop has achieved maximum canopy and the morning relative humidity is very high with low temperature. As the disease advances, patches of disease plant become more prominent, spreading slowly in the entire field. According to Laha and Grewal (1983) symptoms appeared on leaflets, petioles and growing tips as water soaked lesions. The lesions are brown and limited in size. However, under conditions of high humidity leaflets got blighted and bear abundant fungal fructifications.

On thick, hard stems, the gray mold growth is gradually transformed into a dirty, gray mass containing dark green to black sporodochia. The sclerotia are small, dark bodies and should not be confused with larger, black or dark brown sclerotia embedded in white mycelium of *Sclerotinia sclerotiorum* (Lib.) de Bary (Joshi and Singh, 1969).

3.3.3 DISEASE CYCLE AND EPIDEMIOLOGY

Reports on epiphytotics of botrytis gray mold from different parts of the world indicated the existence of definite and efficient mechanisms of survival of the pathogen from one season to another. The information regarding the survival and epidemiology is scanty so far as the botrytis gray mold of chickpea is concerned. BGM can devastate chickpea, resulting in complete yield loss in years of extensive winter rains and high humidity (Reddy et al., 1993 and Pande et al., 1982).

3.3.4 DISEASE MANAGEMENT

Haware and McDonald (1992) reported that delayed sowings reduced BGM incidence even in susceptible cultivars, but significantly reduced the grain yields. Singh (1997) also observed that the late sown crop (around 20 Nov.) in Punjab, India, showed significantly low incidence of BGM. Combination of wider row spacing, intercropping with linseed and two spray application of carbendazim @ 0.2% significantly reduced BGM severity and increased grain yield of chickpea and linseed. Bakr et al. (1993) reported that seed treatment with bavistin + thiram (1:1), indofil M-45, thiabendazole, ronilan, rovral, bavistin @ 0.3% controls seed borne inoculum of *B. cinerea*. Foliar spray with ronilan, bavistin + thiram combination @ 0.1% or bavistin alone @ 0.2% provided complete protection to chickpea plants against aerial infection by *B. cinerea* (Grewal and Laha, 1982). Haware et al. (1997) reported that one spray with vinclozolin (0.2%)

at the time of flowering in the integrated management system reduced BGM incidence. Haware et al. (1999) reported biocontrol potential of *T. viride* isolate T-15 (isolated from chickpea rhizosphere) on *B. cinerea* in chickpea under controlled environmental conditions.

3.4   ASCOCHYTA BLIGHT OF GRAM

Ascochyta blight, caused by *Ascochyta rabiei* (Pass.) is the most important disease in the chickpea (*Cicer arietinum* L.) in many countries. In India the disease occurs in North-western part of Uttar Pradesh, Punjab and Haryana in severe form.

3.4.1   PATHOGEN

Ascochyta rabiei (Pass.) Labrousse is the causal pathogen. The perfect stage of the fungus is *Didymella rabiei* (Kov.) Von Arx. The mycelium of the pathogen is septate. The pycnidia develop on stem, leaves and seedpods are dark brown, globose and measure 140–200 μ. Conidia are formed within pycnidium and remain viable for long period of time. The pycnidia absorb water, swell and release conidia. Several strains of the pathogen are known.

3.4.2   SYMPTOMS

The fungus attacks all above-ground plant parts and infection can occur on leaves, stems and pods at any stage of plant growth, but plants are most susceptible to disease during flowering. Brown, circular spots with brownish red margin appear on leaves and pods of affected plants. On petioles and stem the spots are elongated in shape. The spots on leaves coalesce turning the leaf completely brown. On green pods, the circular lesions have dark margin where black dot like bodies appear known as pycnidia. The pycnidia are arranged in concentric circles. The elongated lesions on stem and petioles also bear black dots and may girdle the stem. The parts above the lesions droop and wilt. If the stem is girdled at the base the whole plant will show wilting. During wet weather the disease spread very fast and may cover the whole field.

3.4.3   DISEASE CYCLE AND EPIDEMIOLOGY

The pathogen overwinters on plant debris left in the field and also on seeds, which serve the source of primary inoculum. Ascospores were also found to play a role in the initiation of the disease epidemics. Further spread of the pathogen is through conidia, which are disseminated by splashing rain, by insects, contact of healthy and diseased leaves and by the movement of man and animals. Other hosts as cowpea and bean also get infection of the pathogen and serve as a source of inoculum. The disease development is favored at 9–24°C with 10 h or more wetness. Wet, windy weather favor rapid disease spread. *A. rabiei* showed variation in morphological, physiological and pathological characters. The disease builds up and spreads fast when night temperatures are around 10°C, day temperatures around 20°C, and rains are accompanied by cloudy days. Excessive canopy development also favors blight development.

3.4.4   MANAGEMENT

Disease free seed is a pre-requisite for effective disease management. Seed treatment with combination of bavistin + thiram (1:2 ratio) @ 2.5 g/kg of seed for eradication of internally and externally seed borne infection of *A. rabiei*. The biocontrol potential of fungal antagonists, *Chaetomium globosum*, *T. viride*, and *Acremonium implicatum* were explored under in vitro and in vivo conditions. *C. globosum* caused 48.6% reduction in colony diameter and 70.9% reduction in pycnidiospores germination under in vitro conditions, whereas, its post-inoculation spray reduced 73.1% disease. Foliar applications of zineb, maneb, captan and daconil also reduced the disease. Desi accessions H00–108, GL 92024 ICC 4475, ICC 6328 and ICC 12004; and Kabuli ILC 3864, ILC 3870 and ILC 4421ILC 200 and ILC 6482 showed resistance to blight. The cultivars Gaurav, GNG 146, GNG 469, PBG 1 and L 551 were found resistant to the disease. Resistance to blight is governed by two complementary dominant genes in GLG 84038 and GL 84099 and one dominant and one recessive independent gene in black-seeded ICC 1468. Resistance in JM 595 and P 1528–1 was different from those in FLIP 91–24C and FL1P 84–92C. Resistant cultivars secreted lesser amount of malic acid and posses more glandular hairs as compared to susceptible ones.

3.5 DRY ROOT ROT

Dry root rot of chickpea has been reported from India, Iran, Australia, Ethiopia, Pakistan, Spain and USA. Wilt and dry root rot diseases alone affecting the productivity of chickpea, 5–20% loss in yields (Singh, 2010). *Rhizoctonia bataticola* is a polyphagous soil borne pathogen infecting over 500 plant species worldwide causing huge losses. Dry root rot disease of chickpea is caused by *Rhizoctonia bataticola* (Taub) Butler=*Macrophomina phaseolina* (Maubl.) Ashby. It has been reported from Australia, Ethopia, California, India, Iran, Lebaonan, Maxico Pakistan, Syria and Turkey (Allen, 1983 and Nene, 1979).

The fungus lacks fruiting bodies and spores. The mycelium is light-brown, thick in which black sclerotia are formed. Sclerotia are variable in form, small and loosely connected by mycelial threads.

3.5.1 SYMPTOMATOLOGY

The first symptom of the disease is yellowing of the leaves within a day or two; such leaves drop and in the course of the next two or three days, and plant showed completely dried symptoms within a week after the appearance of the first symptom. The tissues were weakened and break off easily. The affected roots are dark brown to black and usually dry, unless the soil is wet. The tap root is quite brittle, show shredding of the bark and can be broken easily. If the plant were pulled out from the soil and examined the basal stem and main root system of diseased plant showed extensive rooting with most of the lateral roots destroyed. In advance cases minute dark black sclerotial bodies can be seen on the surface of the root, as well as in the pith. If the plants are pulled from the soil and examined, the basal stem and the main roots may show dry rot symptoms.

3.5.2 DISEASE CYCLE AND EPIDEMIOLOGY

The fungus is seed and soil-borne, facultative parasite and may survive in the soil in the form of sclerotia for long time (Dingra and Sinclair, 1994). Soil borne inoculum is more important in causing infection and disease development. It produces pycnidia when the atmospheric temperature is above 30°C and the pycniospores remain viable for over a year. The fungus is mainly a soil-dweller and spreads from plant to plant through irrigation water, tools and implements and cultural operations. The sclerotia and pycniospores may also become air borne and cause further spread of the pathogen. The disease appears suddenly when ambient temperatures are between 25 and 30°C.

3.5.3 DISEASE MANAGEMENT

Drenching the affected plants and the infested soil with Bordeaux mixture or other effective fungicide may help in reducing the inoculums potential. Field sanitation measures, including cutting down the diseased plants and burning them and deep plowing in summer help to reduce the diseases intensity during the following season. Drought should be avoided. Sowing should always be done on the recommended time. Seed treatment with a mixture of carbendazim 1.5 g and thiram 1.5 g per kg of seed and with *Trichoderma viride* formulation + 3 g thiram per kg seed can reduce the disease incidence. Nagamani et al. (2011) observed that seed treatment with carbendazim @ 2 g/kg of seed+ seed treatment with *T. viride* @ 4 g/kg of seed +soil application of FYM fortified with *T. viride* (T7) was found to control root rot. Hence integrated management of the disease using bio-control agents and chemicals is the best alternative (Ramarethinum et al., 2001).

3.6 CHICKPEA RUST

Uromyces ciceris-arietini has been reported is the causal organism of chickpea (*Cicer arietinum*). Chickpea rust is not known to cause as widespread damage on chickpea as other chickpea diseases, it can occasionally be serious when conditions during the cropping season favor early epidemics. In a particularly favorable year, an epidemic of rust on Bengal gram in Karnataka (32–40%, Singh, 2010), India, caused incidences of up to 90–100%.

3.6.1   PATHOGEN

The fungus produces the uredial and telial stages on chickpea. There is no any alternate host. The uredospores are globose to sub-globose, brownish yellow in color, with minute spines on the walls. They measure 20–28 µ in diameter with 3–4 µ thick cell wall and contain 4–8 germ-pores. The teliospores are also similar to uredospores, but round to ovate, rough and thick walled, with thickened apex. While the uredospore germinates readily, the teliospores are not known to germinate. Its pycnidial and aecial stage are not known.

3.6.2   SYMPTOMATOLOGY

First rust symptoms appear initially on the leaves as small, round or ellipsoidal, cinnamon-brown, powdery pustules. These pustules tend to coalesce. Sometimes a ring of small pustules can be seen around larger pustules, which occur on both leaf surfaces but more frequently on the lower one. Occasionally pustules can be seen on stems and pods especially when infection is severe. Severe infection results in premature defoliation and possible death of the entire plant.

3.6.3   EPIDEMIOLOGY

Cool and moist weather conditions favors rust build up although rain is not essential for its development. The symptoms usually become conspicuous later in the growing season although epiphytotics may occur earlier in the season when conditions are favorable.

3.6.4   MANAGEMENT

Cultural practices such as field sanitation, seed selection, crop rotation and early sown crops help to escape infection. Foliar spray of fungicides such as Mancozeb (0.2% a.i.), Bayleton (0.05% a.i) and Calixin (0.2% a.i.) are found effective against the pathogen.

KEYWORDS

- blight
- chickpea
- fungicide
- rot
- rust
- wilt

REFERENCES

1. Allen, D. J. (1983). The pathology of tropical food legumes. John Wiley and Sons, New York, p413.
2. Asthana, A. N., S. K. Chaturdevi, (1999). A little impetus needed. The Hindu survey of Indian Agriculture, 1999 p61–65.
3. Bakr, M. A., Hossain, M. S., Ahmed, A. U. (1997). Research on Botrytis gray mold of chickpea in Bangladesh. pp. 15–18. In Recent advances in research on Botrytis gray mold of chickpea (Haware, M. P., Lenne, J. M., Gowda, C. L. L., eds.). Patancheru 502324, Andhra Pradesh, India: ICRISAT.
4. Bakr, M. A., Rahman, M. M., Ahmed, F., Kumar, J. (1993). Progress in the management of Botrytis gray mold of chickpea in Bangladesh. Haware, M. P.; Gowda, C. L.L.; McDonald, D. (eds.). International Crops Research Institute for the Semi-Arid Tropics (ICRISAT). Recent advances in research on Botrytis gray mold of chickpea: summary proceedings of the second Working Group Meeting to discuss collaborative research on Botrytis Gray Mold of Chickpea. Patancheru, A. P. (India). ICRISAT. pp.17–18.
5. Chandra, S. et al. (1974). *Indian, J. Genet.* 34, 257–262.
6. Dhingra, O. D., Sinclair, J. B. (1994). Basic plant Pathologt Methods. CRS press, London. p443.
7. Dubey, S. C., S. R. Singh (2008). Virulence analysis and oligonucleotide fingerprinting to detect diversity among Indian isolates of *Fusarium oxysporum* f. sp. *ciceris* causing chickpea wilt. *Mycopathologia*, 165, 389–406.
8. FAO. (2011). FAOSTATS. Food and Agriculture Organization of the United Nations, Rome, Italy. http://faostat.fao.org/site/567/default.aspx#ancor (accessed on 4 Apr. 2013).

9. Grewal, J. S., Laha, S. K. (1982). Chemical control of botrytis gray mold of chickpea. *Indian Phytopath.* 36, 516–520.
10. Gupta, O. (1991). *Legume Res.* 14, 193–194.
11. Halila, M. H., Strange, R. N. (1996). Identification of the causal agent of wilt of chickpea in Tunisia *Fusarium oxysporum* f. sp. *ciceris* race 0. *Phytopath. Medit.* 35, 67–74.
12. Harman, G. E., Howell, C. R., Viterbo, A., Chet, I., Lorito, M. (2004). *Trichoderma* species–opportunistic, avirulent plant symbionts. *Nature Reviews*, 2, 43–56.
13. Haware, M. P. (1998). Diseases of chickpea. *In* 'The pathology of food and pasture legumes' (eds. D. J. Allen, J. M. Lenne) pp. 473–516. (ICARDA, CAB International: Wallingford, UK).
14. Haware, M. P., Jimenez-Diaz, R. M., Amin, K. S., Phillips, J. C., Halila, H. (1990). Integrated management of wilt and root rots of chickpea 129–137. In: Chickpea in the Nineties: Proceedings of the second international work shop on chickpea improvement, Patancheru, India.
15. Haware, M. P., McDonald, D. (1992). Integrated management of botrytis gray mold of chickpea. In 'Botrytis gray mold of chickpea.' (eds. M. P. Haware, D. G. Faris and, C. L.L. Gowda) pp. 3–6. (ICRISAT: Patancheru, AP, India).
16. Haware, M. P., Mukherjee, P. K., Lenne, J. M., Jayanthi, S., Tripathi, H. S., Rathi, Y. P. S. (1999). Integrated biological-chemical control of Botrytis gray mold of chickpea. *India Phytopath.* 52, 174–176.
17. Haware, M. P., Nene, Y. L. (1980). Influence of wilt at different stages on the yield loss in chickpea. *Trop. Grain Legume Bull.* 19, 38–40.
18. Haware, M. P., Nene, Y. L. (1982). *Plant Dis.* 66, 250–251.
19. Haware, M. P., Nene, Y. L., Natarajan, M. (1986). Survival of *Fusariunz oxysporum* f. sp. *ciceri* in soil in the absence of chickpea. National seminar on management of soil-borne diseases of crop plants. TNAU, Coimbatore, 8–10 Jan. (1986). Pl (Abstract).
20. Haware, M. P., Nene, Y. L., Rajeshwari, R. (1978). Eradication of *Fusariutn oxysporum* f. sp. *ciceri* transmitted in chickpea seed. Phytopathology 68, 1364–1367.
21. Haware, M. P., Tripathi, H. S., Rathi, Y. P. S., Lenne, J. M., Jayanthi, S. (1997). Integrated management of botrytis gray mold of chickpea: cultural, chemical, biological, and resistance options. Pages 9–12. *In* Recent advances in research on Botrytis gray mold of chickpea (eds. Haware, M. P., Lenne, J. M., Gowda, C. L. L.). Patancheru 502324, Andhra Pradesh, India: ICRISAT.
22. Jalali, B. L., Chand, H. (1992). Chickpea wilt. Pages 429–444 in: Plant Diseases of International Importance. Vol. 1. Diseases of Cereals and Pulses. U. S. Singh, A. N. Mukhopadhayay, J. Kumar, and, H. S. Chaube, eds. Prentice Hall, Englewood Cliffs, NJ.
23. Joshi, M. M., Singh, R. S. (1969). A Botrytis gray mold of gram. *Indian Phytopathology*, 22, 125–126.
24. Jukanti, A. K., P. M. Gaur, C. L.L. Gowda, and R. N. Chibbar. 2012. Nutritional quality and health benefits of chickpea (*Cicer arietinum,* L.): A review. Br. J. Nutr. 108(Suppl. 1), S11.
25. Kaur, N. P., Mukhopadhayay, A. N. (1992). Integrated control of chickpea wilt complex by *Trichoderma* spp. and chemical methods in India. *Trop. Pest Management.* 38, 372–375.

26. Kolte, S. O., Thakre, K. G., Gupta, M., Lokhande, V. V. (1998). Biocontrol of *Fusarium* wilt of chickpea (*Cicer arietinum*) under wilt sick field condition. Paper submitted, ISOPP at National Symposium on management of soil and soil borne diseases. 9–10th Feb., (1998). p22.

27. Laha, S. K., Grewal, J. S. (1983). Botrytis blight of chickpea and its perpetuation through seed. *Indian Phytopatholgoy*, 36, 630–634.

28. Mahendra, P. (1998). Diseases of pulse crops, their relative importance and management. *J. Mycol. Plant. Pathol.* 28(2), 114–122.

29. Mane, S. S. (1995). Studies on *Fusarium oxysporum f. sp. ciceri* causing chickpea wilt with special reference to its management by bio-agents. PhD thesis, IARI, New Delhi. p72.

30. Nagamani, P., Viswanath, K., Kiran Babu, T. (2011). Management of dry root rot caused by Rhizoctonia bataticola (Taub.) Butler in Chickpea. *Current Biotica*, 5(3), 364–369.

31. Navas-Cortes, J. A., Hau, B., Jimenez-Diaz, R. M. (2000). Yield loss in chickpea in relation to development to Fusarium wilt epidemics. *Phytopathology*, 90, 1269–1278.

32. Nene, Y. L. (1979). Diseases of chickpea. P 171–187 in Proc. International seminar on chickpea improvement, ICRISAT, Hyderabad, India. p298.

33. Nene, Y. L. et al. (1980). In: *Proc. Consultants Group Discussion on the Resistance to Soil Borne Diseases of Legumes*, ICRISAT, Patancheru.

34. Nikam, P. S., Jagtap, G. P., Sontakke, P. L. (2007). Management of chickpea wilt caused by *Fusarium oxysporium* f. sp. *Ciceri*. *African Journal of Agricultural Research*, Vol. 2 (12), p.692–697.

35. Padwick, G. W. (1941). Report of the Imperial Mycologist. Scient. Rept. Agric. Res. Inst., New Delhi, 1937–38; 94–101.

36. Pal, M., Singh, B. (1993). Channe koo Uktha Rog Se Bachayen. *Kheti*, 47(6), 24–25.

37. Pandey, M. P., Beniwal, S. P.S., Arora, P. P. (1982). Field reaction of chickpea varieties to chickpea gray mold. *International Chickpea Newsletter*, 7, 13.

38. Rachana, S. et al. (2002). *Ann Plant Protection Sci.* 10 (1), 156–157.

39. Ramarethinam, S., Morugesan, N., Marimuthu, S. (2001) Compatibility Studies of fungicides with *Trichoderma viride* used in the Commercial formulation- Bio-CURF-F. *Pestology*, (25), 2–6.

40. Reddy, M. V., Ghanekar, A. M., Nene, Y. L., Haware, M. P., Tripathi, H. S., Rathi, Y. P. S. (1993). Effect of vinclozolin spray, plant growth habit and inter-row spacing on botrytis gray mold and yield of chickpea. *Indian Journal of Plant Protection*, 21, 112–113.

41. Reddy, M. V., Haware, M. P., Ghanekar, A. M., Amin, K. S. (1991). Field diagnosis of chickpea diseases and their control. In: *Information Bulletin no. 28.* ed. by International Crops Research Institute for the Semi Arid Tropics, Patancheru, India.

42. Rupela, O. P. (1987). Nodulation and nitrogen fixation in chickpea. In: M. C. Saxena and, K. B. Singh, editors, The chickpea. CAB International, Wallingford, U.K. 191–196. S16. doi: 10.1017/S0007114512000797.

43. Shakir, A. S., Mirza, J. H. (1994). *Pak. J. Phytopathol.* 6, 87–90.

44. Sharma, K. D., Chen, W. D., Muehlbauer, F. J. (2005). Genetics of chickpea resistance to five races of Fusarium wilt and a concise set of race differentials for *Fusarium oxysporum* f. sp. *ciceris*. *Pl. Dis.* 89, 385–390.

45. Singh, B., Dubey, S. C. (2007). Channe Kaa Mallyni Rog – Bachaw Ke Uppaya. *Kheti* 60(8), 18–20.
46. Singh, G. (1997). Epidemiology of Botrytis gray mold of chickpea. pp. 47–50. In: recent advances in research on Botrytis gray mold of chickpea (Haware, M. P., Lenne, J. M., Gowda, C. L. L., Eds.). Patancheru 502324, Andhra Pradesh, India: International Crop Research Institute for Semi Arid Tropics.
47. Singh, K. B., Dahiya, B. S. (1973). Breeding for wilt resistance in Chickpea. Symposium on problem and breeding for wilt resistance in Bengal gram. Sept. 1973 at IARI, New Delhi, 13–14.
48. Singh, N. P. (2010). Annual Group Meet, August 29–31, 2010, Project Coordinator's Report, All India Coordinated Research Projects on Chickpea, IIPR, Kanpur, 20–21.
49. Sugha, S. K. et al. (1994b). *Indian, J. Mycol. Plant Pathol.* 24, 97–102.
50. Tullu, A., Muehlbauer, F. J. Simon, C. J. Mayer, M. S. Kumar, J. Kaiser, W. J., Kraft, J. M. (1998). Inheritance and linkage of a gene for resistance to race 4 of Fusarium wilt and RAPD markers in chickpea. *Euphytica*, 102, 227–232.
51. Upadhyaya H. D. et al. (2013) Mini Core Collection as a Resource to Identify New Sources of Variation, Crop Science, vol. 53, pp. 2506–2517, November–December 2013.
52. Van Emden, H. F., Ball, S. L., Rao, R. (1988). Pest diseases and weed problems in pea lentil and faba bean and chickpea. In: (ed. R. J. Summerfield) World Crops: Cool season Food Legumes. ISBN 90–247–3641–2. Kluwer Academic Publishers. Dordrecht, The Netherlands. 519–534.
53. Vishwa, D., Gurha, S. N. (1998). Integrated Management of Chickipea Diseases. Integrated Pest and disease Management. Rajeev, K., Upadhyay, K. G., Mukerji, B. P., Chamola and Dubey, O. P. (eds.) APH Publishing Co., New Delhi. (India). p.249.

CHAPTER 4

DISEASES OF URD/MUNG BEAN CROPS AND THEIR MANAGEMENT

SANGITA SAHNI,[1] BISHUN D. PRASAD,[2] and SUNITA KUMARI[3]

[1]Department of Plant Pathology, T.C.A., Dholi, Muzaffarpur, Bihar, India

[2]Department of Plant Breeding and Genetics, B.A.C., Sabour, Bihar, India

[3]Krishi Vigyan Kendra, Kishanganj, BAU, Sabour, Bihar, India

CONTENTS

4.1   INTRODUCTION

Mungbean [*Vigna radiate* (L.) Wilczek] and Urdbean [*V. mungo* (L.) Hepper] are the important pulse crops in India after chickpea and pigeon-pea. These are also widely cultivated throughout Southern Asia like Pakistan, Sri Lanka, Bangladesh, Thailand, Laos, Vietnam, Indonesia, China and Taiwan. In India these crops are cultivated in three different seasons, viz., *kharif, rabi* and summer. It is grown as sole relay crop in rice fallows during *rabi* season in Andhra Pradesh, Tamilnadu, Karnataka and Orissa and sole catch crop during spring/summer season in Uttar Pradesh, Bihar, West Bengal, Jharkhand, Punjab, Haryana and Rajasthan. However, maximum area of its cultivation is under *kharif*, where intercropping with sorghum, pearl-millet, maize, cotton, castor, pigeonpea, etc., are popular. Short maturity duration (<60 days) make the crop ideal for catch cropping, intercropping and relay cropping. These crops are grown principally for its high protein seeds that are used as human food, that can be prepared by cooking, fermenting, milling or sprouting, they are utilized in making soups, curries, bread, sweets, noodles, salads, boiled dahl, sprouts, bean cake, confectionery, to fortify wheat flour in making vermicelli and many other culinary products like sabut dhal, dhal, papad, namkeen, halwah, and vari, etc. (Singh et al., 1988). The protein is comparatively rich in lysine, an amino acid that is deficient in cereal grains. They complement each other and hence enhance the food quality. Besides being a rich source of protein, these are also important for sustainable agriculture and enriching soil organic matter through biological nitrogen fixation. India is the largest producer of mungbean and account 54% of the world production and covers 65% of the world acreage. Mungbean is grown on about 3.43 million hectares with annual production of 1.71 million tons. Similarly, Urdbean is grown on about 3.30 million hectares with annual production of 1.83 million tons (AICRP, 2012–2013). The average yield fluctuates between 300 to 500 kg/ha for a decade in India. The yield losses (5–100%) reported due to various biotic stresses, which is responsible for the fluctuation in the average yield. The biotic stresses like diseases incited by fungi, bacteria, viruses, and nematodes are major limiting factors for high yield. Therefore, there is a need to correct identification, diagnosis and adaptation of suitable management strategies against different diseases of these crops. Since mungbean and urdbean are infested by similar bacterial, fungal pathogens and viruses, they have been dealt together. A brief account of the most important diseases of these crops in India, including the causes, symptoms, management of these destructive diseases, are discussed here. These diseases are responsible for reducing overall production as well as quality of the crop produce.

4.2 MUNGBEAN YELLOW MOSAIC

4.2.1 CAUSAL ORGANISM

Yellow mosaic disease is caused by Mungbean yellow mosaic virus (MYMV), a member of Gemini virus group transmitted through whitefly (*Bemisia tabaci* Gen.), is a most destructive disease of mungbean and urdbean in India as well as in other countries in Asia which are growing these crops. MYMV incidence was first reported from the fields of IARI, New Delhi by Nariani (1960). Nene (1968) named it mungbean yellow mosaic virus. The paired particles of the causal virus measure 30 x 15 nm having ssDNA (Honda et al., 1981). The MYMV is not transmitted through sap (Nariani, 1960), seed or soil (Nair, 1971). The whitefly is a very efficient vector as it can acquire and inoculate the virus in certain hosts within 10 to 15 minutes. For 100% transmission, 10 viruliferous whiteflies per plant are required (Nair, 1971; Nene, 1973). This viral disease is found on several alternate and collateral host which act as primary sources of inoculums. Rathi and Nene (1974) found the host range of MYMV to be restricted to species belonging to the families *Leguminosae*, *Compositae* and *Gramineae*. In India, this virus cause more severe yellow mosaic disease in urdbean than mungbean (Williams et al., 1968) (Figure 4.1).

4.2.2 SYMPTOMS

The first visible sign of the disease is the appearance of yellow spots scattered on young leaves, which increase with time leading to complete yellowing. The next trifoliate leaf emerging from the growing apex

FIGURE 4.1 Symptomatology of MYMV disease in mungbean. (a) Field view showing severity of disease. (b) Infected plant, (c) Infected pods.

showed irregular alternating yellow and green patches, which also turn yellow. These color changes in affected plants are so conspicuous that the disease can be detected in the field from a distance. The leaves showed slight puckering with reduction in size. The infected plants usually mature late and bear very few flowers and pods. The pods are deformed and contain shriveled, undersized seeds. Reduction in number of pods/plant seeds/pod and seed weight are the main contributing factors for yield reduction (Nene, 1973; Dhingra and Chenulu, 1985). The infection not only drastically reduces yield but also severely impairs the grain size and quality (Singh and Shrivastva, 1985). Yield losses due to this disease vary from 5 to 100% depending upon disease severity, susceptibility of cultivars and population of whitefly (Nene, 1972; Singh, 1980; Rathi, 2002).

4.2.3 MANAGEMENT

1. Cultivation of resistant varieties:
    Mungbean: Narendra Mung1, Pant Mung 3, PDM 139 (Samrat), PDM-11 (Spring Season), ML 131, ML 267, ML 337, Pusa 105 and MUM 2.
    Urdbean: Narendra Urd1, IPU 94–1 (Uttara), PS 1, Pant U 19, Pant U 30, UG 218,
    WBU 108, KU 92–1 (Spring season) and KU 300 (Spring Season).
2. Inter/mixed cropping of mungbean and urdbean with non-host crops like sorghum, pearl millet and maize.
3. Diseased plants should be rogued out to prevent further spread of the disease.
4. Foliar application of metasystox or triazophos 40 EC @ 2.0 ml/L or malathion 50 EC @ 2.0 ml/L or oxydemeton methyl 25 EC @ 2.0 ml/L at 10–15 days intervals if required for effective management of the disease by reducing vector control.

4.3 LEAF CRINKLE

4.3.1 CAUSAL ORGANISM

Leaf crinkle disease caused by urdbean leaf crinkle virus (ULCV) belonging to Tospovirus. It is the second important viral disease with incidence of 5 to 28%, but is more serious in mashbean than mungbean (Kadian, 1980; Rishi, 1990). The disease was first reported from India (Williams et al., 1968). ULCV is transmitted through sap inoculation, grafting, seeds and insects (Nene, 1972; Kadian, 1980). According to Ahmad et al. (1997) ULCV is transmitted through seed at the rate of 2.7 to 46%. Leaf feeding beetle (*Henosepilachna dodecastigma* (Wied), whitefly

(*Bemisia tabaci* Glov.) and two aphid species (*Aphis craccivora* and *A. gossypii*) have been reported to be putative vectors of ULCV (Beniwal and Bharathan1980, Narayansamy and Jaganthan1973, Dhingra, 1975). For effective transmission a very short acquisition-feeding period of 30 seconds to 2 minutes preceded by a pre- acquisition fasting was found necessary (Figure 4.2).

4.3.2 SYMPTOMS

The disease affects both the vegetative growth and yield components of these plants (Beniwal and Chaubey, 1979; Kadian, 1982; Kolte and Nene, 1973 and Ilyas et al., 1992). The disease is characterized by the appearance of extreme crinkling, curling, puckering and rugosity of leaves, stunting of plants and malformation of floral organs (Kolte and Nene, 1973). The crinkling is observed on some branches while others

FIGURE 4.2   Typical Leaf crinkle symptom caused by ULCV in mungbean.

remain apparently healthy (Brar and Ratual, 1986). Pollen production, fertility and subsequent pod formation is severely reduced with affect on seed weight and size of seeds in infected plants leading to decrease in yield (Nene, 1972). The virus has been reported to decrease grain yield from 35 to 81% depending upon genotype and time of infection (Bashir et al., 1991).

4.3.3 MANAGEMENT

1. Seeds from diseased crops should not be used.
2. Treat the seeds with imidacloprid 70 WS@ 5 mL/kg or give solar Seed treatment by soaking seed in water for 3–4 hours and then exposure to solar heat from 12 to 4 p.m. in May and June.
3. Cultivation of resistant varieties:
   Mungbean: D-3–9, K 12, ML 26, RI 59 and T44 RII
   Urdbean: HUP 27, 102, 164 and HUP 315.
4. Rogue out the infected plants to avoid contact between healthy and diseased plants during intercultural operations.
5. Application of one foliar spray of insecticide (dimethoate 30 EC @ 1.7 mL/ha) on 30 days after sowing.

4.4 MOSAIC MOTTLE

4.4.1 CAUSAL ORGANISM

This disease is caused by Bean Common Mosaic Virus, which belongs to potyvirus group. The mosaic mottle of urdbean and mungbean is common in India as well as Southeast Asian countries (Tsuchizaki et al., 1986). It can be transmitted by sap, mechanically and by seed (Shahare and Raychaudhary, 1963; Nene, 1972). Singh and Nene (1978) also reported its transmission by aphids, *Aphis craccivora* and *A. gossypii*. Host range of the virus is confined to the family *Leguminosae* (Srivastava et al., 1969). However, Urdbean is more susceptible than mungbean (Figure 4.3).

FIGURE 4.3 Symptoms of Bean Common Mosaic potyvirus in mungbean. (a) Infected plants showing green mosaic areas and downward cupping along the main vein of each leaflet. (b) Advance stage of Bean Common Mosaic potyvirus infection. Infected leaves showing green vein banding, blistering and malformation. Picture were taken from http://vegetablemdonline.ppath.cornell.edu/PhotoPages/Bean/Viruses/BeanVirus1.htm.

4.4.2 SYMPTOMS

The disease is characterized by a mosaic pattern of irregular broad patches of light and dark green areas and blistering and puckering of leaf blade. The size of the leaf gets reduced and margins show upward rolling. The leaves become rough and brittle. Affected plants show reduction in overall growth and often display excessive branching. In cases of severe infection the whole inflorescence is changed into leaf like structures, thereby causing 100% loss in seed yield (Nene, 1972).

4.4.3 MANAGEMENT

1. Use diseased free seeds.
2. Rogue out the infected plants to avoid contact between healthy and diseased plants during intercultural operations.
3. Foliar spray of insecticide (dimethoate 30 EC @ 1.7 mL/ha) on 30 days after sowing.

4.5 LEAF CURL

4.5.1 CAUSAL ORGANISM

This disease is caused by Tomato Spotted Wilt Virus, is an important potential killer of mungbean and urdbean plants (Nene, 1972). The virus is transmitted by sap, grafting and the thrip, *Frankliniella schultzei* (Amin et al., 1985).

4.5.2 SYMPTOMS

Nene described the symptoms of this disease for the first time in 1968. Chlorosis will develop around some lateral veins and their branches near the margin of the youngest trifoliate leaf. The leaves show downward curling of margin, sometimes rolling and twisting of young leaves can also be observed. If plants infected early after sowing, they remain stunted and majority of these die due to top necrosis within two weeks, however, plants infected in late stages of growth do not show severe curling and twisting of the leaves but show conspicuous veinal chlorosis. The infected plants produce few pods which contain small seeds.

4.5.3 MANAGEMENT

1. Foliar spray of rogor (0.05%) on first appearance of the disease and subsequently at 10 days interval.

2. Grow resistant/tolerant varieties
   Urdbean: N 212 and Khargone 3
   Mungbean: Pant mung 3

4.6 CERCOSPORA LEAF SPOT

4.6.1 CAUSAL ORGANISM

Cercospora leaf spot (CLS) is caused by several species dominated by *Cercospora canesens* and *Cercospora cruenta* which may cause severe losses of yield under warm and humid weather conditions. Mungbean is more susceptible to this disease than urdbean. The fungus survives on the infected seeds and crop debris (Grewal, 1978). Rath and Grewal (1973) observed heavy sporulation at 27°C temperature and 96% relative humidity.

4.6.2 SYMPTOMS

Leaf spots develop with a somewhat circular to irregular shape. The central area will turn tan or gray with reddish brown or brown to dark brown margin. Lesions vary in size depending on the isolate and the host (Ilag, 1978). The petioles, stems and pods also get affected by the pathogen. During favorable condition the spots increase in size during flowering and increase is most rapid at the pod-filling stage lead to defoliation. The size of pods and seeds is reduced and thus also the yield (Grewal, 1978). Singh et al. (2000) reported yield losses to the tune of 50% in severely diseased field (Figure 4.4).

4.6.3 MANAGEMENT

1. Destruction of infected crop debris and avoiding the collateral hosts in the vicinity of the crop would greatly help in reducing the incidence of the disease.
2. Crop rotation with non leguminous crops.
3. Treat the seeds with thiram or captan @ 2.5g/kg of seed.

A

B

FIGURE 4.4 Symptomatology of Cercospora leaf spot disease in mungbean. (a) Field view showing severity of disease. (b) Cercospora leaf spot on upper leaf surface of mungbean.

4. Cultivation of resistant varieties:

Mungbean: LM 113, LM 168, LM 170 and JM 171

Urdbean: Naveen, Jawahar, Urd-3, Gujarat Urd-1 and Barkha

5. On appearance of the symptoms spray with carbendazim 50 WP @1.0 g/L or mancozeb 45 WP @ 2.0 g/L or copper oxychloride @ 3 to 4 g/L. Subsequent spray should be done after 10 to 15 days, if required.

4.7 POWDERY MILDEW

4.7.1 CAUSAL ORGANISM

Powdery mildew (PM), caused by the pathogen *Erysiphe polygoni* DC, is one the most destructive and wide spread disease of mungbean and urdbean in India and south east Asia countries. It is a serious problem in all the areas having rice based cropping systems of the country (Abbaiah, 1993). It occurs almost every year causing considerable yield loss due to reduction in photosynthetic activity and physiological changes (Legapsi et al., 1978). The fungus is obligate, ectophytic, and spreading on the surface of the host and sending haustoria into the epidermal cells to obtain nutrients. Host range of this fungus restricted to species belonging to the family *Leguminosae*. The fungus survives in its conidial form or cleistothecia form on the host tissues, which become source of primary infection. The secondary spread is through air borne conidia. Severe infection by the fungus occurs in the cool, dry months where the yield losses owing to PM have been estimated to be around 20–40% (Reddy et al., 1994). In India, the losses due to powdery mildew in winter-sown urdbean and mungbean are more as compared to rainy season crop.

4.7.2 SYMPTOMS

The disease appears on all the part of plants above soil surface. Disease initiates as faint dark spots, which develop into small white powdery spots, coalescing to form white powdery coating on leaves, stems and pods. At the advance stages, the color of the powdery mass turns dirty white. In case of severe infection, defoliation takes place. The disease induces forced maturity of the infected plant causing heavy yield losses.

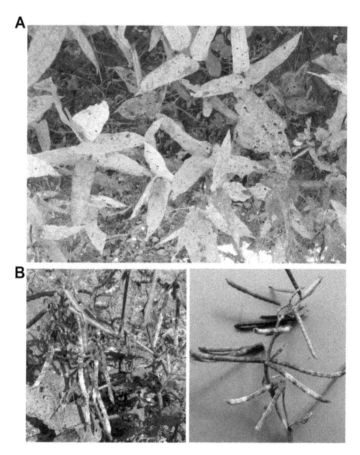

FIGURE 4.5   Symptomatology of Powdery mildew in urdbean. (a) Infected leaf of urdbean. (b) Infected pods in urdbean.

4.7.3 MANAGEMENT

1. The diseased plants should be detected and destroyed.
2. Delayed sowing of mungbean and urdbean with wider spacings considerably reduce the disease severity.
3. Cultivation of resistant varieties:
   Mungbean: LM 223, LM 24, P115, ML 131, MI 322, ML 337, ML 395 SS1, JRUM 1, TARM 1 and AVRDC 1381
   Urdbean: COBG10, LBG 648, 17, Prabha, IPU 02–43, AKU 15 and UG 301

4. Spray with NSKE @ 50 g/L or neem oil 3000 ppm @ 20 ml/L twice at 10 days interval from initial disease appearance or Spray with eucalyptus leaf extract 10% at initiation of the disease and 10 days later also if necessary or Spray with water soluble sulfur 80 wp @ 4 k g/L or carbendazin 50 WP @ 1 g/L (0.05%), benlate (0.05%) and topsin-M (0.15%) and rotate chemicals with different modes of action.

4.8 ANTHRACNOSE

4.8.1 CAUSAL ORGANISM

Five species of *Colletotrichum* are known to attack mungbean and urdbean but *C. lindemuthianum* and *C. capsici* are wide spread and cause severe infections under favorable environment conditions. The pathogen survives from one crop season to the next on infected seeds and crop residue (Singh et al., 1981). Primary leaves and the hypocotyls are foci of secondary infections. Intermittent rains at frequent intervals favor the epidemic development of the disease. The optimum temperature and relative humidity for disease development is 17–24°C and 100%, respectively.

4.8.2 SYMPTOMS

The disease appears on the above ground parts of plant, for example, foliage, stems and pods. The characteristic symptoms of this disease are circular brown sunken spots with dark centers and bright red orange margins on leaves and pods. In severe infection, affected part withers off. Infection just after germination causes seedling blight.

4.8.3 MANAGEMENT

1. Hot water seed treatment at 58°C for 15 minutes has been found effective in checking the seed borne infection and increasing proportion of seed germination.

2. Seed treatment with thiram 80% WP @ 2 g/L or captan 75 WP @ 2.5 g/L helps in eliminating the seed borne infection.
3. Spray the crop with 0.2% zineb 80% WP @ 2 g/L or ziram 80% WP @ 2 g/L with first appearance of symptoms on the crop and repeat after 15 days (if necessary).

4.9   MACROPHOMINA BLIGHT

4.9.1   CAUSAL ORGANISM

It is caused by the fungus *Macrophomina phaseolina* causing root rot, collar rot, seedling blight, stem rot, leaf blight, pod and seed infection. In pre-emergence stage, the fungus causes seed rot and mortality of germinating seedlings. In post-emergence stage, seedling blight disease appears due to soil, water or seed-borne infection. Fungus produces numerous jet black color sclerotia that survive in soil and host residue for long time and become source of primary infection. The pathogen is also carried through the infected seeds. The fungus has wide host range, it perpetuates freely and become virulent when optimum pre-disposing conditions in the host exist. Dark brown to black pycnidia are formed on the diseased spots and pycniospores coming out of pycnidia may contribute to aerial spread of the disease. The pathogen is most favored at a temperature of 30°C and 15% moisture.

4.9.2   SYMPTOMS

The disease is difficult to identify in initial stages. However, dark lesions are formed on the main stalk near soil level, forming localized dark green patches. The tissues of the affected portions become weak and shredded easily. Decay of secondary roots and shredding of the cortex region of the tap root are prominent symptoms. If the plants will pull out, the basal stem and root may show dry rot symptoms. Black dot like sclerotia are formed on the surface and below the epidermis on the outer tissue of the stem and root. The disease develops rapidly and causes severe infestation under high temperature and water stress conditions.

4.9.3 MANAGEMENT

1. Basal application of zinc sulfate @ 25 kg/ha or neem cake @ 150 kg/ha or soil application *Pseudomonas fluorescens* (1 x 10^{10}Cfu/g) or *Trichoderma viride* (1 x 10–^{8} cfu/g) @ 2.5 kg/ha + 50 kg of well decomposed FYM at the time of sowing helps in prevention of the disease.

2. Seeds treated with *Trichoderma* (1 x 10^{8} cfu/g) 5–10 g/kg of seed or captan 75 WP @ 2.5 g/L and thiram 80% WP @ 2 g/L before sowing provides significant protection.

3. The diseased plants should be uprooted and destroyed so that the sclerotia do not form or survive.

4. Spray with carbendazim 50 WP @ 1.0 g/L at an interval of 15 days with the appearance of the symptoms.

4.10   WEB BLIGHT

4.10.1 CAUSAL ORGANISM

Web blight caused by *Rhizoctonia solani kuhn.* (Teleomorph: *Thanatephorous cucumeris*) is one of the most important fungal disease. It causes considerable damage by reducing seed quality and yield. It is reported in Punjab, Haryana, Bihar, Rajasthan, Uttaranchal, Uttar Pradesh, West Bengal, Himachal Pradesh and Jammu and Kashmir states (Saksena and Dwivedi, 1973). The intensive crop cultivation and modified agro-practices have increased the populations of *R. solani* in soil and gradually built up new disease problems. It is a soil and seed borne pathogen that has many hosts, forms sclerotia in/on soil and survives for a long period in the absence of a host either as sclerotia or thick walled brown hyphae in plants debris. It was reported that 26–28°C temperatures and 90–100% relative humidity favored maximum disease development. The pathogen causes considerable yield loss in mungbean and urdbean in India (Dubey, 2003). Yield loss up to 57% in mungbean was reported from Iran (Kaiser, 1970).

4.10.2 SYMPTOMS

The symptoms of web blight occur on roots, stems, petioles and pods, but the disease is the most destructive on foliage. It causes seedling mortality during second and third week of plants growth. Seed decay, pre-and post-emergence mortality occurs. The first symptoms appear as small circular brown spots on the primary young leaves. These spots enlarge, often show concentric banding and surrounded by irregular conspicuous water soaked areas. The lesion expands and coalesces and white mycelial fungal growth can be seen under surface of infected leaves and young branches. The mycelium on infected leaves appears as spider web thus suggested the name web blight disease.

4.10.3   MANAGEMENT

1. Proper sanitation and burying the infected leaves immediately after harvest will reduce the primary inoculums.
2. Crop rotation, which help in controlling the disease to a greater extent.
3. Planting at a time to avoid rainy season during the susceptible crop stage.
4. Avoid the thick canopy of the crop by using proper seed rate.
5. Cultivation of resistant varieties.
6. Seed treatment with Carbendazim and thiophanate methyl were found best controlling seedling mortality of mung bean caused by *R. solani*.
7. Foliar spray of bavistin 0.05% along with seed treatment with bavistin (0.2%) is highly effective in reducing web blight.

4.11   BACTERIAL LEAF SPOT

4.11.1   CAUSAL ORGANISM

This disease is caused by *Xanthomonas phaseoli* (Smith) Dowson, is a gram negative, rod shape bacteria. The bacteria survive in the seeds, plant debris and on the other host plants during off-season. Warm and humid weather is favorable for disease development. The optimum temperature for the growth of the bacterium is 30–33°C.

4.11.2   SYMPTOMS

The disease is characterized by small, brown and dry raised spots develop on leaves and stem. Leaf spots first appear as superficial eruption and gradually invade the tissues, giving corky or rough appearance. Leaves become yellow with advancement of disease and premature defoliation occurs. The stem and pods also get infected.

4.11.3 MANAGEMENT

1. Seed treatment with streptomycin sulfate @ 500 ppm or captan @ 0.3% or bleaching power @0.025%.
2. Three protective spray of streptocycline @ 100 ppm or zineb @ 0.3% or benomyl @ 0.2% or three spray of streptomycin @ 0.025% + 0.1% carbendazim is effective in managing the disease.

KEYWORDS

- bean
- blight
- crinkle
- curl
- mildew
- mosaic

REFERENCES

1. Amin, P. W. (1985). Apparent resistance of groundnut cultivar Robut 33–1 to bud necrosis disease. Pant Dis. 69, 718–719.
2. Abbaiah, K. (1993). Development of powdery mildew epidemics in urd bean in relation to weather factors. *Indian Journal of Pulses Res.*, 6, 186–188.
3. Ahmad, Z., M. Bashir, T. Mtsueda. (1997). Evaluation of legume germplasm for seed-borne viruses. *In:* Harmonizing Agricultural Productivity and Conservation of Biodiversity: Breeding and Ecology. *Proc. 8th SABRAO J. Cong. Annu. Meeting Korean Breeding Soc., Seoul, Korea,* pp. 117–120.
4. AICRP (2012–2013). All India Coordinated Research Project on MULLaRP, Project Coordinator's Report in Annual Group Meet at TNAU, Coimbatore, 2012–13.
5. Bashir, M., S. M. Mughal, B. A. Malik. (1991). Assessment of yield losses due to leaf crinkle virus in urdbean (*Vigna mungo* (L) Hepper). *Pak. J. Bot.*, 23, 140–142.

6. Beniwal, S. P. S., N. Bharathan. (1980). Beetle transmission of urdbean leaf crinkle virus. *Indian Phytopathol.*, 33, 600–601.

7. Beniwal, S. P. S., Chaubey, S. N. (1979). Urdbean leaf crinkle virus: Effect on yield contributing factors, total yield and seed characters of urdbean (*Vigna mungo*). *Seed Res.* 7, 125–181.

8. Brar, J. S., Rataul, H. S. (1987). Evidence against the transmission of urdbean leaf crinkle virus (ULCV) in mash, Vigna mungo (L.) through insects- a field approach. *Indian, J. Ent.*, 49, 57–63.

9. Dhingra, K. L. (1975), Transmission of urdbean leaf crinkle virus by two aphid species. *Indian Phytopathol.* 28, 80–82.

10. Dhingra, K. L., Chenulu, V. V. (1985) Effect of yellow mosaic on yield and nodulation of soybean. *Indian Phytopathology*, 38, 248–251.

11. Dhingra, K. L., Chenula, V. V. (1981). Studies on the transmission of urdbean leaf crinkle and chickpea leaf reduction viruses by *Aphis crassivora* Koch. *Indian Phytopath.*, 34, 38–42.

12. Dubey, S. C. (2003). Integrated management of web blight of urd/mung bean by bio-seed treatment. *Indian Phytopath.*, 56, 34–38.

13. Grewal, J. S. (1978), Diseases of mungbean in India. First Intl. Mungbean Symp. Proc., Univ. Philippines, 1977, pp. 165–168.

14. Honda, Y. M., Twaki, Y., Saito, P. Thangmeearkom, P. Kittisak, K. Deema, N. (1981) Mechanical transmission, purification and some properties of whitefly transmitted mungbean yellow mosaic virus in Thailand. *Plant Disease.*, 65, 801–04.

15. Ilag, L. L. (1978), Fungal diseases of mungbean in the Philippines. First Intl. Mungbean Symp. Proc., Univ. Philippines, 1977, pp. 154–156.

16. Ilyas, M. B., Haq, M. A., Iftikhar, K. (1992). Studies on the responses of growth components of urdbean against leaf crinkle virus. *Pakphyton.*, 4, 51–56.

17. Kadian, O. P. (1980). Studies on leaf crinkle disease of urdbean (*Vigna mungo* (L.) Hepper), mung bean (*V. radiata* (L.) Wilczek) and its control. PhD Thesis, Dept. Plant Pathology, Haryana Agric. Univ., Hisar. 177 pp.

18. Kadian, O. P. (1982). Yield loss in mungobean and urdbean due to leaf crinkle disease. *Indian Phytopath.*, 35, 642–644.

19. Kaiser, N. J. (1970). Rhizoctonia stem canker disease of mungbean in Iran. *Plant Dis Rep.*, 54, 240–50.

20. Kolte, S. J., Nene, Y. L. (1973). Studies on the symptoms and mode of transmission of leaf crinkle virus of urdbean (Phaseolus mungo) *Indian Phytopath.*, 25, 401–404.

21. Lagapsi, B. M., Capiton, E. M. and Hubbell, J. N. (1978). AVRDC., Phillipine, program studies. First International Symposium on Mungbean.

22. Nair, N. G. (1971). Studies on the yellow mosaic of urdbean caused by mungbean yellow mosaic virus PhD thesis, U.P. Agric. Univ., Pantnagar, India.

23. Narayansamy, P., T. Jaganthan. (1973). Vector transmission of black gram leaf crinkle virus. *Madras Agric. J.*, 60, 651–652.

24. Nariani, T. K. (1960). Yellow mosaic of mung (*Phaseolus aureus*). *Indian Phytopathology*, 13, 24–29.

25. Nene, Y. L. (1968). Annual report (No. 1) project, FG-In-358, U.P. Agric. Univ., India.

26. Nene, Y. L. (1972). A survey of viral diseases of pulse crops in Uttar Pradesh, G. B. Pant Univ. Agric. Tech., Pantnagar, *Res. Bull.* 4.

27. Nene, Y. L. (1973). Viral diseases of some warm weather pulse crops in India. *Plant Disease Reporter*, 57, 463–467.

28. Nene, Y. L. (1973). Viral diseases of some warm weather pulse crops in India. *Plant Dis. Rep.* 5, 463–467.

29. Nene, Y. L., Srivastava, S. K., Naresh, J. S. (1972). Evaluation of urdbean (*Phaseolus mungo*, L.) and mungbean (*Phaseolus aureus*, L.) germplasms and cultivars against yellow mosaic virus. *Indian, J. Agric. Sci.* 42, 251–254.

30. Rath, G. C., Grewal, J. S. (1973). A note on *Cercospora* leaf spot of *Phaseolus aureus. Indian, J. Mycol. Plant Pathol.* 3, 204–207.

31. Rathi, Y. P. S. (2002). Epidemiology, yield losses and management of major diseases of Kharif pulses in India. In: Plant Pathology and Asian Congress of Mycology and Plant Pathology, Oct.-1–4, 2002. University of Mysore, Mysore, India.

32. Rathi, Y. P. S., Nene, Y. L. (1974). The additional hosts of mungbean yellow mosaic virus. *Indian Phytopathol.* 27, 429–430.

33. Reddy, K. S., S. E. Pawar, C. R. Bhatia. (1994). Inheritance of powdery mildew (*Erysiphe polygoni* DC) resistance in mungbean (*Vigna radiate,* L. Wilczek). *Theoretical & Applied Genetics.* 88, 945–948.

34. Rishi, N. (1990). Seed and crop improvement of northern Indian pulses (*Pisum* and *Vigna*) through control of seed- borne mosaic viruses *Final Technical Report, (US-India Fund) Dept. Plant Pathology, CCS Haryana Agric. Univ. Hisar, India:*122 pp.

35. Saksena, H. K., Dwivedi, R. P. (1973). Indian, J. *Farms* Sci. 1, 58–61.

36. Shahare, K. C., Raychaudhary, S. P. (1963). Mosaic disease of urd (*Phaseolus mungo,* L.) *Indian Phytopathol.* 16, 316–318.

37. Singh, J. P. (1980). Effect of virus diseases on growth component and yield of mungbean and Urdbean. *Indian Phytopatholo.* 8, 405–08.

38. Singh, A., A. Sirohi, K. S. Panwar. (1998). Inheritance of mungbean yellow mosaic virus resistance in urdbean (*Vigna mungo*). *Indian, J. Virol.* 14, 89–90.

39. Singh, A. K., Srivastava, S. K. (1985). Nodular physiology of urdbean as affected by urdbean mosaic virus. V. Effect on some enzymematic activity-*Phyton* (Austria) 25, 213–217.

40. Singh, D. P. (1980). Inheritance of resistance to yellow mosaic virus in blackgram (*Vigna mungo* (L.) Hepper). *Theor. Appl. Genet.* 57, 233–235.

41. Singh, D. P. (1981). Breeding for resistance to diseases in greengram and blackgram. *Theor. Appl. Genet.* 59, 1–10.

42. Singh, G., S. Kapoor, K. Singh, (1988). Multiple disease resistance in mungbean with special emphasis on mungbean yellow mosaic virus. In: Shanmugasundaram, S. (Ed.) Mungbean, Proceedings of the second International Symposium on Mungbean, Shanhua, Asian Vegetable Research and Development.

43. Singh, R. A., De, R. K., Gurha, S. N., Ghosh. A. (2000). Yellow mosaic disease of mungbean and urdbean. In: Advances in Plant Disease Management, (Eds.) U. Narain, K. Kumar, M. Srivastave. Advance Publishing Concept, New Delhi, India, pp. 337–348.

44. Singh, R. N., Nene, Y. L. (1978). Further studies on the mosaic mottle disease of urdbean, *Indian Phytopath.* 31, 159–162.

45. Srivastava, K. M., Verma, G. S., Verma, H. N. (1969). A mosaic disease of blackgram (*Phaseolus mungo*). *Sci. Cult.* 35, 475–476.
46. Tsuchaizaki, T., Iwaki, M., Thongameearkom, P., Sarindu, N., Deema, N. (1996). Bean common mosaic virus isolated from mungbean (*Vigna radiata*) in Thailand. *Technical Bulletin of the Tropical Agriculture Research Centre*, No. 21, 184–188.
47. Williams, F. J., Grewal, J. S., Amin, K. S. (1968). Serious and new diseases of pulse crops in India in 1966. *Plant Dis. Rep.* 52, 300–304.

PART II

OIL SEED CROPS

CHAPTER 5

GROUNDNUT DISEASES AND THEIR MANAGEMENT

SANTOSH KUMAR,[1] MANOJ KUMAR,[2] AMARENDRA KUMAR,[1] and GIREESH CHAND[1]

[1]Department of Plant Pathology, Bihar Agricultural University, Sabour, Bhagalpur 813210, Bihar, India

[2]Department of Genetics and Plant Breeding, Bihar Agricultural University, Sabour, Bhagalpur 813210, Bihar, India

CONTENTS

5.1   INTRODUCTION

Groundnut is one of the major oilseed crops in the world. It is a valuable cash crop cultivated by millions of small farmers, because of its economic and nutritional value. Under commercial cultivation, it is grown mainly as a sole crop with high levels of inputs whereas under subsistence conditions both sole crop and mixed or intercropping can be seen. The low productivity in groundnut is attributed to many production constraints. Among these, biotic factors particularly diseases play a major role in limiting the yield of groundnut. Among these, biotic factors particularly diseases play a major role in limiting the yield of groundnut. The crop is known to be attacked by a number of fungal and bacterial diseases. Diseases like leaf spot, rust, collar rot, stem rot, bud necrosis, rosette, etc., are very important.

5.2 TIKKA DISEASE OR LEAF SPOT

Tikka disease is reported from all groundnut growing countries of the world such as Africa, Australia, China, India, Indonesia, Malaysia, Philippines, Sri Lanka and USA. The disease is caused by two fungal plant pathogens, for example, *Cercospora arachidicola* and *Cercosporidium personatum*. The perfect stages of both these fungal pathogens (*Mycosphaerella arachidicola* and *M. berkeleyii*) play important role in primary infection and pathogen survival. The yield loss from tikka disease has been reported from 20–50% but may be increased with association other diseases. The all groundnut varieties grown in India are susceptible to tikka disease.

5.2.1 EARLY LEAF SPOT

It is caused by *Cercospora arachidicola* Hori. It develops small necrotic flecks that usually have light to dark-brown centers, and a yellow halo. The spots may range from 1 mm in diameter. Sporulation is on the adaxial (upper) surface of leaflets.

5.2.2 LATE LEAF SPOT

It is caused by *Phaeoisariopsis personata* (Berk & Curt). It develops small necrotic flecks that enlarge and become light to dark brown. The yellow halo is either absent or less conspicuous in late leaf spot. Sporulation is common on the abaxial (lower) surface of leaves. Comparisons of early and late leaf spots are listed in Table 5.1.

TABLE 5.1 Comparisons of Early and Late Leaf Spots

S. No.	Early leaf spot	Late Leaf Spot
1.	It is caused by *Cercospora arachidicola*	It is caused by *Phaeoisariopsis personata*
2.	Spots are brown with yellow halos	Spots are dark brown to black with dense spores forming ring patterns
3.	Sporulation is common on the upper surface of leaves	Sporulation is common on the lower surface of leaves
4.	Appear during early stage of the plant	Appear during late stage of the plant

5.2.3 SYMPTOMS

The primary symptoms of the disease are appearing in 35–60 days old plants. The tikka disease occurs as two distinct types of lea spots caused by two species of *Cercospora*. *C. personatum*causes small (1–6 mm), almost circular and dark colored spots on the leaves, stipules, petioles and stem which may coalesce to form a large dark brown to black irregular patch. There may be few to many spots on each leaf. The severe infection or spotting on the leaves causes premature dropping. The disease is more severe at the time between flowering and harvesting, when the climatic conditions are favorable. The leaf spots caused by *Cercospora arachidicola* are almost circular to irregular, large (1–10 mm), surrounded by bright yellow haloes and dark brown center. The conidia are formed on upper surface of leaf while *C. personatum* produced conidia on lower surface of leaves with concentric rings.

5.2.4 PATHOGEN

The causal organism of tikka disease are *Cercospora arachidicola* Hori (perfect stage of the pathogen: *Mycosphaerella arachidicola* W. A. Jenkins) and *Cercosporidium personatum* (Berk and Curt) Deighton (perfect stage of the pathogen: *Mycosphaerella berkeleyii* W. A. Jenkins). The mycelium of *C. personatum* is intercellular, brown, septate, branched and slender with haustoria. The conidia are hyaline, 18–60 × 6–11 μm, 2–7 septate and borne singly on short, 26–54 × 5–8 μm conidiophores. The conidiophores are produced in bunches from the hymenial layer of subepidermal region. The mycelium of *C. arachidicola* is inter and intracellular, brown, septate, branched and without haustoria. The conidiophores are 22–45 × 3–5 μm, yellowish brown, septate and conidia are hyaline or pale yellow, obclavate, 4–12 septate measuring 38–108 × 3–6 μm.

5.2.5 DISEASE CYCLE

The tikka disease of groundnut is soil borne. The pathogen *C. arachidicola* and *C. personatum* disseminated by wind, which is blown from leaf to leaf. The primary infection of disease is caused by conidia found on the plant debris in the soil. The spores remain viable in the soil for a long time and infect the succeeding crop under favorable environmental conditions. High humidity and relatively low temperature is essential for initiating the fungal infection. It is observed that the high nitrogen fertilizer increases disease intensity.

5.2.6   DISEASE MANAGEMENT

- The disease can be controlled by long crop rotation and sanitation practices.
- The intercropping with pigeon pea and use of phosphatic fertilizers also reduced the disease incidence.
- The early sowing crop varieties to avoid the disease.
- Adjust the date of sowing to reduce the disease conditions, which is favorable for rapid disease development.
- The use of Dithane Z-78 (0.2%), Dithane M-45 (0.2%), Cosan, Breston (0.1%) and copper sulfate mixture (15–25 kg/ha) effectively controlled the disease. Some other effective systemic fungicides are benomyl, bavistin, brestanol and cercobin.

5.3   RUST

The rust of groundnut is distributed in Central and South America, China, India, West Indies and USSR. In India, the disease is found in Andhra Pradesh, Punjab, Tamil Nadu, West Bengal and Uttar Pradesh. Rust of groundnut is an economic disease and causing 14–32% yield loss.

5.3.1   PATHOGEN

The rust of groundnut is caused by *Puccinia arachidis* Speg. The pathogen produces both uredial and telial stages. Uredial stages are produced abundant in groundnut and production of telia is limited. The uredospores are one celled, subglobose, ovoid to round, light brown, thin walled, 2–3 germ pores and measuring 24 × 21 μm with short and hyaline pedicels. Teliospores are dark brown with two cells. Pycnial and aecial stages have not been recorded and there is no information available about the role of alternate host.

5.3.2   SYMPTOMS

The disease attacks all aerial parts of the plant. The disease is usually found when the plants are about 6 weeks old. Small brown to chestnut dusty pustules (uredosori) appear on the lower surface of leaves. At later stages, these pustules may appear on upper leaf surface and other aerial parts of the plant except flower. The epidermis ruptures and exposes a powdery mass of uredospores. Corresponding to the sori, small, necrotic, brown spots appear on the upper surface of leaves. The severely infected leaves wither and drop prematurely. The rust pustules may be seen on petioles and stem. Late in the season, brown teliosori, as dark pustules, appear among the necrotic patches. In severe infection lower leaves dry and drop prematurely. The severe infection leads to production of small and shriveled seeds. The seeds formed on infected plants are small and shriveled.

5.3.3   EPIDEMIOLOGY AND FAVORABLE CONDITIONS

High relative humidity (above 85%), heavy rainfall and low temperature (20–25°C) favor disease development.

5.3.4   DISEASE CYCLE

The pathogen survives as uredospores on volunteer groundnut plants. The fungus also survives in infected plant debris in soil. The spread is mainly through wind borne inoculum of uredospores. The uredospores also spread as contamination of seeds and pods. The continuous cultivation of the crop in India without any significant break may perpetuate the disease. Rain splash and implements also help in dissemination. The fungus also survives on the collateral hosts like *Arachis marginata, A. nambyquarae*. The uredospores found in southern India may act as potential source of disease in northern India blown by wind during monsoon season.

5.3.5   MANAGEMENT

- Avoid monoculturing of groundnut.
- Grow moderately resistant varieties like ALR 1.

- Remove volunteer groundnut plants and reservoir hosts to reduce the primary source of inoculum.
- The application of a mixture of Carbendazim (0.5%) and Mancozeb (0.25%) at 2–3 weeks interval on 4–5 weeks old plants effectively controlled the disease.

5.4   STEM ROT

The disease is distributed throughout the world and prevalent particularly in warm dry climates. It was first reported by McClintock (1917) in Virginia. The loss of yield caused by the pathogen is 25%, but sometimes it reaches 80–90% (Grichar and Bosweel, 1987). Similarly, yield losses over 25% have been reported by Mayee and Datar (1988). Stem rot causes pod yield losses of 10–25%, but under severe diseased conditions yield losses may range to up 80% (Rodriguez Kabana et al., 1975). Patil and Rane (1982) reported yield loss up to 10 to 50% due to this disease. Adiver (2003) reported the yield loss of 15–70% in groundnut is due to leaf spot, rust and stem rot singly or in combination.

5.4.1   PATHOGENS

Stem rot caused by *Sclerotium rolfsii* Sacc is an important pathogen which causes wide spread and serious losses. *Sclerotium rolfsii* was first reported by Rolfs (1892) later the pathogen was named as *Sclerotium rolfsii* by Saccardo (1911). Higgins (1927) worked in detail on physiology and parasitism of *S. rolfsii*. This was the first detailed and comprehensive study in USA. The pathogen *Sclerotium rolfsii* Sacc., is a soil borne in nature survived for years by producing sclerotial bodies and causing the disease on various hosts Weber (1931) and Garret (1956). Scleorotia which are very well organized compact structures, built of three layers, the rind, composed of empty melanised cells; the cortex cells, filled with vesicles and the medulla (Chet, 1975). Sclerotia may be spherical or irregular in shape and at maturity resemble the mustard seed (Barnett and Hunter, 1972). Sclerotial size was reported to be varied from 0.1 mm to 3.0 mm (Ansari and Agnihotri, 2000 and Anahosur, 2001).

5.4.2   SYMPTOMS

Wilson (1953) described the symptoms of stem rot as, mycelium covering the plant stem near the soil surface and produced organic acids, which were toxic to living plant tissue. This followed the necrosis of plant cells. The mycelium invaded the stem, gynophores and also pods causing rotting of the tissues. The production of abundant white mycelium, and small brown spherical sclerotia on the infected parts were characteristic symptoms of the disease. Beattle (1954) also observed same symptoms on infected plants. Sclerotia developed on the surface of soil and infected stem (Baruah et al., 1980). Mehrotra and Aneja (1990) noticed the cortical decay of stem base at ground level and appearance of conspicuous white mycelium which extended into the soil and on organic debris. The mycelial mat may extend several centimeters up to the stem above the soil line. Numerous tan to brown, spherical sclerotia of about mustard seed size formed on infected plant material which was found on the soil surface (Nyvall, 1989 and Aken and Dashiell, 1991) also reported the similar symptom. Mayee and Datar, 1988 and Narain and Kar, 1990 found that the pathogen causing seedling blight, collar rot, wilt, root rot, stem rot and pod rot.

5.4.3   MANAGEMENT

- Cultural practices such as field sanitation, seed selection, crop rotation and early sown crops help to escape infection.
- Soil solarisation during the hot dry season, also helps to control nematodes.
- Eradication of volunteer groundnut and alternate host plants is important in reducing the primary source of inoculums.
- Seed (4 g per kg of seed) and soil application (2.5–3.5 kg/ha one week after transplanting) with talc based formulation of *Trichoderma harzianum* and *T. viride* reduce disease incidence. Kulkarni (1994) showed that, seed and soil treatment with *T. viride* and *T. harzianum* were the most effective in reducing the mortality percentage of groundnut incited by *S. rolfsii*.
- Foliar spray of fungicides such as thiram (0.1%) and carbendazim (0.1%) are found effective against the diseases.

5.5 GROUNDNUT BUD NECROSIS

Bud necrosis disease (BND) is caused by two serologically distinct viruses, bud necrosis virus (BNV) and tomato spotted wilt virus (TSWV). BND was first recorded in Brazil in 1941, and significant crop losses by this disease have been reported from Australia, India, and the USA (Reddy, 1984a).

5.5.1 SYMPTOMS

Initial symptoms are concentric rings or chlorotic spots on young leaflets. Subsequently terminal bud necrosis occurs especially when day temperatures exceed 30°C. Plants infected at early stages are severely stunted. Occasionally, necrosis may spread to the petioles and then to the stem leading to death of the plant. Later infected plants ma y only show bud necrosis on a few branches and axillary shoot proliferations may be restricted to the terminal portion (Reddy et al., 1991). In early infection, pods are seldom produced. In late infections, pod size is reduced, shriveled, and mottled with discolored testa.

5.5.2 TRANSMISSION

The virus is not transmitted by seed; it is transmitted by thrips.

5.5.3 MANAGEMENT

- Use resistant/tolerant cultivars: ICGS 11, ICGS 4–4, ICGV 87141, ICGV 87187, ICGV 87119, ICGV 87121, ICGV 87160, ICGV 8–7 1–5 7, or ICGV 86590.
- Control of vector (thrips).
- Adjust date of sowing to avoid the peak disease incidence.
- Sow groundnut at a high plant density and maintain a good plant stand.
- Intercropping of groundnut with cereals, for example, pearl millet will restrict spread of the virus.

- Avoid groundnut cultivation adjacent to the crops that are susceptible to BNV, such as green gram or black gram.

5.6   GROUNDNUT ROSETTE

Three rosette diseases have been recognized. They are "groundnut chlorotic rosette" (G C R) "groundnut green rosette" (G G R), and "groundnut mosaic rosette" (G M R). GCR and GMR are predominant in eastern and southern Africa, whereas GGR appears to be restricted to western Africa (Reddy, 1984 b).

5.6.1   SYMPTOMS

Groundnut chlorotic rosette (GCR) is characterized by general chlorosis, with a few green islands on young leaflets. Early infected plants are stunted, progressively producing small chlorotic, curled, and puckered leaflets. Older leaflets are bright-yellow with dark-green patches. Plants infected late, show typical leaf symptoms without the marked stunting and bushy appearance (Reddy, 1984b). Groundnut green rosette (GGR) infected plants show mild and narrow chlorotic streaks on young leaflets. The older leaflets are darkgreen and reduced in size with their margins rolled outward. Early infected plants are stunted and bushy, whereas on late infected plants a proliferation of axillary shoots may be observed (Reddy, 1984b).

5.6.2   MANAGEMENT

- Several long-duration cultivars with resistance to rosette are currently available. These include RG 1, RMP 1–2, RMP 91, KH 14–9 A, M 25-M 6–8, and M 6–9-M 101. Short duration rosette resistant cultivars are being developed.
- Aphis craccivora is mainly responsible for the spread of rosette disease. Spray of endosulfan 4% dust with 1 kg a.i. per ha or demeton-s-methyl 72–96 ml a.i. per ha provide effective control for aphids. It is essential to know the peak period of aphid migration before application of insecticides.

- Eradication of volunteer groundnut plants is helpful to prevent perpetuation of virus inoculum during the off-season.
- Early sowing and maintenance of a good plant stand are helpful in reducing the disease incidence.

5.7 ROOT KNOT NEMATODE

The root-knot nematodes (*Meloidogyne* spp.) are the most important nematode species causing damage ranging from 2–0% to 9–0% in infested fields of groundnut (Rodriguez-Kabana, 1984). Root galls contain white swollen adult females. The body tapers anteriorly to a narrow neck and mobile head with stylet, massive median bulb and large esophageal glands. An egg sac often protrudes posteriorly from the female to the exterior of the gall. It contains several hundred eggs. Often one or more elongate males are present in an egg sac. The females are 0.5 mm to 0.8 mm long. At the center of its posterior region, the female cuticle has a pattern of cuticular markings surrounding the anus and vulva. The second stage of juveniles invades roots at or close to the tip and migrates to the site of differentiating vascular tissues. Consequently several giant cells farm around the nematodes head. The complete life cycle takes 3 weeks or more, depending on host and temperature. Males average about 1.1 mm in length. The posterior is characteristically twisted through 90° or more. Larvae are about 400 μm long and have a delicate stylet (Dropkin, 1980).

5.7.1 SYMPTOMS

The symptoms of damage caused by *Meloidogyne hapla* are similar to those caused by *M. arenaria*. Root-knot nematodes enter and damage groundnut roots, pegs, and pods. Infected plants develop enlarged roots and pegs. Galls develop into various sizes resulting from an internal swelling from the root tissue. Infected pods develop knobs, protuberances, or small warts. Infected plants with root-knot nematodes may show various degrees of stunting and chlorosis. Root development is reduced, and vascular systems of infected tissues are disrupted, resulting in the poor flow of water and nutrients from the roots (or pegs) to the shoot. Infected plants tend to wilt under drought conditions.

5.7.2 CONTROL MEASURES

- A crop rotation of cereal-cereal-groundnut can significantly decrease the level of root-knot nematode infestation in soils.
- Nematicides used in groundnut are fumigant and nonfumigant types with contact or systemic properties. Application of a fumigant nematicide like ethylene dibromide (EDB) is made 18 cm deep at a soil temperature between 15 and 21°C @ 18 or 19 L ha^{-1}. Nonfumigant nematicides are aldicarb, carbofuran, and phenamiphos. These nematicides are effective when applied at sowing @ 2–3 kg a.i. ha^{-1}. The best results are obtained when applications of nematicides are made in a band 17–25 cm wide and incorporated 2–4 cm into the soil (Rodriguez-Kabana, 1984a).
- Soil solarization during the hot dry season, also helps to control nematodes.
- Grow resistant cultivars: NC 343, NC 3033, NCAC 17090, or ICGS 2.

KEYWORDS

- groundnut
- necrosis
- nematode
- rosette
- rust
- spot

REFERENCES

1 Adiver, S. S. (2003). Influence of Organic Amendments and Biological Components on Stem Rot of Groundnut," *National Seminar on Stress Management in Oilseeds For Attaining Self Reliance in Vegetable Oil* Indian Society of Oilseeds Research, Directorate of Oilseed Research, Hyderabad Form January 28–30, p15–17.

2. Aken, C. N., Dasiell, K. E. (1991). First Report of Southern Blight Caused By *Sclerotiumrolfsii* on Soybean in Nigeria. *Plant Disease*, Vol. 75, p537.

3. Anahosur, K. H. (2001). Integrated Management of Potato Sclerotium Wilt Caused By *Sclerotium rolfsii. Indian Phytopathology*, 54, 158–166.

4. Ansari, M. M., Agnihotri, S. K. (2000). Morphological, Physiological and Pathological Variations Among *Sclerotium Rolfsii* Isolates of Soybean. *Indian Phytopathology*, 53, 65–67.

5. Barnett, H. L., Barry, B. Hunter (1972). Illustrated Genera of Imperfect Fungi. Burgess Publishing Company, Minnesota.

6. Baruah, H. K., Baruah, P., Baruah, A. (1980). *Textbook of Plant Pathology*, Oxford and IBH, Publishing Co, New Delhi, p498.

7. Beattle, J. H. (1954). "Growing Peanuts United States Department Agriculture," *Farmers Bulletin*, p552.

8. Chet, I. (1975). Ultra Structural Basis of Sclerotial Survival in Soil. *Microbial Ecology*, 2, 194–200.

9. Dropkin, V. H. (1980). Introduction to plant nematology. New York, USA: John Wiley.

10. Gerrett, S. D. (1956). *Biology of Root Infecting Fungi*, p.293, Cambridge University Press, London.

11. Grichar, V. J., Bosweel, T. E. (1987). "Comparison of Lorsban and Tilt With Terrachlor For Control of *Southern Blight* on Peanut the Texas," Agriculture Experiment Station Pr-4534.

12. Higgiens, B. B. (1927). Physiology and Parasitism of *Sclerotium Rolfsii* (Sacc). *Phytopathology*, 17, 417–448.

13. Kulkarni, S. A., Anahosur, K. H. (1994). Effect of Age of Groundnut Plant to Infection of *Sclerotium Rolfsii* Sacc A Causal Agent of Stem Rot Disease. *Karnataka Journal of Agricultural Sciences*, 7, 367–368.

14. Mayee, C. D., Datar, V. V. (1988). Diseases of Groundnut in the Tropics. *Review of Tropical Plant Pathology*, 5, 85–118.

15. Mcclintock, J. A. (1917). "Peanut Wilt Caused By *Sclerotium Rolfsii*," *Journal of Agricultural Research*, 8, 441–448.

16. Narain and Kar, A. K. (1990). Wilt of Groundnut Caused By *Sclerotium Rolfsii, Fusarium Sp.* and *Aspergillus Niger. Crop Research*, 3, 257–262.

17. Nyvall, R. F. (1989). *Field Crop Diseases* Hand Book, Second Edition Published By Van Nat Rand Reinhold, New York, Vol. 13, 31–32.

18. Patil, M. B., Rane, M. S. (1982). Incidence and Control of *Sclerotium Wilt* Groundnut. *Pesticides*: 16, 23–24.

19. Ready, D. V. R. (1984a). Tomato spotted wilt virus, pages 48–49 in Compendium of peanut diseases (Porter, D. M., Smith, D. H., Rodriguez-Kabana, R. eds.) St. Paul, MN, USA: American Phytopathological Society.

20. Ready, D. V. R. (1984b). Tomato spotted wilt virus, pages 49–50 in Compendium of peanut diseases (Porter, D. M., Smith, D. H., Rodriguez-Kabana, R. eds.) St. Paul, MN, USA: American Phytopathological Society.

21. Ready, D. V. R., Nightman, J. A., Beshear, R. J., Highland, B., Black, M., Sreenivasulu, P., Dwivedi, S. L., Demaski, J. W., Mcdonald, D., Smith, J. W., Smith, D. H. (1991). Bud necrosis: A disease of groundnut caused by tomato spotted wilt virus. Information Bullet in no. 31, Patancheru, A.P. 502324, India: International Crops Research Institute for the Semi Arid Tropics.

22. Rodriguez-Kabana (1984). Root knot nematodes. Compendium of peanut diseases. P38–41 In: Porter, D. M., Smith, D. H., Rodriguez-Kabana (Eds.), St. Paul, M N, USA: American Phytopathological Society.
23. Rodriguez-kabana, R, Backman, P. A., Williams, J. C. (1975). Determination of Yield Losses Due to *Sclerotium Rolfsii* in Peanut Fields, *Plant Dis Rept*, 59, 855–858.
24. Saccardo, P. A. (1911). Notae Mycologicae, *Annals Mycologici*, 9, 249–257.
25. Weber, G. F. (1931). Blights of Carrots Caused By *Sclerotium Rolfsii With Geographic* Distribution and Host Range of the Fungi. *Phytopathology*, 21, 103–109.

CHAPTER 6

DISEASES OF SOYBEAN AND THEIR MANAGEMENT

SUNIL KUMAR

AICRP on Soybean, School of Agricultural Sciences and Rural Development, Nagaland University, Medziphema – 797106, Nagaland, India

CONTENTS

6.1　INTRODUCTION

Soybean (*Glycine max* L. Merrill) is a leguminous crop; it belongs to the family Leguminocae. It is rich in high quality protein (40–42%), oil (18–20%) and other nutrients like calcium, iron and glycines. It is a good source of isoflavones. Soybean helps in preventing heart diseases, cancer, HIV, etc. (Kumar, 2007). Soybean protein is rich in the valuable amino acid lysine (5%) in which most of the cereals is deficient. In addition, it contains good amount of minerals, salts and vitamins (thiamine and riboflavin). Its sprouting grains contain a considerable amount of vitamin C, minerals, salts and vitamins (thiamine and riboflavin) (Singh et al., 2003). Soybean is the richest, cheapest and easiest source of best quality protein and fat. Hence, it is called as vegetarian meat and wonder crop. This crop is severely affected by a number of diseases and causes much yield losses.

6.2 FUNGAL DISEASES

6.2.1 COLLAR ROT

6.2.1.1 Causal Organism

The causal organism of this disease is *Sclerotium rolfsii*. *S. rolfsii* is a well-known polyphagous and most destructive soil borne fungus. This was first reported by Rolfs (1892) as a cause of tomato blight in Florida. Later, Saccardo (1911) named the fungus as *Sclerotium rolfsii*. But, in India, Shaw and Ajrekar (1915) isolated the fungus from rotted potatoes and identified as *Rhizoctonia destruens* Tassi. However, later, studies showed that, the fungus involved was *S. rolfsii* (Ramakrishnan, 1930). Higgins (1927) worked detail physiology and parasitism of *S. rolfsii*. However, its perfect stage was first studied by Cruzi (1931) and proposed generic name as *Corticium*. Mundkur (1934) successfully isolated the perfect stage of *S. rolfsii*. Sclerotium is soil inhabitant basidiomycetes, produces abundant white fluffy, branched mycelium that forms numerous sclerotia but is usually sterile (does not produce spores) and cause serious diseases on many hosts by affecting the roots, stems, tubers, corns and other plant parts that develop in or on the ground. The perfect stage of the fungus is *Aethalium rolfsii*. In fact, the fungus' growth is so fast, Rolfs mentioned that "if the temperature is 80–90°F, in 48 hours you will have a growth that will in appearance rival swan down." Both in culture and in plant tissue, a fan-shaped mycelial expanse may be observed growing outward and branching acutely. The fungus produces two types of hyphae. Coarse, straight, large cells (2–9 μm × 150–250 μm) have two clamp connections at each septation, but may exhibit branching in place of one of the clamps. Branching is common in the slender hyphae (1.5–2.5 μm in diameter), which tend to grow irregularly and lack clamp connections. Slender hyphae are often observed penetrating the substrate. Sclerotia (0.5–2.0 mm diameter) begin to develop after 4–7 days of mycelial growth. Initially a felty white appearance sclerotia quickly melanise to a dark brown coloration. Townsend and Willetts (1954) recognize four zones in the mature Sclerotium: (i) thick skin, (ii) rind of thickened cells, (iii) cortex of thin walled cells, and

(iv) medulla containing filamentous hyphae. Sclerotia forming on a host tend to have a smooth texture, whereas those produced in culture may be pitted or folded. Serving as a protective structure, sclerotia contain viable hyphae and serve as primary inoculum for disease development.

6.2.1.2 Symptoms

The infected plants gradually lose their color and turn pale, followed by drooping. The affected roots, particularly the collar portion turn yellow-ish-brown. Affected plants can be easily pulled out from the soil. White to tan-brown mustard seed like sclerotia are seen around the infected roots. The symptoms may be extended on stem, causing shriveling of the stem (the fungus can also be seen naturally causing water-soaked spots on leaves) and finally result in the death of the plants (Kolte, 1985).

6.2.1.3 Disease Cycle

The fungus overwinters mainly as sclerotia. Pathogen is spread by con-taminated tools, infected transplants seedling, moving water, infested soil, infected vegetables and fruits and in some hosts as sclerotia mixed with the seed. The fungus attacks tissues directly. However, the mass of myce-lium it produces secretes oxalic acid and also pectinolytic, cellulolytic and other enzymes and it kills and disintegrates tissues before it actually penetrates the host. Fungus once establishes in the plants, advances and produces mycelium and sclerotia quite rapidly, especially at high moisture and high temperature from 30 to 35°C.

6.2.1.4 Disease Management

Management of collar rot disease is difficult. Crop rotation provides only partial control. Cultural practices, for example, deep summer plowing to bury the fungal sclerotia in surface debris, ammonia fertilizations, and calcium compounds application are effective in controlling the diseases. Soil solarisation and use of Pentachloronitrobenzene (PCNB) which is sold under the name of Brassicol, Quintozone or Terrachlor are very effective for controlling this disease. The control is attributed to the hydrolysis products of glucosinolates in to allyl and butenyl isothicyanates which are toxic to *Pythium aphanidermatum, Sclerotium rolfsii, Sclerotinia sclerotiorum* and *Phytophthora capsici* (Singh, 2009).

6.2.2 CERCOSPORA LEAF SPOT

6.2.2.1 Causal Organism

Cercospora leaf spot is caused by *Cercospora kikuchii* (Teleomorph – *Mycosphaerella*). The fungus produces long, slender and colorless to dark, straight to slightly curved, multicellular conidia on short dark conidiophores. Conidiophores arise from the plant surface in the clusters through stomata and form conidia successively on new growing tips. Conidia are detached easily and are often blown long distances by the wind. The fungus is favored by high temperatures and therefore is most destructive in the summer months and in warmer climates. Fungus produces non specific toxin cercosporin which acts as a photosensitizing agent in the plant cells, for example, it kills cells only in light. The pathogen remains over seasons in or on seed and as small black stomata in plant debris.

6.2.2.2 Symptoms

Foliar symptoms usually are seen at the beginning of seed set and occur in the uppermost canopy on leaves exposed to the sun. Affected leaves are discolored, with symptoms ranging from light purple, pinpoint spots to larger, irregularly shaped patches typically only on the upper leaf surface. As disease develops, affected leaves may become leathery and dark purple with bronze highlights. Symptoms may be confused with sunburn. Discoloration may extend to the upper stems, petioles and pods. Infection of petioles and severe symptoms may lead to defoliation of the uppermost leaves and give the appearance of a maturing crop. However, petioles of fallen leaves remain attached to the stem, and lower leaves of the plant remain green. Symptoms of purple seed stain are distinct pink to dark purple discolorations of seed. Discolored areas vary in size from small spots to the entire surface of the seed coat; however, infected seeds may not show symptoms.

6.2.2.3   Disease Cycle

The fungus survives winter in infected crop residue and infected seed. Mostly early season infections do not cause symptoms but contribute to infection of foliage and pods later in the season. Warm and wet weather is favorable for infection. Foliar symptoms are the result of an interaction between a toxin produced by the fungus and sunlight. Weather conditions during flowering and plant maturity will affect the incidence of purple seed stain. Despite being caused by the same organism, there is no consistent relationship between the occurrence of Cercospora leaf blight and purple seed stain.

6.2.2.4   Management

Use the disease free seeds and resistant varieties to control this disease. Seed treatment is essential to eliminate the seed-borne inoculum. Disinfection of seeds by dip in 0.5% copper sulfate solution for 30 minutes. Foliar application of fungicides namely hexaconazole @ (0.025%), bavistin (0.025%) and chlorothalonil (0.2%) are economic and effective to control this disease. Applications made during pod-filling stages can reduce the incidence of purple seed stain, but may not affect soybean yield. Rotation to non-host crops such as alfalfa, corn and small grains and tillage to bury infested crop residue will reduce pathogen levels. If considering tillage, use proven conservation practices to maintain soil quality.

6.2.3   *DOWNY MILDEW*

6.2.3.1   Causal Organism

The causal organism of this disease is *Peronospora manshurica*. Downy mildew is a very common foliar disease of soybeans, but it seldom causes serious yield loss. The pathogen may also infect seed and reduce seed quality. Diseased plants are usually widespread within a field.

6.2.3.2 Symptoms

Seedlings that are infected from oospores on the seed can develop large chlorotic areas on the first and second pairs of true leaves. The disease is more common in late vegetative and reproductive growth stages. Lesions occur on upper surfaces of leaves as irregularly shaped, pale green to light yellow spots that enlarge into pale to bright yellow spots. Older lesions turn brown with yellow-green margins. Young leaves are more susceptible than older leaves, so disease is often found in the upper canopy. Lesion size varies with the age of the leaf affected. On the underside of the leaf, fuzzy, gray tufts may be seen growing from each lesion, particularly when humidity is high or leaves are wet, for example, early in the morning. Infected pods show no external symptoms, but the inside of the pod and seed may be covered with a dried, whitish fungal mass that appears crusty and contains spores. Infected seed can be smaller, appear dull white and have cracks in the seed coat.

6.2.3.3 Disease Cycle

The pathogen is primarily soil borne through oospores lying in the diseased plant debris. *Peronospora manshurica* survives in leaves and on the surface of seed. Extended periods of leaf wetness are favorable for movement of the pathogen. High humidity and moderate temperatures favor infection. The increased resistance of older leaves and higher temperatures midseason usually stop disease development before extensive damage occurs.

6.2.3.4 Management

Use only resistant and certified seeds for sowing. However, many races of the pathogen have been identified, and varieties that are resistant to all known races have not yet been developed. Crop rotation and burial of infested crop residue using conservation tillage practices can reduce pathogen levels. Two- to three-foliar spray of fungicide such as sulfur fungicide should be done at the disease initiation and after that 15 days interval.

6.2.4   FROGEYE LEAF SPOT

6.2.4.1   Causal Organism

Frogeye leaf spot has become more prevalent in north hills zone of India. The causal organism of this disease is *Cercospora sojina*. It is especially problematic in continuous soybean fields. Diseased plants are usually widespread within a field.

6.2.4.2   Symptoms

Early season infections from infected seed result in stunted seedlings. On leaves, lesions are small, irregular to circular and gray with reddish-brown borders that most commonly occur on the upper leaf surface. Lesions start as dark, water-soaked spots that vary in size, and as lesions age, the central area becomes gray to light brown with dark, red-brown margins. In severe cases, disease can cause premature leaf drop and will spread to stems and pods. Symptoms on stems are not as common or distinctive as foliar symptoms and appear as narrow, red brown lesions that turn light gray with dark margins as they mature. Lesions on pods are circular or oval shaped and are initially red-brown and turn to light gray with a dark brown margin. Seed close to lesions on pods can be infected. Infected seeds have light to dark gray discolored blotches that vary in size and cover the entire seed in severe cases. The seed coat often cracks.

6.2.4.3   Disease Cycle

The fungus survives in infested crop residue and infected seed. Early season infections contribute to infection of foliage and pods later in the season. Warm, humid weather promotes spore production, infection and disease development. Young leaves are more susceptible to infection than older leaves, but visible lesions are not seen on young, expanding leaves because the lesions take two weeks to develop after infection. It is common for disease to be layered within the canopy. This is a result of little to no infection during dry periods and higher levels of infection during wet or humid weather.

6.2.4.4   Management

Resistant varieties are available and should be used where disease is a potential problem. Several races of the pathogen have been identified, and varieties with resistance to all known races are available. Crop rotation and tillage will reduce survival of *Cercospora sojina*. Crops not susceptible to this pathogen are alfalfa, corn and small grains. If tillage is considered to promote decay of crop residue, great care should be taken to minimize soil erosion and maintain soil quality. Foliar fungicides applied during late flowering and early pod set to pod-filling stages can reduce the incidence of frogeye leaf spot and improve seed quality and yield.

6.2.5   SEPTORIA BROWN SPOT

6.2.5.1   Causal Organism

Brown spot is the most common foliar disease of soybean. The pathogen of this disease is *Septoria glycines*. Disease develops soon after planting and is usually present throughout the growing season. Yield losses depend on how far up the canopy the disease progresses during grain fill. Diseased plants are usually widespread within a field.

6.2.5.2   Symptoms

Symptoms are typically mild during vegetative growth stages of the crop and progress upward from lower leaves during grain fill. Infected young plants have purple lesions on the unifoliate leaves. Lesions on later leaves are small, irregularly shaped and dark brown, and are found on both leaf surfaces. Adjacent lesions can grow together and form larger, irregularly shaped blotches. Infected leaves quickly turn yellow and drop. Disease starts in the lower canopy and, if favorable conditions continue, will progress to the upper canopy. Lesions on stems, petioles and pods are not as common, but appear as brown, irregularly shaped spots ranging from small specks to 1/2 inch in diameter.

6.2.5.3   Disease Cycle

Warm and wet weather favors the disease development. The fungus survives on infected leaf and stem residue. Disease usually stops developing during hot and dry weather, but may become active again near maturity or when conditions are more favorable.

6.2.5.4   Management

Use of resistant variety is good source of managing this disease but there are no known sources of resistance, but differences in susceptibility occur among soybean varieties. The host range of *Septoria glycines* includes other legume species and common weeds such as velvet leaf. Crop rotation with non-host crops such as alfalfa, corn and small grains and incorporation of infested crop residue into the soil will reduce the survival of *Septoria glycines*. If tillage is an option, use conservation tillage practices to maintain soil quality. Foliar fungicides labeled for brown spot control are available. Applications made after appearance of the disease may slow the rate of disease development into the middle and upper canopy and protect yield.

6.2.6   SOYBEAN RUST

6.2.6.1   Causal Organism

The causal organism of soybean rust is *Phakopsora pachyrhizi*. Soybean rust is an aggressive disease capable of causing defoliation and significant yield loss. Soybean rust is an endemic to India and found in most soybean growing areas of the world.

6.2.6.2 Symptoms

Soybean plants are susceptible at any stage of development, but symptoms are most common after flowering. Early symptoms of rust infection begin on lower leaves. Lesions begin to form on lower leaf surfaces, starting as small, gray spots and changing to tan or reddish-brown. Lesions are scattered within yellow areas that appear translucent if the affected leaves are held up to the sun. Mature lesions contain one to more small pustules that usually occur on lower leaf surfaces. These pustules produce uredospores and spore production may continue for weeks. Premature defoliation and early maturity occurs while an infection is severe.

6.2.6.3 Disease Cycle

The rust pathogen can only survive on green tissue; thus, the pathogen is unable to survive in areas where killing frosts eliminate susceptible hosts. The movement of rust depends on rust spores increasing at sites where the pathogen has survived the winter, dispersal of the spores to new areas and establishment of the disease in those areas. These steps need to be repeated several times within a growing season in order for rust to cause an epidemic in the country. When spores land in new areas, infection takes place only when prolonged periods of leaf wetness (6 to 12 hours) and moderate temperatures occur in those areas. Cool, wet weather or high humidity favor soybean rust epidemics. Dense canopies also can provide ideal conditions that encourage disease development. Infection can spread rapidly to middle and upper leaves once the canopy closes.

6.2.6.4 Management

A limited number of resistant breeding lines have been identified; however, there is currently some commercially available soybean rust resistant variety (DSb 21) in India. Resistant varieties have been released in other countries, but none are resistant to all known races of the pathogen. Currently, foliar fungicides are the only viable option for managing soybean rust. To manage the disease effectively and profitably, fungicides need to be sprayed prior to infection or, at the latest, very soon after initial infection, for example, hexaconazole @ 0.1% or propiconazole @ 0.1% (Singh, 2009). National and local spread of soybean rust can be tracked to help gage if/when to start scouting or initiate fungicide applications.

6.2.7 CHARCOAL ROT

6.2.7.1 Causal Organism

Charcoal rot can be an important disease and is most yield-limiting when weather conditions are hot and dry. This disease is caused by *Macrophomina phaseolina*. This disease is more common is southern and North Eastern part of the India and causes huge losses.

6.2.7.2 Symptoms

Symptoms of charcoal rot usually appear after flowering. Initial symptoms are patches of stunted or wilted plants. Leaves remain attached after plant death. The lower stem and taproots of these plants are discolored light gray or silver. When stems are split, black streaks are evident in the woody portion of the stem. In addition, the fungus produces numerous tiny, black fungal structures called microsclerotia that are scattered throughout the pith and on the surface of taproots and lower stems. These microsclerotia give the tissue a charcoal-like appearance. Infected seed either show no symptoms or having microsclerotia embedded in seed coat cracks or on the seed surface. Infected seed have lower germination, and if seed germinates, the seedlings usually die within a few days.

6.2.7.3 Disease Cycle

The fungus survives in soil or soybean residue as microsclerotia. Microsclerotia infect roots of soybean plants, sometimes very early in the season. Many environmental factors like temperature, humidity, rainfall, etc. affect microsclerotia survival, root infection and disease development. The fungus is more abundant in soil when pH is very acidic or alkaline. Charcoal rot is most prevalent during hot, dry weather, especially when it occurs during the flowering/pod formation stages.

6.2.7.4 Management

In summer crops, irrigation lowers soil temperature and increases soil moisture. These conditions are unfavorable for the disease. Most efforts on control of *M. phaseolina* involve management of populations of microsclerotia. Growing small grains, such as wheat or barley, can reduce microsclerotia numbers. Corn is also a host of *M. phaseolina* so it will not reduce levels of the fungus when planted in rotation with soybeans. The fungus is less damaging to corn than to soybeans. Fields with minimal or no tillage may have fewer symptoms because of lower soil temperatures and greater water-holding capacity. Avoid excessive seeding rates so that plants do not compete for moisture, which increases disease risk during a dry season.

6.2.8 FUSARIUM WILT AND ROOT ROT

6.2.8.1 Causal Organism

Fusarium is a very common soil fungus, and more than 10 different species are known to infect soybean roots and cause root rot. The species *Fusarium oxysporum* is responsible for causing Fusarium wilt. Although Fusarium root rot is a widespread disease in the country, the economic impact on yield is not well documented.

6.2.8.2 Symptoms

Symptoms of Fusarium wilt are more noticeable under reduced moisture and hot conditions and are often misdiagnosed as those of Phytophthora root rot. Infected plants have brown vascular tissue in the roots and stems and show wilting of the stem tips. However, external decay or stem lesions are not seen above the soil line. Foliar symptoms include scorching of the upper leaves, while middle and lower canopy leaves can turn chlorotic and later wither and drop from the plant. Young plants are at the greatest risk to root rots caused by *Fusarium* species. Infected plants may exhibit poor or slow emergence, and seedlings are often stunted and weak. Seedlings with root rot have reddish-brown to dark brown discolored roots. Infected plants may have poor root systems and poor nodulation, which may cause the plants to wilt and finally die.

6.2.8.3   Disease Cycle

The fungus survives in the soil either as spores or as mycelium in plant residue. Certain weeds may serve as hosts to some pathogenic *Fusarium* species. The fungi can infect plants at any stage of soybean development but infection is particularly favored when plants are weakened. Stresses such as herbicide injury, high soil pH, iron chlorosis, nematode feeding and nutritional disorders can all predispose plants to infection. After infection, damage to plants can be worsened if soil moisture is limited because of the compromised root systems.

6.2.8.4   Management

Varieties have varying levels of susceptibility, but no resistant varieties have been described. Reducing or eliminating stress factors, such as use of herbicides that cause injury to soybeans, wet soils and soybean cyst nematode, can help reduce root rot problems. Growing of tolerant varieties to iron deficiency chlorosis should be considered if the root rot seems associated with iron deficiency chlorosis. If *Fusarium* is a problem in a field, seed treatments with bavistin @ 2g/kg seed may protect seedlings in subsequent years.

6.2.9   POWDERY MILDEW

6.2.9.1   Causal Organism

The powdery of soybean is caused by *Microsphaera diffusa*. The disease is more prevalent in cooler than normal seasons. While this disease is uncommon, when it does show up in fields, there can be noticeable yield loss.

6.2.9.2 Symptoms

The most common and characteristic sign of powdery mildew is white, powdery fungal growth appeared on aboveground plant parts, particularly the upper surface of leaves. Powdery mildew usually does not appear until mid- to late reproductive stages. Initially, small fungal colonies form and grow together as they enlarge. Eventually, entire surfaces of infected plant parts are covered with white fungal growth. Advanced symptoms include yellowing of plant tissues and premature defoliation.

6.2.9.3 Disease Cycle

Microsphaera diffusa is a biotrophic parasite. The fungus survives in infested crop residue. In general belief had been that the pathogen survives between crop seasons through cleistothecia in soil. The favorable conditions for the disease development are cool, cloudy weather and low humidity. Powdery mildew of soybean is severely affect when crop sown in late season.

6.2.9.4 Management

Planting of resistant varieties to minimize the disease and early sowing to escape the disease. Chemicals such as sulfur fungicide effectively manage the powdery mildew; however, there are limited situations where fungicide use will be profitable. Efficacy of some plant extracts and plant products against the pathogen has been experimentally demonstrated Nemadole (a neem product) and *Allium cepa* (onion), *Allium sativum* (garlic), rhizome of ginger and neem leaves (*Azadiracta indica*) are non phytotoxic but fungicidal and at par with Karathane in the suspension of powdery mildew of pea. Several fungi such as *Ampelomyces*, *Tilletiopsis* and *Verticillium* and insects (*Thrips tabaci*) are natural biocontrol agents of the powdery mildew.

6.2.10   ANTHRACNOSE STEM BLIGHT

Colletotrichum truncatum (hemibiotrophoc fungus) causes the anthracnose stem blight of soybean. Anthracnose is generally a late season disease that is prevalent on maturing soybean stems throughout the world. Soybean, however, is susceptible to infection throughout the growing season. Diseased plants are usually widespread within a field.

6.2.10.1   Symptoms

Infected seed may or may not show symptoms. When seed symptoms do occur, they appear as brown discoloration or small gray areas with black specks. Foliar symptoms include reddish veins, leaf rolling and premature defoliation. On stems and petioles, symptoms typically appear as irregularly shaped red to dark brown blotches during early reproductive stages. Damping off may occur if infected seed is planted. Leaves, pods and stems may also be infected without showing symptoms. Petiole infection may result in a shepherd's crook. Early infection of leaf petioles may cause premature defoliation and yield loss. Infection of young pods results in seedless pods at maturity while pods infected later contain seeds that are infected. Near maturity, black fungal bodies that produce small, black spines and spores are evident on infected stems, petioles and pods.

6.2.10.2   Disease Cycle

The fungus overwinters as mycelium in crop residue or infected seed. Although plant stand may be affected by early season infection, most infection occurs during the reproductive stage of the crop. Spores produced by the fungus are sensitive to drying; thus, free moisture for 12 hours or longer is necessary for successful infection. Warm, wet weather favors infection and disease development. The most important factors affecting the infection are temperature and moisture. Moderate temperatures between 13° and 26°C favor infection. No infection occurs at temperatures above

27° and at 13°C also the disease is considerably reduced. A relative humidity of above 92% is necessary for infection, the optimum being close to 100%. A 10 hour wet period is reported to be necessary for conidial infection and new lesions usually appear in 3–7 days depending on prevailing temperature.

6.2.10.3   Management

There are no known sources of resistance to anthracnose, but soybean varieties differ in susceptibility. The seed must be disease free hence it should be collected from only healthy pods. Usually, seed produced in dry areas or free from infection. Crop rotation and tillage will reduce survival of *Colletotrichum* species. Non-legume crops such as corn are not susceptible to this pathogen. If tillage is considered, great care should be taken to minimize soil erosion and maintain soil quality. Foliar fungicides labeled for anthracnose are available. Benlate, Ziram, Vitavax, Ferbam and lime sulfur, in order listed had been recommended for foliar sprays. Bavistin, Vitavax and Agroson GN were recommended for seed treatment. Applications should be made during the early to mid-reproductive growth stages of the crop, although there are limited situations where fungicide use will be profitable. There are many reports of biological control of the anthracnose of bean through seed bacterization and through inoculation with avirulent strains of the pathogen (Sticher et al., 1997; Van Loon et al., 1998).

6.2.11   PHYTOPHTHORA ROOT AND STEM ROT

6.2.11.1   Causal Organism

Phytophthora sojae causes the Phytophthora root and stem rot of soybean. Phytophthora root and stem rot is an economically important disease of soybeans that is most severe in poorly drained soils. Diseased plants often occur singly or in patches in low-lying areas of the field that are prone to flooding.

6.2.11.2   Symptoms

The most characteristic symptom of Phytophthora root rot, however, is a dark brown lesion on the lower stem that extends up from the taproot of the plant. *Phytophthora sojae* can infect soybeans at any growth stage from seed to maturity. Early season symptoms include seed rot and pre- and post-emergence damping off. Stems of infected seedlings appear water-soaked, while leaves may become chlorotic and plants may wilt and die. On older plants, symptoms vary depending on the variety. For susceptible plants, leaves become chlorotic between the veins and plants wilt and die, with the withered leaves remaining attached. Varieties that are not fully susceptible may appear stunted, but plants are typically not killed. The lesion often reaches as high as several nodes and will girdle the stem and stunt or kill the plant.

6.2.11.3   Disease Cycle

Phytophthora sojae survives on crop residue or in the soil as oospores. Optimum soil moisture is 15 to 20% is needed for oospores germinate to produce structures that release swimming spores, called zoospores, under saturated soil conditions. The zoospores are attracted to soybean roots. Infection occurs via the roots, and from there the pathogen colonizes the roots and stems. Disease is most common in poorly drained soils, but may occur in other soils as well.

6.2.11.4   Management

Management of Phytophthora root rot is by planting resistant varieties. Many race-specific resistance genes (called *Rps* genes) to *Phytophthora sojae* have been identified in soybean breeding lines. Some of these genes have been incorporated in commercial soybean varieties; thus, there are soybean varieties available that have complete resistance to a specific race of *Phytophthora sojae*. There are numerous races (now called pathotypes) of *Phytophthora sojae*, and many pathotypes can exist in a single field. Furthermore, new pathotypes can develop that can infect varieties with specific *Rps* genes. Partial resistance is available to *Phytophthora sojae*. Partial resistance is effective against all races of *Phytophthora sojae*; however, it is only expressed after the first true leaves emerge, not in very young seedlings. Continuous soybean production may increase disease severity. But rotation to non-hosts may reduce disease severity because oospores can survive in soil for long periods of time. Disease is more severe in no-till fields because these fields can be wetter. If tillage is considered to improve drainage, use proven conservation tillage practices to maintain soil quality. Where *Phytophthora sojae* is a serious problem, seed treatments with metalaxyl as an active ingredient can provide some protection. Seed treatments are especially helpful with poor quality seed and in fields with a history of this problem.

6.2.12 POD AND STEM BLIGHT AND PHOMOPSIS SEED DECAY

6.2.12.1 Causal Organism

The causal organism of these diseases is *Diaporthe phaseolorum* var. *sojae* and *Phomopsis longicolla.* Pod and stem blight is one of three diseases that make up the *Diaporthe-Phomopsis* complex. Other diseases in this complex include seed decay and stem canker. Stems, petioles, pods and seeds are severely affected by this disease.

6.2.12.2 Symptoms

The most characteristic symptoms of pod and stem blight are linear rows of black specks on mature stems of soybeans. The specks, which are flask-shaped fruiting structures of the fungus known as pycnidia, can be seen during the season on prematurely killed petioles or stems. Poor seed quality may result from infection. Seed infection occurs only if pods become infected. Pod infection can occur from flowering onwards, but extensive seed infection does not occur until plants have pods that are beginning to mature. Insect damage to pods favors development of seed infections.

Phomopsis-infected seed are cracked and shriveled and are often covered with chalky, white mold. Infected seedlings have reddish-brown, pinpoint lesions on the cotyledons or reddish-brown streaks on the stem near the soil line. If infected seeds are planted, emergence may be low due to seed rot or seedling blight.

6.2.12.3　Disease Cycle

The fungi survive winter in infected seed and infested crop residue. Certain weeds may serve as hosts to some pathogenic *Diaporthe* and *Phomopsis* species. Infection can occur early in the growing season without causing symptoms. Disease is favored by warm, humid weather, when soybean plants are maturing. Also, disease is more severe if harvest is delayed.

6.2.12.4　Management

Sources of resistance have been identified, and variation in seed infection has been reported among commercial soybean varieties. Unfortunately, there currently are no resistant varieties or lists of seed reactions of current varieties available. Varieties with an earlier relative maturity for a region are at greater risk of Phomopsis seed decay and pod and stem blight than fuller-season varieties. Do not plant seed with a high incidence of infection. Crop rotation and tillage will reduce survival of *Diaporthe* and *Phomopsis* species. Non-host crops include corn. If tillage is considered to promote decay of pathogen-infested residue, be careful to minimize soil erosion and maintain soil quality. Application of foliar fungicides near R5 stage can protect seed quality, but may not affect yield. Harvest early maturity varieties first to lower the incidence of seed rot. Fungicidal seed treatments with Thiram, Ziram and Apron are effective against *Phomopsis* species. Treating *Phomopsis*-infected seed lots may increase germination and improve plant establishment.

6.2.13 PYTHIUM ROOT ROT

6.2.13.1 Causal Organism

Several species of *Pythium* are reported to cause this disease. Early planting dates increase the risk of disease in the major soybean growing areas. Diseased plants often occur singly or in small patches in low-lying areas of the field that are prone to flooding.

6.2.13.2 Symptoms

Pythium species cause pre- or post-emergence damping off. Infected seed appear rotted and soil sticks to them. Infected seedlings have water-soaked lesions on the hypocotyl or cotyledons that develop into a brown soft rot. Diseased plants are easily pulled from the soil because of rotted roots. Older plants become resistant to soft rot, but root rot may cause plants to become yellow, stunted or wilted if infection is severe.

6.2.13.3 Disease Cycle

The pathogen survives either in plant residue or in soil as oospores. Severity of disease depends on the amount of the pathogen in the soil, plant age and environmental conditions at the time of infection. Saturated soil is critical for infection for all *Pythium* species. As *Phytophthora*, *Pythium* produces zoospores that swim in free water and infect the roots of plants. In general, *Pythium* species that are prevalent in the north infect plants at lower temperatures (10–15°C), and *Pythium* species in the south infect plants at warmer temperatures (30–35°F), although there are exceptions.

6.2.13.4 Management

Planting in cold, wet soils should be avoided to reduce infection by *Pythium* species that infect at low temperatures. Where *Pythium* is a problem, seed treatments with Apron, metalaxyl or strobilurins as active ingredients can provide some protection. Resistance to metalaxyl/mefenoxam has been accepted; however, they are generally considered more effective than strobilurins. Soil application of metalaxyl at transplanting time followed by weekly sprays of potassium phosphonate (1 g/L) plus acibenzolar-S-methyl (0.025 g/L) also significantly reduced root rot infection. No-till soils often have higher soil moisture and lower soil temperatures, factors that increase the risk of *Pythium* infection. If tillage is considered to improve drainage, use conservation tillage practices to maintain soil quality.

6.2.14 RHIZOCTONIA ROOT ROT

6.2.14.1 Causal Organism

The pathogen of Rhizoctonia root rot is *Rhizoctonia solani.* Rhizoctonia root rot is one of the most common soil borne diseases of soybeans. Diseased plants usually occur singly or in patches in the field. Disease is typically more common on the slopes of fields.

6.2.14.2 Symptoms

Rhizoctonia infects young seedlings, causing pre- and post-emergence damping off. Infected seedlings have reddish-brown lesions on the hypocotyls at the soil line. These lesions are sunken, remain firm and dry and are limited to the outer layer of tissue. If seedlings survive the damping off phase, infections may expand to the root system, causing a root rot. The root rot phase may persist into late vegetative to early reproductive growth stages. Older infected plants may be stunted, yellow and have poor root systems.

6.2.14.3 Disease Cycle

The fungus survives on plant residue or in soils as sclerotia. When soils warm, the fungus becomes active and infection may occur soon after seed is planted. The fungus grows better in aerated soils; thus, disease is more severe on light and sandy soils. Symptoms may disappear if infected plants grow out of the root rot problems although plants may remain stunted.

6.2.14.4   Management

Resistance has been reported in some varieties; however, there are no varieties being developed for resistance to Rhizoctonia root rot. Unfortunately, many strains of *Rhizoctonia* can infect corn, alfalfa, dry bean and some cereal crops. Eliminating stress factors, such as use of herbicides that cause injury to soybean roots, can help reduce root rot problems. Most fungicide seed treatments such as Bavistin or Benlate (2 g/kg) are effective against *Rhizoctonia* and same fungicide can be used as foliar sprays 2–3 times gives good control.

6.3   BACTERIAL DISEASE

6.3.1   BACTERIAL PUSTULE

6.3.1.1   Causal Organism

This disease caused by *Xanthomonas axonopodis* pv. *glycines*. Bacterial pustule occurs mid- to late season when temperatures are warmer and more favorable for disease development. Symptoms may be mistaken for bacterial blight, Septoria brown spot or soybean rust. Diseased plants are usually widespread within a field.

6.3.1.2   Symptoms

Lesions are found on outer leaves in the mid- to upper canopy. Lesions start as small, pale green specks with elevated centers and develop into large, irregularly shaped infected areas. Unlike bacterial blight, no water soaking is associated with lesions, but each lesion is surrounded by a greenish-yellow halo. A pustule may form in the center of some lesions, usually on the lower leaf surfaces. Pustules crack open and release bacteria. Bacterial pustule will not cause leaves to tatter like bacterial blight.

6.3.1.3　Disease Cycle

Bacteria survive winter in crop residue and seeds and are spread by rain and wind. Infection occurs through leaf stomata or wounds. Rainy weather favors disease development. Unlike bacterial blight, high temperatures do not slow disease development.

6.3.1.4　Management

Avoid planting extremely susceptible varieties. Some varieties are marketed as resistant to this disease. Rotation and tillage reduce survival of *Xanthomonas axonopodis* pv. *glycines*. Other legume crops may be hosts; non-hosts include alfalfa, corn and small grains. If tillage is considered, use proven conservation tillage practices to maintain soil quality.

6.4　VIRAL DISEASE

6.4.1　*SOYBEAN YELLOW MOSAIC*

This viral disease is the most destructive disease of soybean in India. It was first reported in 1960 and is now known to occur throughout the country. The loss of yield depends upon the stage at which the crop is infected. If the infection is early in the season there may be total loss of seed yield.

6.4.1.1　Causal Organism

Four viruses causing yellow mosaic disease of legumes across the South Asia have been identified as bipartite begomoviruses (genus *Begomovirus*, family *Geminiviridae*). The soybean strain of MYMV occurring in north India is distinct from the strain occurring in southern and western India (Usharani et al., 2004). A strain of MYMIV, designated as MYMIV-Cp causes golden mosaic of cowpea. It has restricted host range and transmission by *Bemisia tabaci*. These viruses have evolved independently of the begomoviruses in plant species of other families. The paired particles of the virus measure 30×18 nm. The particle contains two circular ssDNA molecules, which account for 20% of the particle weight. The coat protein contains one polypeptide with MW of 28.5 kDa.

6.4.1.2   Symptoms

Disease appeared in the field when the crop is about one month old. Two types of symptoms appeared depending upon the host response. The general pattern of development of both symptoms is the same. The first visible sign of the disease is the appearance of yellow spots scattered on the lamina surface. They are mostly round in the shape. In yellow mottle, the spots are diffuse and expand rapidly. The leaves show yellow patches alternating with green areas that also turn yellow. Such completely yellow leaves gradually change to a whitish shade and ultimately become necrotic. These color changes of affected plants are so conspicuous that the disease can be spotted in the field from a distance. In case of necrotic mottle, the center of yellow spots develops necrosis, which is demarcated by finer veins. The virus becomes systemic in the plant and all newly formed leaves show signs of mottle from the very beginning. Number of size of spots per plant and seeds per pod are greatly reduced.

6.4.1.3   Management

Certified and healthy seeds use for sowing. Cultivar PK 21–22 of soybean is tolerant to the disease. Control of the disease through prevention of population build up of the vector has also been recommended. Sprays of 0.1% metasystox, starting when the crop is about a month old or as soon as a single diseased plant is seen in the field, can give relief from severe incidence of the disease. Anthio is effective at 0.2% when used as spray 3 times.

6.5   NEMATODE DISEASE

6.5.1   *SOYBEAN CYST NEMATODE*

In India the most important pathogen of soybean is soybean cyst nematode (SCN). In high-yielding production fields or during years when soil moisture is plentiful, damage from SCN may not be obvious. However, yield losses up to 40% on susceptible varieties are still possible. When symptoms are associated with damage, infected plants usually occur in patches within a field.

6.5.1.1   Causal Organism

The soybean cyst nematode that causes the disease is known as *Heterodera glycines.* The body of females is swollen, pearly white and lemon shaped and usually varies between 0.6–0.8 mm in length and 0.3–0.5 mm in diameter. The male is wormlike about 1.3 mm long and 30–40 μm in diameter. The males remain in the root for a few days during which they may or may not fertilize the females and then they move into the soil and soon die. Cysts are typically lemon shaped. Mature cysts of Indian populations measures 470–1010 × 370–730 μm. Each females produces 300–600 eggs most of which remain inside her body when the females die. Eggs in the gelatinous matrix may hatch immediately and the emerging second stage juveniles may cause new infestation.

6.5.1.2   Symptoms

Obvious symptoms may not develop, even though yield loss occurs. Noticeable symptoms of SCN include stunting, slow or no canopy closure and chlorotic foliage. Infected plants have poorly developed root systems. Soybean cyst nematode infection also may reduce the number of nodules formed by the beneficial nitrogen-fixing bacteria necessary for optimum soybean growth. Signs of SCN include white females that are most readily seen in the field starting about six weeks after crop emergence. To see them, roots must be dug and soil carefully removed. However, the only way to get a reliable diagnosis as to the amount of SCN in the soil is through analysis of a properly collected soil sample by a diagnostic laboratory. Plant damage is not just limited to direct and indirect effects of feeding by the nematodes. Wounds caused by infecting nematodes and by maturing females serve as entry points for other soil borne pathogens. Diseases such as brown stem rot, Rhizoctonia root rot, sudden death syndrome and charcoal rot are more severe in the presence of SCN.

6.5.1.3   Disease Cycle

SCN survives in the soil as eggs within dead females called cysts. These eggs can survive several years in the absence of a soybean crop. The second-stage juvenile (J2) hatches from the eggs and infects soybean plants. After infection, these juveniles migrate to the vascular system before setting up specialized feeding cells within the root. As they feed, the nematodes become immobile. The juveniles molt three more times before maturing into adults, with females becoming so large they burst through the outer surface of the roots. A female will produce 200 to 300 eggs that are deposited in an external egg mass or are retained within her body. Soybean cyst nematode can complete four or more generations during the growing season, depending on planting date, soil temperature and length of the growing season, host suitability, geographic location and maturity group of the soybeans. Conditions that favor soybean growth are also favorable for SCN development. High soil pH may be used to predict where SCN is more problematic. Areas of fields with soil pH levels of 7.0 to 8.0 typically have more SCN compared to areas with soil pH 5.9 to 6.5.

6.5.1.4   Management

The number of SCN in a field can be greatly reduced through proper management, but it is impossible to eliminate SCN from a field once it is established. Soil tests are recommended prior to every third or fourth soybean crop to monitor SCN population densities (numbers). Resistant varieties are available to manage SCN. The three most common sources of resistance are PI 88788 (most common), PI 548402 (Peking) and PI 437654 (also referred to as Hart wig or PUSCN-14). Resistant varieties are not resistant to all SCN populations. Most resistant varieties contain only one source of genetic resistance. Rotating sources of SCN resistance may help prevent the development of more damaging SCN populations. SCN-resistant varieties, even high-yielding varieties, can vary considerably in how well they control nematode population densities. Greater SCN reproduction will result in a higher SCN egg population in the soil at the end of the growing season, and consequently, higher numbers of SCN in subsequent seasons. Thus, growers must consider how SCN-resistant soybean varieties affect SCN population densities, in addition to how well the varieties yield, to maintain the long-term productivity of the land for soybean production. If SCN is a problem, rotation should include non-host crops (usually corn) and resistant soybean varieties. Years of non-host crops may decrease SCN numbers by as much as 90% south, but only 10–40% in the north. Maintaining adequate soil fertility, breaking hardpans, irrigation and controlling weeds, diseases and insects improves soybean plant health. These practices help plants compensate for damage by SCN, but do not decrease SCN numbers. Zero tillage practices may slow SCN movement and lower population densities. Soil that remains on tillage and harvest equipment can move SCN and should be removed before equipment is relocated from an infested to a non-infested field. Seed treatments labeled for use on SCN may provide early season protection. A limited number of nematicides labeled for use on SCN can be applied at planting.

KEYWORDS

- mildew
- nematode
- rot
- rust
- soybean
- spot

REFERENCES

1. Cruzi, M. (1931). Aleumi csidi 'Canorena Pedale' da sclerotium observation in Italia. *Atti Academia Nazionale des* Lincei. Rendiconti, 14, 233–236.
2. Higgins, B. B. (1927). Physiology and parasitism of *Sclerotium rolfsii* Sacc. *Phytopath*, 17(7), 417–448.
3. Kolte, S. J. (1985). Diseases of annual edible oilseed crops. Vol. 3. Rapeseed-Mustard and Sesame Diseases. CRC Press, Inc. Boca Raton, Florida. USA, p. 97.

4. Kumar. (2007). A Study of Consumer Attitudes and Acceptability of Soy Food in Ludhiana. *MBA Research Project Report, Department of Business Management, Punjab Agricultural University, Ludhiana, Punjab.*

5. Mundkur, B. B. (1934). Perfect stage of *Sclerotium rolfsii* Sacc. in pure culture. *J. Agric. Sci.*, 4, 779–781.

6. Ramakrishnan, T. S. (1930). A wilt of zinnia caused by *Sclerotium rolfsii. Madras Agric. J.*, 16, 511–519.

7. Rolfs, P. H. (1892). Tomato blight some hints. *Bulletin of Florida Agricultural Experimental Station*, p. 18.

8. Saccardo, P. A. (1911). Notae mycologicae. *Annales Mycologici*, 9, 249–257.

9. Singh Chhidda, Singh Prem, Singh Rajbir. (2003). *Modern Techniques of Raising Field crops*, Oxford and IBH Publishing Co. Pvt. Ltd. pp. 273.

10. Singh, R. S. (2009). Plant Diseases. Oxford and IBH Publishing Co. Pvt. Ltd. New Delhi.

11. Sticher, L. B. Mauch-Mani, Metroux, (1997). Systemic acquired resistance. *Annual Review of Phytopathology*, 35, 235.

12. Townsend, B. B., H. J. Willetts. (1954). The development of sclerotia of certain fungi. Ann. Bot. 21, 153–166.

13. Usharani, K. S., Surendranath, B., Haq, Q. M. R., Malathi, V. G. (2004). Yellow mosaic virus infecting soybean in northern India is distinct from the species infecting soybean in southern and western India. *Current Science*, 86(6), 845.

14. Van Loon, L. C., Bakker, P. A. H. M., Pieterse, C. M. J. (1988). Systemic resistance induced by rhizosphere bacteria. *Annual Review of Phytopathology*, 36, 453.

CHAPTER 7

DISEASES OF LINSEED AND THEIR MANAGEMENT

SUNIL KUMAR

AICRP on Soybean, School of Agricultural Sciences and Rural Development, Nagaland University, Medziphema – 797106, Nagaland, India

CONTENTS

7.1 INTRODUCTION

Linseed (*Linum usitatissimum* L.) varieties are two types, for example, fiber and seed. Linseed is also known as 'alsi'. These differ considerably in character of the plant growth. The fiber plants having tall and slender stem, produces a high amount of good quality fiber and bears seeds of poor quality oil content. Seed flax varieties develop shorter stems with more tendency to branch and usually bears larger seeds and higher oil content. Flax is grown for its use as a nutritional supplement, edible oil and as an ingredient in many wood-finishing products. Flax is also grown as an ornamental plant in gardens. Flax fibers are used to make linen. The Latin species name *usitatissimum* means *most useful*, pointing to the several traditional uses of the plant and their importance for human life. Flax fibers are taken from the stem of the plant and are two to three times as strong as those of cotton. As well, flax fibers are naturally smooth and straight. Europe and North America depended on flax for vegetable-based cloth until the nineteenth century, when cotton overtook flax as the most common plant used for making linen paper. Flax is grown on the Canadian Prairies for linseed oil, which is used as a drying oil in paints and varnish and in products such as linoleum and printing inks. Linseed grown in the warmer regions of the temperate zones as a winter crop and in the cooler regions as a summer crop. Seed flax varieties are generally more grown in the drier areas, whereas fiber flax is grown in the humid regions.

The cultivation of linseed and flax is greatly handicapped by infecting the diseases in the most of the growing areas of the world. In India the disease is of significance since linseed is a major oilseed crop of the commercial importance. The crop is sown from late October to November and harvested in late March to April. The loss from the diseases is about 10 to 100%. Some diseases of the crop causes huge loss are described below:

7.2 ANTHRACNOSE

This disease occurs on both the seed and flax varieties in humid cool areas throughout the world.

7.2.1 PATHOGEN

Colletotrichum linicolum Pethyb and Laff

7.2.2 SYMPTOMS

The symptoms are characteristic of generally of the anthracnose disease. Cankers appear on the cotyledons are circular zonated sunken brown spots that spread under cool and moist conditions to involve the cotyledons and apex of the stem. Seedling blight occurs either pre- or post-emergence, in the later case usually as a stem canker occurs at soil surface level. In the high moister conditions the leaf spots and stem cankers are widespread during growing season. The central portion of the spots shows pinkish mass of spores generally in moist weather condition. Brown spots formation occurs on the capsules and less conspicuous lesions are found on the seed. The acervuli formation appears on the mature lesions of the stem.

7.2.3   DISEASE CYCLE

Infected seed and crop reduce are the important sources of infection. When infected seeds are sown the lesions develop and produce spore masses on the cotyledons. These initiate the disease in the crop. In cool climate regions decomposition of plant debris is slow and fungus can survive for up to 2 years on the debris if it is not buried in deep soil. Conidia produced on the seedling cankers furnish the primary inoculum. Secondary infection occurs whenever weather conditions are favorable. The percentage of infected seed is an important factor of the severity of the disease during growing season since the secondary inoculum produced will be proportional to the amount of primary inoculum. Stands are reduced and fiber is damaged by the disease (Hirua, 1924, Pethybridge and Lafferty, 1920).

7.2.4   CONTROL

The disease is successfully controlled by the using of disease free seed, seed treatment, crop rotation and resistant varieties. Buda (C.I. 326) and Crystal (C. I. 982) are resistant varieties, whereas Punjab (C. I. 20) is highly susceptible (Ray, 1945). The plant debris should be either removed to a place where linseed is not likely to be grown in the near future or it should be deeply plowed. If the disease assumes serious form the crop should be given fungicidal sprays. In field trials, seed treatment with thiram 75 WP, carbendazim 50 WP, carbendazim + thiram, mancozeb 75 WP, triadimenol 15 DS and metsulfovax 20 WP at 2.5 g/kg seed have given good control of seed borne infection, the best being cabedazim and carbedazim + thiram. Seed bacterization with a strain of *Pseudomonas aeruginosa* (rhizobacteria) and treatment with a derivative of benzothiadiazole also induce systemic resistance against the anthracnose fungus (Bigirimana and Hofte, 2002).

7.3 DAMPING OFF

7.3.1 PATHOGEN

Sclerotinia fuckeliana

7.3.2 SYMPTOMS

The infection of the damping off pathogens first appears on the basal leaf and stem rot when crop is seedling stage. Latter lesions develop on the stem as tan or brown water soaked spot, which may become grayish on drying out. The profuse gray brown sporulation of the fungus occurs on old diseased tissue is a characteristics features. Rotting of plant produce at harvest or in store causes heavy losses. Blight of buds blossom, leaves and stem may also occur and may result into dieback. A sometimes canker formation also occurs on woody plant parts.

7.3.3 DISEASE CYCLE

The fungus overwinters as sclerotia on or within infected tissues that have fallen on ground and as mycelium in dead or living plants. In the spring or early summer sclerotia germinate and produce cylinder stalks terminating at a small disk or cup shaped apothecium. A large number of ascospores are discharged from the apothecia in the air over a period of 2 to 3 weeks.

7.4 FUSARIUM WILT

7.4.1 PATHOGEN

Fusarium oxysporum fsp. *lini*

7.4.2 SYMPTOMS

Linseed plants are infected by wilt at any stage in their development and symptoms vary with varieties and with environmental conditions. Although primarily wilt seedling blight occurs when susceptible seedlings are grown at high temperatures. In typical wilt, the leaves turn yellow or grayish yellow, the apical leaves thicken, growth spots and the plant die and turn light brown. Frequently the plant is only stunted, in which case the leaves turn yellow and fall prematurely or the primary stem dies and new apparently healthy lateral branches develop from the first node. A late infection or a weak attack may be evidenced by premature ripening.

7.4.3 DISEASE CYCLE

The fungus is primarily soil-borne it is persists for several years in the soil to invade the young plants through the roots and develops chiefly in the xylem vessels. High temperatures and low moisture are important factors in the development of the disease and the expression of resistance in most flax varieties (Tisdale, 1916, 1917). Fungus mycelium apparently must be present in the plant tissues to produce wilting (Schuster, 1944). Seed infection occurs and accounts for the spread of the parasite to new areas (Bolly, 1924).

7.4.4 CONTROL

The use of wilt resistant varieties constitutes the chief means of control (Bolly, 1932, Stalkman et al., 1919). Wilt resistance is conditional by several factor pairs and selected strains breed true for different degrees of resistance (Barker, 1923). The nature of wilt resistance has been studied by Nelsan and Dworak (1925) and others. In India, RR 9, NP 12, 21, 124, RR 5B, RR 80 etc. are resistant to wilt disease. Plant inoculated with arbuscular mycorrhizal (AM) fungi show resistant to wilt.

7.5   RUST

Flax rust occurs in the major flax production areas of the world. Specialized races of the rust parasite occur on both the cultivated and wild species of *Linum.*

7.5.1   PATHOGEN

Melampsora lini (Pers.) Lev.

7.5.2   SYMPTOMS

The flax rust parasite is an autoecious, long cycle fungus producing pycnia, aecia and telia on the same plant. The occurrence of pycnia and aecia usually during early part of the growing season and they appear as light yellow to orange yellow sori on the leaves and the stems. The reddish brown uredinium occurs on the both surface of the leaves as well as on the other aerial parts of the plant. The small pustule may be surrounded by the chlorotic zone. The brown to black telia covered by the epidermis occur chiefly on the stems but also on the leaves and the capsules late in the growing season. The telia on the stem do not rupture.

7.5.3   DISEASE CYCLE

The autoecious long cycle rust produces all stages on the flax plant. The teliospores on the crop refuse germinate in the spring to produce the sporidia, which infect the young tissues of the flax plant. The pycnial stage develops, fusion of compatible haploid cells occurs to initiate the binucleate phase and the aecial stage forms from the binucleate fusion hyphae. According to Allen (1934) the fungus is heterothallic. The primary uredial infection develops from the aeciospores throughout the early part of the growing season and secondary infections from urediospores account for much of the later spread. Telia are formed around the uredia and from independent uredial infection as the flax plant matures. The telia persist in the flax straw to renew the cycle. The uredial stage continues development in the regions where the flax plants are growing in both in summer and winter; however the telial material on the crop refuse is the common source of primary inoculum in most flax producing areas. Gold and Stalter (1983) have revealed that the teliospores are formed 10–15 days after inoculation of stems with urediospores.

7.5.4 MANAGEMENT

Crop rotation and removal or plowing under of the flax refuse is important in controlling the epidemic development of the disease. The teliospores on small pieces of infected tissue frequently are carried with the seed therefore careful cleaning of flax seed is important especially when seeding on new hand. Fungicidal sprays have been found useful in controlling the disease but are not economical. Foliar sprays of zineb + copper, mancozeb, ziram benomyl, tridemorph tiradimenol are reported to suppress the disease pressure. Resistant varieties are the best means of the rust control as reviewed by Flor (1941) and Vallega (1944). Varieties namely Jawahar-7, Jawahar-17 and JLS (J)-1 are found resistant against this disease in Madhya Pradesh. Likewise, in Punjab and Himachal Pradesh LC-54, LC-115, K-2 and Himani are found resistant varieties to this disease. Cultivars LC-216, LC-255, LC-256 are resistant to all races of the pathogen prevalent in the hills (Saharan, 1991).

7.6 PASMO DISEASE

Pasmo is a serious disease of flax. This disease causes 50–70% yield losses depending upon its occurrence. Delayed infections cause severe damage to the fiber and a reduction in seed yield; the fiber is weakened and breaks while infected bolls either do not yield seeds or yield undersized and worthless ones.

7.6.1 PATHOGEN

Septoria linicola

7.6.2   SYMPTOMS

The pathogen attack of flax pasmo infects all parts of the plant above the soil surface from the moment of germination to the boll stage, spreading downward along the plant. External symptoms of this disease are apparent at an early stage. Damage is first seen on the cotyledons leaves from which the disease spreads to other leaves, stems, buds and bolls. The initial symptom of the disease is the appearance of patches on the cotyledons leaves, which turn brown, wither, and develop innumerable black dots and drop. Eventually these patches develop on the true leaves at different points on the leaf blade. The patches are more or less rounded and vary in color from greenish yellow in the initial stage to dark brown later. The damaged leaves gradually dry curl and drop, denuding the stalk downward. New patches appear on the upper leaves. The greatest leaf damage occurs at the time of the plant flowering. Damage to the stems usually commences in the lower part. The attack is initially insignificant; the patches are elongated and do not encircle the stem. Later, however, the patches enlarge to few centimeters and encircle the stem and the fruit-bearing branches. At this stage affected sections alternate with green bands of healthy tissue, imparting a variegated appearance to the stems and branches. This pattern of bands on stem and branches is a characteristic external symptom of flax pasmo. As the disease advances, the number of pycnidia increases on the infected portions. Eventually the patches fuse and the entire stems turns brown to brownish gray profusely dotted and drop. Simultaneously with the appearance of patches on the upper leaves and stems, the disease becomes evident on the sepals; when the buds are infected flowers do not develop, the duds wither, become covered with pycnidia and drop. Diseased young bolls do not develop. Infected immature and fully developed bolls yield normal seeds except when the damage is severe. Infected seeds have a bluish tinge and white eruptions cover the seed surface or concentrate on its germinal section. These white eruptions are the sporophores of the flax pasmo pathogen.

7.6.3   MANAGEMENT

The import of flax seeds from regions in which the disease is prevalent is prohibited; the import of seeds for research purposes is permitted in form of samples under the condition that such are free from infection and are grown in quarantine nurseries; flax seeds received for commercial processing should not be used for sowing; should flax pasmo be detected in a farm, quarantine should be imposed and measures implemented in accordance with approved system; all infected flax plants and plant remnants should be burned on the spot, the soil disinfected and reflowed and agricultural implements disinfected; the planting of flax in farms under quarantine should be prohibited for 6 or 7 years; flax seeds should be treated; and flax varieties resistant to pasmo should be developed.

7.7 ALTERNARIA BLIGHT

Alternaria blight of linseed is caused by Alternaria lini. In India this disease was first reported from Kanpur (UP) in 1993 by Dey. Presently this disease is prevalent in all the growing regions of India.

7.7.1 CAUSAL ORGANISM

Alternaria lini

7.7.2 SYMPTOMS

The symptom of this disease occurs on all the above ground portions of the plant. The symptoms of the disease observed as the failure of the bud to open and it is followed by the appearance of minute dark color black spots near the calyx. Sizes of the spots increases and reach up to pedicel. Bud rotting appears when unopened bud is severely infected. If the infection takes place after the fertilization the development of the ovule proceed normally but seeds formed are the under sized, shriveled and capsule may give burnt look. Dark brown spots appears on the young leaves which starts from the base and it's gradually reach on the stems. The severely infected leaves dry up and twisted. In the lower leaves, the tips of which hang downwards come in contact with soil.

If the infection takes place very severe the whole plats dried up before the bud formation.

7.7.3 DISEASE CYCLE

In Indian conditions the pathogen is survive through the contaminated seed as well as diseased plant debris. Pathogen may colonize on the seed via capsule, sepals and walls. The outer layer of the seed coat gets colonized and resting hyphae are formed. Under field condition the survival of the pathogen in summer is not possible however, in lab condition it is possible has reported by the several workers. The temperature between 26 to 33–°C and humid conditions are most favorable for the growth of the fungus and for infection to the plants. The relative humidity below 75% restricts the disease development. The role of toxins has been reported involved in the pathogenicity.

7.7.4 MANAGEMENT

The field should be well drained for the cultivation of linseed. Late sowing varieties should be preferred for the areas where high humidity prevails. Disease free seeds should be used for the sowing purpose. Seed treatment with Bavistin @ 2g/kg or Iprodione 1.5 g/kg seed removes the seed borne inoculum followed by foliar spraying of Dithane M 45 @ 0.2% at the time of flowering and another spray repeated at 15 days intervals are effective to control this disease. Singh et al(2013) reported that the seed treatment with *T. viride* @ 4 g/kg seed followed by two foliar sprays of Mancozeb (0.25 %) decreased the blight intensity and bud damage significantly.

7.8 POWDERY MILDEW

Powdery mildew of linseed is very important next to the rust. Most of the rust resistant varieties of the linseed are highly susceptible to the powdery mildew. If the disease appears at early growth stages its causes very high yield reduction. A severely infected plant produces very poor quality seed and fiber.

7.8.1   CAUSAL ORGANISM

Oidium lini Skoric

7.8.2   SYMPTOMS

The disease appears as small, circular or irregular dirty white powdery patches on few leaves of the plants. It spreads very quickly on the other plants including leaves, stem, branches, flowers and buds. The infected leaves covered by a thick powdery masses shows curling, twisting and drooping symptoms which ultimately dry up. Severely affected plants do not die but the yield is considerably reduced.

7.8.3   DISEASE CYCLE

The pathogen survives in soil on the diseased plant parts through the formation of cleistothecia. In the next season under the favorable environmental conditions asci and ascospores are released which initiate the primary infection. However, in India there is no any report of the formation of cleistothecia. In northern and eastern part of India this disease appears in last week of February and reaches its maximum severity by middle of March when the temperature is between 20–25–°C and humidity is about less than 65%.

7.8.4   MANAGEMENT

Early sowing should escape the infection the disease. In Punjab and Himachal Pradesh regions grow K-2, LC-54 and LC-185 cultivars which are resistant to this disease. Foliar spray of Bavistin (0.2%) or Sulfex (0.3%) or Karathane (0.2%) or Wettable Sulphur (0.25%) at weekly intervals depending upon the intensity of the disease. Proper coverage of the crop with fungicides spray is gave the good control. Thiovit was found highly effective for controlling the disease followed by Calixin and Rovral for this disease (Singh and Singh, 2002).

7.9   NON-PARASITIC DISEASE

7.9.1   HEAT CANKER

The heat injury of the cortical tissues of hypocotyls and stem near the soil surface is common on the flax and other succulent plants. Due high surface temperatures in the dark colored soils frequently are high enough to kill the cells of young plant tissues. The formation of cankers from such injury is common in the semi humid regions and at high altitudes throughout the world. Cortical tissues disintegrate, resulting in the death of the young seedlings or in sunken brown spots on the stems. The stems usually enlarge above the cankers in the plants that survive the initial injury. The cortical rotting saprophytic organisms frequently invade the injured tissues to increase the damage. Preventing excessive soil temperatures by early sowing or by drilling the rows north and south so as to secure maximum shading, by higher rates of seedling, by use of a nurse crop, by mulching the soil surface, and by irrigation are means of reducing this type of damage.

7.9.2   BACTERIAL DISEASES

The occurrence of bacterial diseases of linseed/flax is very few instances. Out of them some are economic importance in the country and reducing the seed yield and straw yield. Among the microorganism chiefly responsible for damage of flax fiber in the countries is:

1. *Pseudomonas herbcola*
2. *Pseudomonas florescence*
3. *Bacillus subtilis*
4. *Clostridium macerans*

These bacterial pathogen of linseed/flax are present throughout the growth period but become acute only at the end of flowering and heavenly damage the fiber during retting process become seed and straw and seed yield. Another bacteria causes bacteriosis in linseed has been isolated by Bushkova (1967) and Lebedeva (1975).

7.9.3   VIRAL DISEASES

The most conspicuous symptoms this viral disease is crinkling of the leaves. This symptoms result in enation of lateral veins on the margin of leaves. Stunted growth of the plants and reduced the flower setting. The color of leaves and flowers appear normal while, the capsule and seed formation is reduced. The causal organism of this disease is *Oat blue dwarf virus*. This virus is transmitted by the six spotted leafhopper *Macrosteles fascifrons* is vector of Aster yellow virus.

KEYWORDS

- **blight**
- **damping off**
- **linseed**
- **mildew**
- **rust**
- **wilt**

REFERENCES

1. Allen, R. F. (1934). A cytological study of heterothallism in flax rust. *Journal Agric. Res.*, 49, 765–791.
2. Barker, H. D. (1923). A study of wilt resistance in flax. *Minn. Agr. Exp. Sta. Tech. Bul.*, 20.
3. Bigirimana, J., Hofte, M. (2002). Induction of systemic resistance to *Colletotrichum lindemuthianum* in bean by a benzothiazole derivative and *Rhizobacteria*. *Phytoparasitica*, 30(2), 159–168.
4. Bolly, H. L. (1924). Fax wilt and flax sick soil. *No. Dak. Agr. Exp. Sta. Bul.*, 174.
5. Bolly, H. L. (1932). Fax wilt and flax sick soil. *No. Dak. Agr. Exp. Sta. Bul.*, 256.
6. Bushkova, L. N. (1967). Gummosis of cotton and other bacteriosis of industrial crops. *Rev. Appl. Mycol.* 46, 2940.
7. Dey, P. K. (1933). An Alternaria blight of linseed plant. *Indian J. Agric. Sci.*, 3, 881–896.
8. Flor, H. H. (1941). Inheritance of rust reaction in a cross between the flax varieties Buda and J. S.W. *Jour. Agr. Res.*, 63, 369–388.

9. Hirua, M. (1924). On the flax anthracnose and its causal fungus, *Colletotrichum lini* (West.) Toch. *Jour. Bot. Japan.* 2, 113–132.

10. Levedeva, M. A. (1975). Bacterial diseases of fiber flax. *Rev. Plant Pathol.* 55, 1816.

11. Nelson, C. I., Dworak, M. (1926). Studies on the nature of wilt resistance in flax. *No. Dak. Agr. Exp. Sta. Bul.*, 202.

12. Pethybridge, G. H., Lafferty, N. A. (1918). A disease of flax seedling caused by a species of *Colletotrichum* and transmitted by infected seed. *Sci. Proc. Roy. Dublin. Soc.*, 15, 359–384.

13. Ray. C. Jr. (1944). Cytological studies on the flax genus, Linum. *Am. Jour. Bot.*, 31, 241–248.

14. Schuster, M. (1944). The nature of resistance of flax to *Fusarium lini. Phytopath.*, 34, 356.

15. Singh, Jyoti, Singh, J. (2002). Effect of fungicidal sprays against foliar diseases of linseed. *Ann. Plant Protec. Sci.*, 10, 169–170.

16. Stalkman, E. C. (1919). Controlling flax wilt by seed selection. *Jour. Am. Soc. Agron.* 2, 291–298.

17. Tisdale, W. H. (1916). Relation of soil temperature to infection of flax by *Fusarium lini. Phytopath.*, 6, 412–413.

18. Vallega, J. (1944). Especialization fisiologia de *Melampsora lini*, en Argentina. Santa Catalina. *Inst. Fitotec. Pub.*, 39.

PART III

HORTICULTURAL CROPS

CHAPTER 8

NUTRITIONAL DISORDERS OF CITRUS AND THEIR MANAGEMENT

J. N. SRIVASTAVA,[1] P. K. SHARMA,[2] UPMA DUTTA,[3] and RAKESH KUMAR[4]

[1]*Bihar Agricultural University, Sabour, Bhagalpur, Bihar, India*

[2]*National Bureau of Agriculturally Important Microorganisms (ICAR), Mau, Uttar Pradesh, India*

[3,4]*Sher-e-Kashmir University of Agriculture Science and Technology, Jammu (J&K), India*

CONTENTS

8.1   INTRODUCTION

Fruit trees and crop plants suffer from nutritional disorders. Nutritional disorders are a result of inadequate supply or excess of desired minerals/ fertilizer. The term macro and micronutrients are used to denote collectively group of mineral/nutritional chemical elements, which are indispensable for optimal growth and which plants absorb primarily through roots. Nitrogen, phosphorous and potassium requiring larger quantity are known as macronutrients while calcium, magnesium, iron, manganese, zinc, boron, copper, molybdenum, sulfur, etc. are known as micronutrients and require in smaller amounts. Deficiency of minerals *viz.* macronutrients and micronutrients, result in disorders in plant metabolism and plants express hunger signs *viz.* chlorosis, leaf spot, leaf blotch, leaf blight, dieback, reduced growth of tree, poor fruit quality and decreased number of fruits in citrus tree. Excess of mineral disturbs nutritional balance, which is most necessary for the proper metabolism in citrus tree. If the supply of minerals is high, trees show toxicity symptoms. Deficiencies and excesses of these minerals also reduce resistance of plants to fungal, bacterial and other diseases (Singh, 1983).

Apart from visual diagnosis, the analysis of plant and leaf samples is also helpful in identification of nutritional disorders in plant, which can be supplemented through soil analysis. Twelve such minerals and disorders caused by them in citrus are discussed as under.

8.2   NITROGEN (N)

Being an important constituent of protein amino acid, enzymes, hormones, vitamins and chlorophyll, N is more important than other minerals. It is integral part of plant tissues (0.2–4.1% N on dry matter basis), which require for proper photosynthesis by green tissue. Nitrogen greatly influences important tree functions such as growth, leaf protection, flower initiation, fruits setting and fruit development and quality (Mooney et al., 1991).

The main cause of N deficiency is lack of available N in soil, which can be due to many factors. Nitrogen leaching is caused by the combination of heavy summer rainfall or over irrigation in highly porous soil. Water logging in soil can cause N loss through denitrification that may lead to a temporary N deficiency that can be relived by dry weather (Zekri, 1995 I).

8.2.1 DEFICIENCY SYMPTOMS

In citrus tree N deficiency symptoms include yellowing of the foliage that begins with older leaves, and then appears on younger leaf flush. Leaves become progressively more yellow, with no distinct pattern, but some-times-mature green leaves slowly bleach to a mottled irregular green and yellow pattern, become entirely yellow and are shed (Zekri and Obreza, 2003). N deficiency often occurs in winter or early spring because of low tree N reverse, low soil temperature and/or lack of root activity. The N deficient trees are stunted with thin canopy, no fruit load and can be highly erratic in bearing habit. They bloom sparsely and flushes emerge irregularly, and produce limited twig and leaf growth (Zekri, 1995 I, II) (Figure 8.1).

The color of citrus fruits peel tends to be pale and smooth, and the juice has lower soluble solids and acid concentration. If N is deficient during summer and fall seasons, when the fruit is expending and matur-ing, some of the green leaves will turn yellow and may shed. Trees that are constantly stunted with irregular and very short twig growth, twig

FIGURE 8.1 Whole-leaf chlorosis caused by nitrogen deficiency.

dieback can occur, and crop production is greatly reduced. (Zekri, and Obreza, 2004).

8.2.2 MANAGEMENT

Citrus trees deficient in N can be improved by applying supplementary N fertilizer in frequent application according to the age of plants. The use of low biuret urea as foliar spray is very effective and rapid way to correct N-Deficiency (Zekri, 2003).

8.3 PHOSPHORUS (P)

Phosphorus is important for normal plant growth and reproduction. It is a constituent of phospholipids, nucleic acid and many proteins. Phosphorus is involved in carbohydrate and fat transformation and protein metabolism and also respiration. It plays important role in high energy bonding (i.e., ATP $\Leftrightarrow$ ADP) and in respiration. It is essential for timely differentiation and maturation of plant tissues. (Mehrotra, 1999).

The cause of phosphorus deficiency is the poor availability of P in soil. Phosphorus deficiency may occur in area of high rainfall due to leaching and erosion. In strongly acid soils, P becomes quickly unavailable. Phosphorus availability is also reduced in calcareous soil.

8.3.1 DEFICIENCY SYMPTOMS

Phosphorus deficiency symptoms appear first on older leaves, then on younger tissues, which lose their deep green color (Zekri, 1995 I). Citrus trees deficient in P have reduced growth and leaves are small and narrow with purplish or bronze discoloration. This type chlorosis spread inwards from the midrib, some time leaving areas of healthy green tissues. Necrosis of tissue leads to withering of leaves and breaking petioles at the pseudo stem. Some leaves may later develop necrotic areas and young leaves will show reduced growth rate. Leaves shed prematurely and fruit drop before normal harvesting time (Zekri and Obreza, 2003). Citrus trees show limited flower development with reduced fruit set and yield.

The fruit will be coarse and rough in texture, and have thick rind and a hollow core. The fruit will also have a high acidity in proportion to total soluble solids. In such trees, fruit maturity is delayed. Usually roots are stunted and poorly branched (Zekri and Obreza, 2004).

8.3.2 MANAGEMENT

Phosphorus deficiency can be improved by applying water soluble P fertilizer to soil after confirmation of P deficiency by leaf and soil analysis (Zekri and Obreza, 2003). Foliar concentration ranges between 0.10 and 0.60% because healthy citrus leaves are capable of tolerating wide variation in P content (Zekri, 1999).

8.4   POTASSIUM (K)

Potassium is important for normal growth and development of plant, in addition it also influences the ghost reaction (tolerance) to various pest/ pathogen. Potassium is needed by plant in large amount. Potassium plays important role in carbohydrate and protein synthesis, osmo-regulation, and also stomatal movement. It is essential as a catalyst of many reactions. But it is not effective without co-nutrients such as N & P. (Mehrotra, 1999)

Potassium deficiency symptoms usually result from an insufficient K supply in the soil. Lack of soil moisture also reduces K uptake and may lead to K deficiency. If the supply of N and P is high relative to that of K, growth may be rapid at first, but the K concentration in the plant may be ultimately decreased to K deficiency (Zekri, 1995 I). K deficiency may occur in sandy acid soils where leaching takes place. The supply of K to plants may be decreased in soil that have very high concentration of Ca and Mg or by heavy application of N. Decreased K uptake occurs typically in some calcareous soils (Zekri and Obreza, 2003).

8.4.1 DEFICIENCY SYMPTOMS

Deficiency of K in citrus tree causes a general reduction in growth and dropping of leaves without visual deficiency symptoms. The K deficiency results in premature yellow brown discoloration of leaves from the tips and margins, which then gets border. Necrotic areas and spotting can develop on leaves. Purplish brown patches may appear at the base of petiole and in severe case water soaked areas may be seen (Zekri, 1996) (Figure 8.2).

Symptoms appear first on older leaves because K tends to concentrate in the rapidly growing tissues. Potassium deficiency causes compact tree appearance, slow growth, small leaves, smaller fruit size with inferior flavor, very thin peel and smooth texture, premature shedding of fruit, lower acid concentration in the fruit, an increase in susceptibility to drought and cold (Camp and Fudge, 1939).

8.4.2   MANAGEMENT

Potassium deficiency can be improved by applying K fertilizer viz. potassium chloride or potassium sulfate to the soil. Foliar application of potassium nitrate or mono-potassium phosphate can be very effective and is a rapid control measure to improve K deficiency (Zekri and Obreza, 2003).

8.5   CALCIUM (CA)

Calcium is involved in cell wall (calcium pectate) formation which gives turgidity of cell and indirectly in cell division, highly required in telophase

FIGURE 8.2   Whole-leaf chlorosis caused by Potassium deficiency.

for cell plate formation. It is essential in activated the growing point especially root tip. The activity of several enzymes is influenced by this mineral (Mehrotra, 1999). In acid soil of hill regions calcium becomes unavailable and its deficiency takes place (Saha, 2002).

8.5.1 DEFICIENCY SYMPTOMS

In affected citrus trees, young leaves become distorted; tip of leaves hook back and their margins become curled. Leaves may be irregular in shape, some times brown scorching or spots may appear on leaves. Terminal buds are also affected or die (Zekri, 1995 II). Trees are extremely stunted having inadequate root system (Bhargava et al., 2000).

8.5.2 MANAGEMENT

Quick lime (bujha chuna) is major source for calcium. The quick lime should be powdered and mixed with soil around the trunk in a radius of 1.5–2 meter. Apply lime @ 4–6 Kg/tree/year to manage the deficiency of this mineral (Saha, 2002).

8.6 MAGNESIUM (MG)

Magnesium is the constituent of chlorophyll and plays an important role in the structural material of certain enzymes involved in carbohydrates synthesis (Mehrotra, 1999). Magnesium deficiency is the result of imbalanced availability of calcium or imbalanced use of potassium fertilizer or calcium containing fertilizers.

8.6.1 DEFICIENCY SYMPTOMS

The symptoms occur first on mature/ older leaves, with younger leaves become mottled or chlorotic, then reddish some time necrotic spot appear. Leaves yellowing are on both sides of the midrib and extend from the base to the apex of the leaf (Saha, 2002). The green portion tapers towards the tip of the leaf so that inverted "V" shape is formed. Affected leaves fall prematurely in late summer and autumn (Camp and Fudge, 1939) (Figure 8.3).

8.6.2   MANAGEMENT

- Drench tree basin with magnesium sulfate @ 100g + Quick line @ 200 g/100 L water (Saha, 2002).
- Spraying magnesium nitrate @ 1 g/L water also control the disorder (Glendining, 1999).

8.7   IRON (FE)

Iron is constituent of certain enzymes and proteins. It is essential for synthesis of chlorophyll, and seems to play a catalytic role. It is an electron carrier in oxidation-reduction during respiration of plants (Mehtrotra, 1999). Iron deficiency often appears in winter due to low soil temperature and root inactivity. High soil pH can cause iron deficiency, especially in trees on trifoliate hybrid rootstocks or trifoliate root stocks. Iron deficiency can also occur in poorly drained soil and also in alkaline soil (Brown, 1956).

8.7.1   DEFICIENCY SYMPTOMS

Young leaves are worst affected and show the symptoms, while older leaves may remain green. Leaves of deficient plants become chlorotic,

FIGURE 8.3   Leaf chlorosis caused by magnesium deficiency.

cream colored nearly too white with the main veins remaining green. In severe cases, even the veins may turn yellow (Zekri, 1995 III). The affected plants show poor growth, and have small fruit. Die back of braches takes place. In a tree single branch or in orchard a single or few trees may be affected (Wallace and Lunt, 1960).

8.7.2 MANAGEMENT

- Spraying with the mixture of ferrous sulfate 400 g and Lime 400 g in 100 L of water gives the best remedy of deficiency of iron (Saha, 2002).
- Sparing ferrous sulfate @ 0.5–0.9% alleviates the deficiency symptoms of iron (Alvs and Tucker, 1992).
- Apply ferrous Sulphate to the acidic soil @ 20g/tree, and alkaline soil @ 50g/tree in a year (Tisdale and Nelsoman, 1975).

8.8   ZINC (ZN)

Zinc is a component of many enzymes involved in auxin and carbohydrates synthesis. It plays important role in chlorophyll formation and photosynthesis activity (Mehrotra, 1999).

Zinc deficiency in citrus is described as "little leaf," "mottle leaf" and "resetting leaf" because of the distinctive leaf pattern produced on most citrus species. In case of severe deficiency, it is also called "mottling." Excessive phosphate or nitrogen has been shown to induce zinc deficiency (Saha, 2002). It is most acute in alkaline or acid coastal soils (Buckman and Brady, 1969).

8.8.1 DEFICIENCY SYMPTOMS

Symptoms of zinc deficiency are more noticeable on north side of tree. Leaves show interveinal chlorosis. Leaf symptoms include small, narrow leaves (little leaf) and whitish yellow areas between the veins (mottle leaf). Leaves are few and small, internodes are short and shoots from rosettes. The leaves especially the terminal growth develops "mottle leaf" symptoms, the growth becomes yellowish and bright creamy and unthrifty (Saha, 2002) (Figure 8.4).

FIGURE 8.4   Leaf chlorosis caused by zinc deficiency (most often on young foliage).

"Mottling" become more pronounced with severe deficiency. Leaves are reduced in size and have pale color. Necrosis may occur beginning at leaf tip and margins, and terminal growth is affected. Affected twigs are erect and bushy. Chlorotic leaves drop off early leading to die-back of twigs. Trees remain stunted. Zinc deficiency is most severe in spring growth in citrus (Zekri, 1995 III).

Reduced vigor, low fruit production, smaller fruit size, poor fruit quality results from deficiency of zinc in the citrus tree. In lemon, the yield may be reduced even through without evident symptoms. In case of orange, mild deficiency may have little effect on yield.

8.8.2   MANAGEMENT

- Zinc deficiency can be corrected by foliar spray of 0.5–1.2% zinc sulfate twice at weekly intervals on tender foliage followed by soil applications of zinc sulfate @ 100g-500g/tree, according age of tree. (Zekri, 2003).
- Spraying by of Ziram or Dithane Z-78 @ 2g/ liter water can also correct the deficiency of Zinc (Saha, 2002).

8.9   MANGANESE (MN)

Manganese is a cofactor of enzymes of cellular respiration, nitrogen metabolism and photosynthesis (Mehrotra, 1999). Manganese deficiency is particularly evident in the spring after a cold winter, because it frequently occurs in combination with deficiencies of zinc or iron or both, and its symptoms may be over shadowed.

8.9.1 SYMPTOMS

Symptoms of manganese deficiency are usually more noticeable on the South side of the tree and are more pronounced in the spring growth flush. Symptoms are evident on both young and mature leaves (Sprague, 1964).

Leaves exhibit interveinal yellowing with a darker green band along the midrib and veins and interveinal chlorosis on the new foliage. Necrotic spot may appear scattered on leaf, leave affected leaves turn brown. Leaf size is not reduced due to the Mn. deficiency. In case of mild deficiency in citrus tree, a slight reduction of vigor and yield occurs. While severe manganese deficiency can induce defoliation, loss of vigor and lower fruit yield. In extreme cases, symptoms are accompanied by premature leaf drop (Zekri, 1995 III).

8.9.2 MANAGEMENT

Manganese deficiency can be corrected by the spraying 0.2–0.5% manganese sulfate twice at weekly interval or manganese sulfate @ 5–10 kg/ha (Zekri, 2002).

8.10 BORON (B)

Boron plays an important role in many activities of plants *viz.* cell division, protein synthesis, pollination of flower, flower formation, fruit setting and seed production. Boron is essential for plant growth too (Mehrotra, 1999). Boron deficiency usually occurs in calcareous soil or irrigation with alkaline water. Hard fruit disease of citrus is cased by boron deficiency (Zekri, 2004).

8.10.1 DEFICIENCY SYMPTOMS

Symptoms of Boron deficiency are seen on young leaves and fruit. The first visual symptoms of B deficiency are generally the death of the terminal growing points of the main stem (Smith and Reuther, 1949) (Figure 8.5).

In leaf, characteristic symptom is the discoloration and downward curling of leaves. The margin of leaves turn brown and the veins turn yellow in color. The veins on the upper leaf surface are enlarged, become thickened, corky and split (Zekri, 1995 IV).

Fruits are brownish in color, mis-shaped and become hard and dry due to lumps in the rind caused by gum impregnations. So boron deficiency is also known as "hard fruit." Cracking of fruits can be also seen (Hass, 1945; Zekri obreza, 2003).

Boron deficiency is caused growing point die, growth is reduced and twigs may split. Stems are brittle, internodes are shortened, roots are thick and stunted (Zekri and obreza, 2003).

8.10.2 MANAGEMENT

- Boron deficiency can be corrected by soil application of borax @ 250g per tree or 10 kg/ha in a year.
- Spraying 0.2–0.5% solution of borax or boric acid on the foliage once or twice a year after the formation of new flushes is helpful in alleviating boron deficiency (Saha, 2002).

FIGURE 8.5 Raised veins and discoloration on leaf caused by boron deficiency.

8.11   COPPER (CU)

Copper is a co-factor of several oxidation enzymes. It is acts as catalyst in certain reaction of respiration (Mehrotra, 1999).

Copper deficiency is also known as "dieback," "ammonization" and "exanthema." These names are derived from the dying back of twigs, frequent association with heavy application of N (ammonia) and gum exudes/excrescences on the surface of twigs and fruit (Floyd, 1977; Zekri and Obreza, 2003).

8.11.1   DEFICIENCY SYMPTOMS

Copper deficiency is characterized by dark green young leaves, development of multiple buds, general stunting and bushy appearance. The leaves become mottled, narrowed and reduced in size (Saha, 2002). Thin cut strip appears on bark of stem and gum exudes. Gum exudation can be seen on the rind also (Dickey et al., 1948). In severe cases, shoots become dry leading to die-back of twigs and become distorted in irregular shapes generally "S" shaped. The fruits are small in size and discolored. Sometimes fruits split and drop before maturity. Fruit show gum pockets around central pith (Zekri, 1995 IV).

8.11.2   MANAGEMENT

- One spraying of Bordeaux mixture (1%) is recommended for correcting the deficiency of copper.
- Spray of Copper sulfate @ 0.5–0.9 (Saha, 2002).
- Copper sulfate is also recommended @ 4–6 Kg/tree in a year for copper deficiency.

8.12   MOLYBDENUM (MO)

Molybdenum is a constituent of enzymes that are responsible for the reduction of nitrates to nitrites. It is essential for protein synthesis. Molybdenum deficiency is most common in acidic soil (Mehrotra, 1999).

8.12.1   DEFICIENCY SYMPTOMS

The symptoms appear on mature leaves, while young leaves appear normal. Leaves become pale yellow with marginal chlorosis. The leaf lamina becomes thin and dry. Yellow spots can be seen on leaves, these spots initially appear as water soaked areas and gradually develop into yellow spots. Gum formation occurs on the underside of the leaf. In severe cases, the necrotic yellow spots enlarge and extend to margins. Affected tree becomes almost defoliated during the winter (Zekri, 1995 IV) (Figure 8.6).

Large irregular brown spots surrounded with yellow may develop on the fruit (Sprague, 1964).

8.12.2   MANAGEMENT

Foliar spray of Sodium molybdate @ 0.2% is useful for correcting molybdenum deficiency in the trees (Zekri, 2002).

8.13   SULPHUR (S)

The amount of sulfur in plant ranges between 0.1 and 1.0% on dry weight basis. A total S in plant mobile and its moves through phloem. Sulphur is a component of amino acids, tripeptides, protein, vitamins (thiamine and biotin) and enzymes. Mostly soil organic matter is held sulfur and deficiency is not common. Sulphur availability is reduced by high soil pH (Knorr, 1973; Welter et al., 1958).

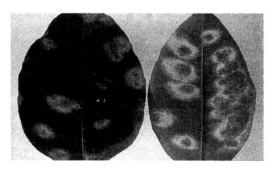

FIGURE 8.6   Molybdenum deficiency.

8.13.1 DEFICIENCY SYMPTOMS

Symptoms always start with the newest foliage. Visual symptoms are small pale green or light yellow on leaves with lighter veins without any spots. The symptoms resemble those of nitrogen deficiency (Zekri, 2002).

8.13.2 MANAGEMENT

Additional dose applied with recommended dose of organic fertilizer.

8.14 GENERAL MANAGEMENT OF DEFICIENCY DISORDERS

In view of overlapping symptoms of different nutritional orders, combination of following nutrients can be sprayed on flushes of the trees.

Zinc Sulphate	500 g
Ferrous Sulphate	200 g
Copper Sulphate	500 g
Manganese Sulphate	200 g
Borex	100 g
Urea	450 g
Lime	400 g
Water	100 L

8.14.1 PREPARATION OF SOLUTION WITH ABOVE MINERALS

The salts should be dissolved in small quantity of water separately, the quick lime is mixed with half quantity of water in a separate container and then the salt solution added to quick lime solution for better mixing (Saha, 2002; Embleton et al., 1967).

8.15 NUTRIENTS/FERTILIZER REQUIREMENTS OF YOUNG CITRUS TREES

Fertilizer application should be divided in two parts; first part is applied in the beginning of September and second part by the end of February. The approximate quantities of nutrients required by young tree listed in Table 8.1.

TABLE 8.1   Estimated Nutrient/Fertilizer Requirement (g/tree) for Young Citrus Trees

Tree age (years)	Fertilizer requirement			
	FYM (kg/tree)	Urea (g/tree)	Diammonium phosphate (g/tree)	Muriate of potash (g/tree)
1	5	95	50	35
2	5	190	105	65
3	10	285	155	100
4	15	380	205	135
5	20	475	260	165
6	25	570	310	200
7	30	715	360	235
8	35	865	415	2265
9	40	1010	465	300
10	45	1175	515	335

Source: Package and practices for horticultural crops SKUAST-Jammu.

8.15.1   FOLIAR SPRAYS FOR TREATING MICRONUTRIENT DEFICIENCIES IN CITRUS (TABLE 8.2)

TABLE 8.2   Nutritional Foliar Sprays for Correcting Deficiencies in Young Citrus Tree

Nutrient	Treatment	Application Rate	Comments	Timing
Magnesium	Magnesium	1 kg/100 L	Mix. Magnesium Sulphate in half – full vat, then add calcium nitrate separately while agitator is running. Then fill vat.	When spring flush leaves are ½ or 2/3 expanded
	Magnesium Sulphate	1 kg/100 L		
	+ Calcium nitrate	1 kg/100 L		
Zinc	Zinc Sulphate heptahydrate (23% Zn)	150 g/100 L	–	As above
	Zinc sulfate (23% (Zn)	500 g/100 L	–	As above
	+ hydrate line	250 g/100 L		

TABLE 8.2   Continued

Nutrient	Treatment	Application Rate	Comments	Timing
Manganese	Manganese Sulphate	100 g/100 L	500g of urea can be added to improve uptake of manganese	As above
Iron	Ferrous Sulphate 0.5–0.9%	–	–	As above
Zinc and Manganese	Zinc Sulphate + Manganese Sulphate	150 g/100 L 100 g/100 L	500 g–750 g urea can be added to improve uptake	As above
	Zink sulfate 23% (Zn)	500 g/100 L	Zinc and manganese deficiencies often occur together; a combined spray correcting both	
	+ Manganese Sulphate	300 g/100 L		
	+ Hydrate lime	250 g/100 L		
Copper	Copper Sulphate	As per label	–	Spring or autumn to suit fungicide program
	Copper Oxchloride	As per label	–	
	Copper hydroxide	As per label	–	

Source: Koo (1983); Embleton (1973); Weir and Sarooshi (1991).

8.15.2   NUTRIENT RECOMMENDATION FOR CITRUS TREES

Observation of visual symptoms for nutrient deficiency and their correction is adequate for average production of citrus fruit. For optimum production and quality of citrus fruit develop a range of tools such as soil analysis and leaf analysis can be used to assess the nutrient requirements of trees. Nutrient analysis is essential to correct potential problem before it becomes a limiting factor in production (Rajput and Haribabu, 1985; Zakri, 2002).

8.15.3 SOIL ANALYSIS

Analytical procedures used in soil testing vary considerably among labs, as do the results they obtain. None are capable of reporting available nutrient levels in a sample; they can only report the chemically extractable levels. Moreover, no soil testing extraction procedure has yet been calibrated to correlate the extractable value of any nutrient element with citrus production levels of citrus fruit quality (Embelton, 1967; Zakri and Obreza, 2003).

It is simple to verify pH and available P and certain exchangeable cations notably Ca and mg by soil analysis. It is usually more difficult to assess the N and K status in the soil because both these elements are subject to leaching in humid regions (Embleton, 1967).

8.15.4 LEAF ANALYSIS

Leaf analysis is an effective technique for monitoring the nutrient status of citrus trees. It has been the most extensively researched tool for determining the needs of citrus trees (Smith, 1966). The leaf sample consists of at least 100 leaves that are 4–6 months old, taken from non-fruiting twigs or terminals of the previous spring's growth flush. Thus, sampling should be conducted from July to September. The orchard to be represented by the leaf sample should consist of only one rootstock/scion combination of uniform-aged trees, within a single soil type (Smith, 1966; Zakri, 1994).

Select 15–20 trees randomly across the orchard from which to collect 5 or 6 leaves each. Leaves should be free of damage from insects or disease. Leaves with obvious chlorosis should be excluded, unless the sample is being taken specially to ascertain a potential cause of the chlorosis. While still fresh, the leaves should be washed to remove soil or dust, then air dried, packaged and submitted to the laboratory for analysis (Jorgenson and Price, 1978; Embleton et al., 1967).

When the results are obtained from the laboratory, compare them with the standard shown in Table 8.3. The laboratory will report the levels of major element as percentage of dry weight, while microelements will be reported as parts per million.

TABLE 8.3 Leaf Analysis Standards for Citrus as Percentage of Dry Matter of Leaf

Minerals	Symbol	Unit	Deficient	Low	Optimum	High	Excess
Nitrogen	N	%	<2.20	2.20–2.40	2.50–2.70	2.80–3.00	>3.00
Phosphorus	P	%	<0.09	0.09–0.11	0.12–0.16	0.17–0.29	>0.30
Potassium	K	%	<0.70	0.70–1.10	1.20–1.70	1.80–2.30	>2.40
Calcium	Ca	%	<1.50	1.50–2.90	3.00–4.90	5.00–7.00	>7.00
Magnesium	Mg	%	<0.20	0.20–0.29	0.30–0.49	0.50–0.70	>0.80
Sulphur	S	%	<0.14	0.14–0.19	0.20–0.39	0.40–0.60	>0.60
Iron	Fe	ppm	<35	36–59	60–120	121–200	>200
Zinc	Zn	ppm	<17	18–24	25–100	101–300	>500
Manganese	Mn	ppm	<17	18–24	25–100	101–300	>500
Boron	Br	ppm	<20	21–35	36–100	101–200	>250
Copper	Cu	ppm	<3	3–4	5–16	17–20	>20
Molybdenum	Mo	ppm	<0.05	0.06–0.09	0.10–1.0	2.0–5.0	>5.0

Source: Smith (1966); Koo (1984); Malavolta and Nettof (1989).

KEYWORDS

- **citrus**
- **deficiency**
- **disorder**
- **management**
- **minerals**
- **symptoms**

REFERENCES

1. Alvs, A. K., Tucker, D. P. H. (1992). Foliar application of various sources of iron, manganese and zinc to citrus. *Proc. Flo. State Hort. Soc.* 105, 70–74.
2. Anonymous (2004). Package of practices for horticultural crops. Directorate of Extension Education. SKUAST-Jammu (J&K).
3. Bhargava, B. S., Raghupathi, H. B. Dohroo, N. P., Bharat, N. K., Sushma Nayar (2000). Nutrient Deficiency and toxicity disorders in tropical and subtropical fruits. In: V. K. Gupta and Satish, K. Sharma (Ed) *Diseases of Fruit Crops.* Kalyani Publishers, New Delhi. pp. 283–286.
4. Brown, J. C. (1956). Iron Chlorois. *Ann. Rev. Pl. Physiol.* 7, 171–190.
5. Buckman, H. O., Brady, N. C. (1969). The nature and properties of soil. MacMillan Co. New York. pp. 395–397 and 504–523.
6. Camp, A. P., Fudge, B. R. (1939). Some symptoms of citrus malnutrition in Florida. *Florida Agr. Sta. Bull:* 335.
7. Dickey, R. D. Drosdoff, M., Hamilton, J. (1948). Copper deficiency of tung in Florida. *Flo. Agric. Exp. Stn. Bull.* 447, 1–32.
8. Embleton, T., Jones, W., Reiz, H. (1967). Citrus fertilization. In: *Citrus Industry,* Vol.3
9. (Reuther, W. Webber, H. J., Batchelor, L. D. Eds.) University of California, U.S.A., pp. 122–182.
10. Embleton, T., Jones, W. Labanauskas, C., Reuther, W. (1967). Leaf analysis as a diagnostic tool and guide to fertilization. In: *Citrus Industry,* Vol.3 (Reather, W., Webber, H. J., Batchelo, L. D. (Eds.) University of California. USA., pp. 183–210.
11. Embleton, T. W. H., Reitz, H. J., Jones, W. W. (1973). Citrus fertilization. In Citrus Industry, Vol. 3 (Recther, W. (ed.) University of California Dev. of Agric. Science Berkeley, C. A. USA.
12. Floyd, B. F. (1977). Dieback or exanthema of citrus tree. Proc. Fla.Fr. Gr. Assoc. (Jan) as cited by, H. H. Hume (1900). *Fla. Agric Exp. Stn. Bull.* 53, 157.
13. Glendining, J. S. (1999). Australian soil fertility manual. CSIRO.

14. Haas, A. R. C. (1945). Boron in citrus tree. *Plant Phsiol.* 20, 323–343.
15. Jorgensen, K. R., Price, G. H. (1978). The Citrus leaf and soil analysis system in queens land. *Proceeding of the International Society of Citriculture* pp. 297.
16. Hume, H. H. (1900). Some citrus troubles. *Fla. Agric. Exp. Stn. Bull.* 53, 157–161.
17. Koo, R. C. J. (1983). Recommended fertilizers and Nutritional spray for citrus. *Florida Agriculture Exp. Stn. Bull.*, 5–36.
18. Knorr, L. C. (1973) Citrus Diseases and disorders. University press of Florida. Gainesville. pp. 26–27.
19. Malavolta, E., Netto, A. V. (1989). Nutricao mineral, calagem, cessageme adubcao dos citros. Associacoa Brasileria para pesquisa da potasa do fosfato. Piracicada-S. P. Brazil.
20. Mehrotra, R. S. (1999). Plant Pathology. Tata McGraw-Hill Publishing Co. Ltd. New Delhi. pp. 750–753.
21. Mooney, P. A., Richardson, A., Harty, A. R. (1991). Citrus nitrogen nutrition – A fundamental approach. N. Z. kerikeri Horticultural Research Station, *Citrus Research Seminar*, June-1991, 69–88.
22. Rajput, C. B. S., Haribabu, Sri. R. (1985). Citriculture. Kalyani Publishers, New Delhi. pp. 273–355.
23. Saha, L. R. (2002). Hand book of plant diseases. Kalyani Publishers, Ludhiana. pp. 290–292.
24. Singh, R. S. (1983). Plant diseases. Oxford & IBH Publishing Co. Pvt. Ltd. New Delhi, pp. 552–553.
25. Smith, P. F., Reuther, W. (1949). Observation on boron deficiency in citrus. *Proc. Florida State Hort. Soc.*, 62, 31–37.
26. Smith, P. F. (1966). Leaf analysis of citrus. In: Temperate and Tropical fruit Nutrition (Childers. N. F., Ed) Horticulture Publication, Rutgers, The State University, USA. pp. 208–228.
27. Sprague, H. B. (1964). Hunger signs in crops. (3rd Ed.) Darid McKay Co. Inc. New York. 61pp.
28. Tisdale, S. L., Nelsomn. W. L. (1975) Soil fertility and fertilizer. Mac Million Publishing Co. New York. pp. 300–313.
29. Walter, R., Embleton, T. W., Jones, W. W. (1958). Mineral nutrition in tree crops. *Annual review of Plant Physiology.* University of California, Citrus experiment station Bull.
30. Wallace, A., Lunt, O. R. (1960). Iron Chlorosis in horticulture plants—A Review. *Proc. Am. Soc. Hort. Sci.* 75, 819–841.
31. Weir; R. G., Sarooshi, R. (1991). Citrus Nutrition. Ag fact, H.2.3.11, N SWDPI.
32. Zekri, M. (1994). Assessment of fertilizer needs. *The Citrus Industry* 75(9), 30.
33. Zekri, M. (1995 I). Nutritional deficiencies in citrus trees. I. Nitrogen, phosphorus and potassium. *The Citrus Industry* 76(8), 58–61.
34. Zekri, M. (1995 II). Nutritional deficiencies in citrus trees. II. Calcium, magnesium and sulfur. *The Citrus Industry* 76(9), 19–20.
35. Zekri, M. (1995 III). Nutritional deficiencies in citrus trees. III. Iron, Zinc and manganese. *The Citrus Industry* 76(10), 16–17.
36. Zekri, M. (1995 IV). Nutritional deficiencies in citrus trees. IV. Boron, Copper, Molybdenum and chlorine. *The Citrus Industry* 76(11), 34–36.

37. Zekri, M. (1996). Several disorders in citrus trees. *The Citrus Industry* 77(2), 18–20.53.
38. Zekri, M. (1999). Citrus fertilizer management. *The Citrus and Vegetable Magazine* 63(9), 28–30.
39. Zekri, M. (2002). Fertilizer management and nutrition of citrus trees. *The Citrus Industry* 83(12).
40. Zekri, M. (2002). Micronutrients in citrus nutrition. *The Citrus Industry* 83(40), 21–23, 32.
41. Zekri, M. (2003). Guidelines for fertilizer program in citrus. *Citrus and Vegetable Magazines* 67(8), 56–60.
42. Zekri, M., Obreza, T. A. (2003). Essential micronutrients for citrus. Functions, deficiency. Symptoms and correction. In: Futch, S. H. (ed.) Nutrient management for Optimum Citrus Tree Growth and Yield Short Courses. University of Florida. Lake Alfred. Florida. pp. 17–24.
43. Zekri, M. (2004). Citrus nutrition in relation to soil acidity and calcareous soils. *Citrus Industry* 85(40), 21–24.
44. Zekri, M., Obreza, T. A. (2004). Plant nutrients for citrus trees. *Citrus and Vegetable Magazine* 68 (10), 17–20.

CHAPTER 9

FUNGAL DISEASES OF OKRA (*Abelmoschus esculentus* L.) AND THEIR INTEGRATED DISEASE MANAGEMENT (IDM)

C. P. KHARE,[1] SUSHMA NEMA,[2] J. N. SRIVASTAVA,[3]
V. K. YADAV,[4] and N. D. SHARMA[5]

[1,5]*Division of Plant Pathology Indira Gandhi Agricultural University, Raipur, Chhattisgarh, India*

[2]*Division of Plant Pathology, Jawaharlal Krishi Vishwa Vidyalaya, Jabalpur, Madhya Pradesh, India*

[3]*Department of Plant Pathology, Bihar Agricultural University, Sabour, Bhagalpur, Bihar, India*

[4]*College of Agriculture, Kundeshwar, Tikamgarh, Madhya Pradesh, India*

CONTENTS

9.1 INTRODUCTION

Okra (*Abelmoschus esculentus* (L.) Moench) is an important vegetable crop grown mainly for its tender green fruits in India. The green fruits are rich in vitamins A and C and minerals like Ca, Mg and Fe. In home consumption, India tops the world (Dhankhar and Mishra, 2004). It is a multipurpose crop due its various uses. Okra seeds are also good sources of protein and vegetable oil (Yadav and Dhankhar, 2001). Okra crop is grown throughout the year and is susceptible to many fungal pathogens. Fungal diseases are major constraint next to the YVMV of all area of the country in the production of okra. It is suffered by fungal diseases, which are belonging to 23 genera and 31 species of fungal pathogens.

     Important fungal diseases of okra (*Abelmoschus esculentus* L.) are shown in Table 9.1.

9.2 DAMPING OFF OF SEEDLINGS AND ROOT ROTS

Damping-off kills seedlings before or soon after they emerge. It is observed in severe from in Karnataka, Assam and on early sown crops in Northern

TABLE 9.1   Fungal Diseases of Okra

SN.	Name of Diseases	Causal Organisms
1	Damping-Off	*Pythium* spp., *Rhizoctonia* spp.
2	Fusarium Wilt of Okra	*Fusarium oxysporium* f. sp. *vasinfectum*
3	Verticillium Wilt	*Verticillium albo-atrum/V. dahale*
4	Powdery Mildew	*Erysiphe cichoracearum Sphaerotheca fuliginea* and *Oidium abelmosche*
5	Cercospora Leaf Spots	*Cercospora abelmoschi*, C. malayensis, *C. hibisci, C. hibiscina*
6	Phyllostica Leaf Spots	*Phyllostica hibiscini*
7	Alternaria Leaf Spots	*Alternaria hibiscinum*
8	Seedling Blight/collar rot	*Macrophomina phaseolina* and *Collectotrichum dematium*
9	Die Back	*M. phaseolina, C. dematium*
10	Southern Blight	*Sclerotium rolfsii*
11	Anthracnose	*Colletotrichum capsici, C. hibisci*
12	Root and Stem Rot	*Phytophthora palmivora*
13	Wet Rot	*Choanephora cucurbitarum*
14	Stem canker	*Fusarium chlamydosporum*
15	Rust of Okra	*Uromyces heterogenus*
16	Fruit rot	*Pleospora infectoria*
17	Fruit Rots	*Pythium spp. and Phytopthora spp.*
18	Pod Spots	*Ascochyta spp.*
19	Seed-Rots	*Colletotrichum dematium, Fusarium* spp. (*F. oxysporium, F. moniliforme*
20	Seed-Rots	*Alternaria alternata, Cladosporium cladosporioides, Fusarium spp., Aspergillus spp.* and *Penicillium spp.*

parts of India. It is coupled with others and complete failures occur in such conditions Cool, cloudy weather, high humidity, wet soils, compacted soil, and overcrowding especially favor development of damping-off. Infection before seedling emergence results in poor germination. If the decay is after seedlings emergence, they fall over or die which is referred to as "damp-off." Seedlings that emerge develop a lesion near where the tender stem contacts the soil surface.

Damping off of seedlings caused by *Pythium spp.* May affect okra in the late sown summer crop. *Fusarium solani, Rhizoctonia solani and*

Macrophomina phaseolina proved to be the causal organisms of okra damping off and root rot diseases. *F. solani* proved the most aggressive fungi in okra crops.

9.2.1 MANAGEMENT

Over irrigation should be avoided to reduce humidity around the crop. The efficacy of clean fallow and rotational crop, reduce population densities of *Pythium aphanidermatum, P. myriotylum, and Rhizoctonia solani* in soil (Johnson et al., 1997). The field should be regularly inspected for the disease-affected seedlings. Such seedlings should be removed and destroyed. Anitha and Tripathi (2001) screened fungicides against *R. solani* and *P. aphanidermatum*, which cause seedling mortality in okra, carbendazim, thiophanatemethyl, carboxin, thifluzamide and captan were effective against R. solani, while metalaxyl, captan, carboxin and iprodione were effective against *P. aphanidermatum*. Seed treatment with antagonist fungal culture of *Trichoderma viride* (3–4 g/kg of seed) or Thiram (2–3 g/kg of seed) will be used against the disease. Soil drenches with benomyl, captan and vitavax (carboxin) were effective against R. solani. Fusarium infection was reduced 38% by benomyl. Soil drenching with Dithane M 45 (0.2%) or Bavistin (0.1%) affords protection against the disease. A combination of benomyl + captan was effective against all 3 root rot causing organisms. Seed treatment with carbendazim followed by soil application of *Trichoderma viride* was proved effective controlling the seedling diseases of okra caused by *Rhizoctonia solani* and *Pythium aphanidermatum* (Johnson et al., 1997). Organic amendments increased seed germination and reduced *R. solani* infection. *Trichoderma harzianum* showed complete reduction in growth of *R. solani* and Plant Guard (containing of *T. harzianum*) reduced the growth of all pathogens *F. solani*, *M. phaseolina* and *Fusarium oxysporum*. Rhizo-N (containing *Bacillus subtilis*) and *B. subtilis* reduced the growth of all pathogens. The fungicide Rizolex-T (tolclofos-methyl) caused reduction on the growth of *R. solani*, *M. phaseolina and F. solani*, (Johnson et al., 1997).

Among the plant growth promoting rhizobacteria, *Bacillus pumilus* (SE-34), *B. pasteurii* (T4), *B. subtilis* (IN937-b) and *B. subtilis* (GBO3) strains significantly improved the crop and reduced the incidence of seed mycoflora (Mashooda et al., 2003). Strains of *Rhizobium* and *Bradyrhizobium* species were effective in controlling the soilborne fungi for their biological control potential against soilborne, root-infecting fungi (*Fusarium spp.*, *Macrophomina phaseolina* and *Rhizoctonia solani*) on okra (Shahnaz et al., 2005).

Rhizobium meliloti used as a seed dressing or as a soil drench inhibited growth of the soil borne root infecting fungi *Rhizoctonia solani* and *Fusarium solani*, (Ehteshamul and Ghaffar, 1993). *Rhizobium meliloti* was antagonistic to *Rhizoctonia solani* and *Fusarium spp.* in okra, when applied as seed or soil treatments (Ghaffar, 1993). *Bradyrhizobium sp.* and *R. meliloti* either used as seed dressing or as soil drench significantly suppressed root-rot infection caused by *M. phaseolina*, *F. solani* and *R. solani* in okra (Siddiqui et al., 2000).

Integrated disease management strategy for controlling the seedling diseases of okra caused by *Rhizoctonia solani* and *Pythium aphanidermatum* by using seed treatment with carbendazim, carboxin followed by soil application of *Trichoderma viride* was proved effective (Anitha and Tripathi, 2000). The efficiency of bacterial antagonists was more towards *P. aphanidermatum* than towards *R. solani*. *T. viride* was effective against R. solani (Anitha and Tripathi, 2001).

Some of the lines such as Red Ghana, HH27, IC12096, and IC17252 behaved resistant (Sohi et al., 1974).

9.3 CERCOSPORA LEAF SPOTS

This disease is serous in the month of August when there is high humidity (Jhooty et al., 1977; Sohi and Sokhi, 1974). Several species of *Cercospora* viz, *Cercospora abelmoschi*, *C. hibisci*, *C. hibiscina* and *C. malayensis*. are reported on leaves of okra causing leaf spots and sometimes blight. Leaf spots by *Cercospora* spp. have no definite shape, size or margin. The causal fungus appears as an olivaceous to sooty-colored growth on the lower leaf surface. Injured leaves will often roll, wilt and abscise.

C. abelmoschi causes no definite spots but grows as sooty to dark olivaceous mold on the lower surface of the leaf. Badly affected leaves roll, wilt and fall down. severe defoliation of affected leaves (Sohi and Sokhi, 1972).

The spots caused by *C. hibiscina* produces dark olivaceous patches of moldy growth on lover surface of the leaf. The spots caused by *C. malayensis* are brown, irregular with gray center and darker colored margins.

These fungi survive through conidia and stromata on crop debris in soil and cause maximum infection at 15°–29 °C.

9.3.1   MANAGEMENT

The control of leaf spots requires regular spraying with fungicides such as copper oxychloride, zineb, maneb, ziram or captan. Bavistin (0.1%) applied at 15 days interval had the lowest disease incidence followed by thiovin+bavistin+diazinon at 15 or 30 days interval (Rahman et al., 2000) Kavach, @0.8% and also Bavistin @ 0.5%, was most efficient in controlling the disease by spraying on green leaves (Dharam et al., 2001).

9.3.2   OTHER LEAF SPOTS

1. **Phyllostica leaf spots** caused by *Phyllosticta hibiscini* are sparingly observed on leaves along with Cercospora leaf spots. The spots are large with gray center and later produce shot holes. The pycnidia appear as minute black dots on both leaf surfaces. spores are hyaline and cylindrical.

2. **Alternaria leaf spots** caused by *Alternaria hibiscinum* appear as brown, subcircular spots of varying size and sometimes with concentric rings. Such spots are formed only on senescent leaves or when the plant is weakened.

9.4   POWDERY MILDEW

The disease is known to occur in severe southern parts of India. It is not reported to be severe in northern plain as it normally occurs very late in northern parts where major cultivation of this crop is done.

Disease is characterized by the obvious white coating of fungal mycelium on lower and upper leaf surface. Severe infection will cause to leaf to roll upward and result in leaf scorching. A large part of the talc-like powder on the leaf surface is composed of spores. These spores are easily blown by winds to nearby susceptible plants. Heavily infected leaves become yellow, and then become dry and brown. The disease is found mainly on the older leaves and stems of plants. Yields of many of the infected crops are reduced due to premature foliage loss (Sohi and Sokhi, 1974).

Powdery Mildew is caused by *Erysiphe cichoracearum and Sphaerotheca fuliginea* and also O*idium abelmosche and Leveillula taurica* (Souza and CafeFilho, 2003) found to as casual agent of powdery mildew.

An outbreak of *Erysiphe cichoracearum* on okra was associated with unusually dry weather. However the disease was associated with high rainfall (Diaz, 1999).

9.4.1 MANAGEMENT

Plants under nutritional stress in most cases will develop powdery mildew much sooner than plants the same age grown under a good nutritional program. Hence the plant should be well manured and application of fertilizers should be done on the basis of standard recommendations. Seed dressing with thiophanate and benomyl conferred resistance to the disease for a few days. Application of Wettable Sulphur (0.2%) or Bavistin (0.1%) or at the 1 week interval effectively control the disease. The incidence of disease was reduced by application of Karathane (dinocap), Solbar (barium polysulfide) 1%, Orthophaltan (folpet) 0.2% and also sulfur dust (0.2%). Spray of wettable sulfur (0.5%) at 50 and 65 days after sowing reduce the disease (Prabhu et al., 1971). Through a dimethirimol-resistant *Oidium sp.* appeared on okra during seventies. (Omer, Meh, 1972) but use of systemic fungicides for disease control is still reliable. Spraying with 0.01% Topas recorded no incidence of powdery mildew. However, the efficacy was comparable to that of carbendazim (0.1%). Topas 0.01% can be recommended as an alternative of carbendazim (Naik and Nagaraja, 2000). Three sprays of 0.05% tridemorph or 0.2% sulfur after the appearance of disease symptoms can be recommended for disease control in okra crops (Singh et al., 1998). Four sprays of penconazole (0.05%) and cyproconazole (0.03%) at 15 days interval were the most effective in reducing the disease (Ragupathy et al., 1998).

As an alternative method to control powdery mildew, use of cattle urine at 30% was efficient to control disease (Broek et al., 2002). The lady-bird (*Psyllobora bisoctonotata*) is a mycopredator of powdery mildew. Third and fourth larval stages were found to be the most efficient feeding stages the insect ate vegetative (mycelium) and reproductive structures, such as conidia and conidiophores, produced by the fungus on the surface of leaves (Soylu and Yigit, 2002).

Disease tolerance hybrids are Vijaya, JNDOH-1, NOH-15, JNDOH-1, AROH-47 and HYOH-1 able to reduce the disease (Neeraja et al., 2004).

9.5 WILT OF OKRA

The disease is found wherever okra is gown intensively. In the recent past occurrence of wilt disease, with some severity has appeared. In field surveys (Turki) about 60% of okra plants were found to be infected with okra wilt pathogen, *Verticillium dahliae*. (Esentepe et al., 1972).

The disease can appear at any stage of plant growth. Crops sowing in May-June suffer more than the crops sowing in February-March. Often, 20–30% plants die due to its attack. Younger plants are more susceptible than the maturing plants.

The conspicuous symptom appears as yellowing and stunting of the plant followed by wilting and rolling of the leaves. Finally, the plant dies. However, before appearance of typical wilting, the leaves showed vein clearing. They lose turgor. Often the leaves hang down in day time to recover again in the night but ultimately they wilt and the plant dies. Sometimes, the plant may look healthy but the apical buds and fruits dry. If a diseased stem or root is cut longitudinally, the vascular bundles appear as dark streaks. In severe attacks the whole stem is blackened.

The causal organism *Fusarium oxysporum f. sp. Vasinfectum* (Atkinson) Synder and Hansen is known to cause the cotton wilt. The race may be difference one. The *F. solani f.sp. hibisci*, are seed-transmissible infects okra only (Ribeiro et al., 1971). The *V. dahliae* infect cacao crop also (Emechebe, et al., 1972). While *R. solani* isolates from winged beans

(*Psophocarpus tetragonolobus*) was found to infect okra (Singh and Malhotra, 1994).

It is mainly a soil-borne fungus. But there is one time reported that fungus is found on seeds also. It survives for sometime as a saprophyte on colonized roots and then as chlamydospores in the soil. In contact with host roots the chlamydospores or conidia germinate and penetrate to roots. After some growth in the cortex the fungus reaches the xylem where it multiplies very rapidly. Dissemination of the fungus can occur by any method that can transfer the soil from one place to another (Jadhv et al., 2000).

9.5.1   MANAGEMENT

Use of certified seed and choice of date of sowing in affected fields is very important. February-March sowing in affected fields enables the plants to escape infection. Apply crop rotation for removal of roots of diseased plants, and deep summer plowing reduce disease incidence. Mulching with transparent polyethylene sheets reduced populations of *R. solani*, and *Fusarium* spp. to zero in soil to a depth of 15–20 cm and soil fumigation with 2,3 dibromopropionitrile and trichloronitroethylene are recommended. Seed treatment with bavistin at 2 g/kg seeds was the most effective seed treatment followed by Agrozim, Derosal and Pausin-M at the same rate for chemical control of okra wilt (*Fusarium solani*). (Patel et al., 2004).

The effectiveness of plant growth promoting rhizobacteria isolates was tested among them, *Bacillus pumilus* (SE-34), *B. pasteurii* (T4), *B. subtilis* (IN937-b) and *B. subtilis* (GBO3) strains significant against some seed-borne fungal diseases of okra. (Mashooda et al., 2003).

The efficacy of plant extracts in inhibiting the growth of *F. solani* (causing wilt in okra) was investigated. Garlic extract (unsterilized) produced the maximum inhibition while extracts of *Allium cepa* (bulb), *Celsia coromandeliana, Ipomoea fistula* [I. carnea], *Jatropha curcas* and *Ocimum sanctum* [*O. tenuiflorum*] showed slight inhibition (Patel and Vala, 2004). The growth of the fungal species from rhizospheric soil and rhizoplane okra was also remarkably reduced by the garlic extract. (Muhsin et al., 2001).

Maximum germination was recorded when seeds were treated with *Trichoderma viride* (@ 25 g/kg seeds) high antagonistic potential against *Rhizoctonia solani*. Treating seeds with high doses of biofungicide (50 or 100 g/kg seeds) did not inhibit germination (Mathivanan et al., 2000). In vitro, *Trichoderma hamatum* and *T. harzianum* were antagonistic to *R. solani*. The highest germination of treated seeds with *T. viride* were recorded against *Fusarium pallidoroseum*, followed by *F. oxysoporum*, and *F. moniliforme* (Gurjar et al., 2004).

Seed dressings with antagonists, using gum arabic as a sticker, reduced infection by the root rot fungi. Combined use of antagonists and organic fertilizers were better than their separate use. *Paecilomyces lilacinus* was more effective than several chemical treatments against root rot and root knot disease complex (Ghaffar, 1988).

Varieties/lines Okra I.S. 6653, 7194, 9273, 9857, C.S. 3232, 8899, Pusa Sawani and Pusa Makhamali are resistant to wilt.

9.6   COLLAR ROT DISEASE

Naturally infected seeds of okra (*Abelmoschus esculentus*) with *M. phaseolina* appear brown to black and show die-back, root and collar rot diseases. The incidence of the disease ranged from 12.7 to 58.3% (Jha and Dubey, 2000). Infected seeds were symptomatic with or without microsclerotia. In asymptomatic seeds, the mycelium was confined to the seed coat and endosperm only, whereas mycelium and micro-sclerotia occurred in the seed coat, endosperm and embryo of symptomatic seeds. Extraembryonal infection resulted in disease transmission to seedlings whereas intra-embryonal infection mostly caused pre- and post-emergence mortality. (Agrawal and Singh, 2000) *M. phaseolina* was found to the present in the seed coat and embryo, seed infection due to *M. phaseolina* led to both pre- and post-emergence mortality of okra, transmission of the pathogen from seed to seedling occurs. (Shahid et al., 2001).

9.6.1   MANAGEMENT

The later planting was associated with the pre- and post-germination mortality and development lowest incidence of collar rot disease (*Macrophomina phaseolina*) (Dubey and Jha, 1999).

Jha et al. (2000) evaluated in vitro against *Macrophomina phaseolina* Among plant extracts, leaf extract of *Eclypta alba* showed maximum inhibition of mycelial growth at 5% concentration. Whereas, at 10% concentration, leaf extract of *Argemone maxicana* showed maximum inhibition of mycelial growth, followed by *Eclypta alba* leaf extract. Among oil cakes tested, *Brassica juncea* cake exhibited maximum inhibition of mycelial growth at 5% concentration. At 10% concentration, cake of *Pongamia glabra* showed maximum mycelial growth inhibition, followed by *Azadirachta indica* cake. Fungicides were tested against fungal and bacterial antagonists in the laboratory, carboxin and metalaxyl did not inhibit the fungal antagonists *Trichoderma viride*, while little inhibition of *Gliocladium virens* was noticed at 0.1% concentration. However, carbendazim and thiophanate methyl inhibited both the fungal antagonists. The fungicides affected none of the bacterial antagonists. (Anitha and Tripathi, 2001) The antagonism of *Trichoderma viride*, *Pseudomonas fluorescens* and *Bacillus subtilis* were effective to *R. bataticola* (*Macrophomina phaseolina*) (Kaswate et al. 2003).

T. viride combined with neem cake was better in controlling the disease complex of root-knot nematode (*M. incognita*) and root rot fungus (*R. bataticola* [*Macrophomina phaseolina*]) than *T. viride* combined with groundnut cake. (Chaitali et al., 2003).

9.7   STEM CANKER

During the rainy season a severe outbreak of stem canker was observed in okra, Maharashtra, India. This is thought to be the first report of *Fusarium chlamydosporum* causing canker in okra (Fugro, 1999).

(Fugro and Jadhav, 2003) recorded stem canker of okra in Konkan, Maharashtra, India, The pathogen was isolated from the infected stem, branches, calyx, immature pods and leaf petioles of okra showing the typical disease symptoms. A total of 23 plant species belonging to 8 different families were studied to determine the host range of the pathogen.

Dark brown to black, circular to elongated lesions developed on the stem of young okra seedlings. Many infected seedlings girdled at the point of infection, which ultimately led to death of the seedlings. Under high rainfall and humid conditions, many infected plants showed splitting of the bark exposing inner cortex tissues. The leaf petioles, flower buds, calyx and immature pods were also infected showing small, circular to elongated dark brown lesions. (Fugro, 1999)

The fungus was highly pathogenic to plant species belonging to the family Malvaceae. Some of the plant species belonging to Amaranthaceae, Solanaceae and Caricaceae were also susceptible but disease incidence was lower than that observed in Malvaceae. The identified hosts of the pathogen were *Abelmoschus tetraphyllus, Abelmoschus tuberculatus, Abelmoschus ficulenus [Abelmoschus ficulneus], Abelmoschus moschatus*, cotton, ambadi, hollycock (Malvaceae), Amaranthus, tomato, and pawpaws.

9.7.1 MANAGEMENT

Among the cultivars, only KS-404, KS-410 and JNDO-5 showed field resistance to stem canker.

9.8 FRUIT ROTS

Okra fruit rot disease caused by *Pleospora infectoria.* Fruit rots caused by *Pythium*, and *Phytophthora* occur when fruits are mishandled, bruised, packed tightly and transported or stored in humid and ward conditions. *Ascochyta* also attacks on okra causing pod spots (Kumar and Rao, 1976). The fruits develop lesions with ash gray centers bearing minute fructifications of the fungus. They shrivel and dry. Abou (1985) recorded *Alternaria radicina, Botrytis. sp.* and *Fusarium sp.* from okra fruits. In Kuwait, okra were collected from several markets, *Alternaria alternata* and *Sclerotina spp.* were the most common fungi isolated (Abdel, 1988). Premature fruit abortion disease of okra in Nigeria was reported to cause by *Choanephora cucurbitarum.* Presence of seed borne pathogens like *Botryodiplodia theobromae*, showed the suppressive effect on seed germinability and okra seedlings. (Ndzoumba et al., 1990).

9.9 SEED-BORNE FUNGAL INFECTION OF OKRA

Eight seed borne fungal diseases were recorded in okra from Nigeria, none of which has previously been reported as seed borne in okra. The most common fungi were isolated from seeds of okra are *Alternaria alternata, Curvularia lunata [Cochliobolus lunatus], Cladosporium cladosporioides, Rhizopus nigricans [R. stolonifer var. stolonifer], Aspergillus niger, Aspergillus flavus, Penicillium citrinum, Trichoderma harzianum* and a dark sterile mycelium (Esuruoso et al., 1975). The prominent field fungi recorded were *Alternaria alternata, Cladosporium cladosporioides, Fusarium spp., Aspergillus spp.* and *Penicillium spp.* (Asha et al., 2001). While (Jamadar et al., 2001) recorded Twenty-seven fungi associated with the different colored seeds, among which *Aspergillus flavus, A. niger, Colletotrichum gloeosporides [Glomerella cingulata], Fusarium moniliformae [Gibberella fujikuroi], Rhizoctonia solani, Rhizopus nigricans [Rhizopus stolonifer] and Phomopsis sp.* were the predominant fungi. Black colored seeds had the highest percentage of seed mycoflora (15.20%), while white colored seeds had the lowest (8.7%). Vigor and seed weight was also low in black seeds while high in white seeds with low mycoflora association.

9.9.1   MANAGEMENT

Benomyl was the most efficient seed treatment, followed by copper oxychloride + zinc and mancozeb to elucidate the fungi associated with seed (Al-Kassim, 1996). Common fungicides, for example, Dithane M-45 (mancozeb), Bavistin (carbendazim), Agrosan GN (phenylmercury acetate and ethylmercury chloride) and thiram, at 0.1, 0.2 and 0.3% were investigated to reduce the seed mycoflora. (Asha et al., 2001) The incidence of seed-borne five fungal seed-borne diseases of okra, viz. Foot and root rot, Anthracnose and die-back, Cercospora leaf spot, Corynespora leaf spot and leaf blight, respectively caused by *Fusarium oxysporum, Colletotrichum dematium, Cercospora abelmoschi, Corynespora cassiicola* and *Macrophonina phaseolina* diseases have been found to be reduced by the use of clean apparently healthy seeds and seeds treated with Vitavax-200 (Anam et al., 2002) Fungicides combinations, like Anucop+Bavistin, Anucop+Dithane, Bavistin+Dithane, Anucop+Captan+Vitavax, Bavistin+Captan+Vitavax were most effective against the seed-borne fungal diseases the crop both in greenhouse and field conditions. Plant lattices as biopesticide against seed-borne fungi of okra were tested (Agarwal and Singh, 2002). The maximum control of seed borne fungal infection (75%) was observed in diluted *M. champaca* latex against seed borne fungi on okra.

9.9.2 SOME OTHER FUNGAL DISEASES OF OKRA

Colletotrichum capsici and *C. hibisci* cause anthracnose of okra stems, fruits and leaves. Spraying with zineb, captan or mancozeb can reduce their occurrence. Rust of okra is of rare occurrence. It is caused by *Uromyces heterogenus.*

9.10 CONCLUSION

Excessive pesticides load on the crops can be minimized to adopt IDM for fungal disease management. Application of plant/herbal extracts, other alternative of fungicides, and use of trace elements have to be utilized for disease management. Biocontrol agents blended with fungicides are promising to seed and soil borne diseases management. Crop rotation and many other cultural practices seem to have little effect on foliar diseases management. However, mycoparasite like *Ampelomyces quisqualis* and mycopredator like ladybird (*Psyllobora bisoctonotata*) could be utilized for powdery mildew disease management. Non-target effect of pesticides on foliar diseases and resistant material exploitations may be beneficial component of IDM.

KEYWORDS

- canker
- mildew
- mosaic
- okra
- spot
- wilt

REFERENCES

1. Abdel-Rahim, A. M. (1988). Post-harvest fungal diseases of some vegetables in Kuwait. *Arab Journal of Plant Protection* 6(2), 83–87.
2. Abou-Heilah, A. N. (1985). Post-harvest fungal diseases of some vegetables in the two main markets of Riyadh (Saudi Arabia). *Journal of the University of Kuwait, Science* 12(1), 103–111.
3. Agrawal, Sunita and Singh, Tribhuwan (2000). Effect of extra- and intra-embryonal infection of *Macrophomina phaseolina* on disease transmission in okra seeds. *Journal of Mycology and Plant Pathology* 30(3), 355–358.
4. Agarwal, Sunita and Singh, Tribhuwan (2002). Plant latices as biopesticide against seed-borne fungi of okra. *Journal of Mycology and Plant Pathology* 32(1), 135.
5. Al-Kassim, M. Y. (1996). Seed-borne fungi of some vegetables in Saudi Arabia and their chemical control. *Arab Gulf Journal of Scientific Research* 14(3), 705–715.
6. Anam, M. K., Fakir, G. A., Khalequzzaman, K. M., Hoque, M. M., Rahim, A. (2002). Effect of seed treatment on the incidence of seed-borne diseases of okra. *Pakistan Journal of Plant Pathology* vol. 1(1) p. 1–3.
7. Anitha, K., Tripathi, N. N. (2000). Integrated management of seedling diseases of okra caused by Rhizoctonia solani Khun and Pythium aphanidermatum (Edson) Fitzp. *Indian Journal of Plant Protection* 28(2), 127–131.
8. Anitha, K., Tripathi, N. N. (2001). Laboratory screening of fungal and bacterial antagonists against *Rhizoctonia solani* (Khun) and *Pythium aphanidermatum* (Edson) inciting seedling diseases of okra. *Indian Journal of Plant Protection* 29(1/2), 146–148.
9. Anitha, K., Tripathi, N. N. (2001). Screening of fungicides against seedling mortality of okra caused by *Rhizoctonia solani* and *Pythium aphanidermatum*. Plant Disease Research 16(1), 52–56.
10. Broek, R. V. D., Iacovino, G. D., Paradela, A. L., Galli, M. A. (2002). Alternative control of *Erisiphe cichocearum* on Okra crop. *Ecossistema* 27(1/2), 23–26.
11. Chaitali., Singh, Lokendra., Singh, Satyendra and Goswami, B. K. (2003). Effect of cakes with *Trichoderma viride* for the management of disease-complex caused by *Rhizoctonia bataticola* and *Meloidogyne incognita* on okra. *Annals of Plant Protection Sciences* 11(1), 178–180.
12. Dhankhar, B. S., Mishra, J. P. (2004). Objectives of okra breeding. Journal of New Seeds, 6(2/3), 195–209.
13. Dharam, Singh, Maheshwari, V. K., Gupta, Anuja (2001). Fungicidal control of Cercospora leaf spot in seed crop of okra [*Abelmoschus esculentus* (L.) Moench.]. *Seed-Research* 29(2), 254–256.
14. Diaz, F. A. (1999). Okra (*Abelmoschus esculentus*) powdery mildew in Mexico. *Revista-Mexicana-de-Fitopatologia* 17(1), 44–45.
15. Dubey, S. C., Jha, A. K. (1999). Influence of dates of sowing and environmental factors on collar rot of okra. *Indian Phytopathology* 52(3), 291–293.
16. Ehteshamul, Haque, and Ghaffar, S (1993). A Use of rhizobia in the control of root rot diseases of sunflower, okra, soybean and mungbean. *Journal of Phytopathology* 138(2), 157–163.

17. Emechebe, A. M., Leakey, C. L. A., Banage, W. B. (1972). Verticillium wilt of cacao in Uganda: the relationship between Verticillium dahliae and cacao roots. *Annals of Applied Biology* 70(2), 157–162.

18. Esentepe, M., Karcilioglu, A., Sezgin, E. (1972). The first report of Verticillium wilt on sesame and okra. *Journal of Turkish Phytopathology* 1(3), 127–129.

19. Esuruoso, O. F., Ogundiran, S. A., Chheda, H. R., Fatokun, D. O. (1975). Seedborne fungi and some fungal diseases of okra in Nigeria. *Plant Disease Reporter* 59(8), 660–663.

20. Fugro, P. A. (1999). A new disease of okra (*Abelmoschus esculentus, L.*) in India. *Journal of Mycology and Plant Pathology* 29(2), 264.

21. Fugro, P. A., Jadhav, N. V. (2003). Stem canker of okra in Konkan region of Maharashtra. *Journal of Mycology and Plant Pathology* 33(2), 288–289.

22. Ghaffar, A. (1988). Soilborne diseases research center of Bangladesh. Final research report, 1st January–30th June, 102–110.

23. Ghaffar, A. (1993). Rhizobia as biocontrol organisms. *BNF Agriculture Bulletin* 12(2), 6.

24. Gurjar, K. L., Singh, S. D., Rawal, P. (2004). Management of seed borne pathogens of okra with bio-agents. *Plant Disease Research* 19(1), 44–46.

25. Jadhav, N. V., Fugro, P. A., Sawant, G. G. (2000). Effect of media, pH, carbon and nitrogen sources on the growth and sporulation of Fusarium chlamydosporum causing stem canker/wilt of Okra. *Indian Journal of Environment and Toxicology* 10(2), 81–83.

26. Jha, A. K., Dubey, S. C. (2000). Occurrence of collar rots of okra (*Abelmoschus esculentus*) in the plateau region of Bihar. *Journal of Research, Birsa Agricultural University* 12(1), 67–72.

27. Jha, A. K., Dubey, S. C. Jha, D. K. (2000). Evaluation of different leaf extracts and oil cakes against *Macrophomina phaseolina* causing collar rots of okra. *Journal of Research, Birsa Agricultural University* 12(2), 225–228.

28. Jamadar, M. M. Ashok, Sajjan and Jahagirdar, Shamarao (2001). Studies on seed mycoflora and nematodes and their effect on germination and vigor index of color graded okra (*Abelmoschus esculentus, L.*). *Crop Research Hisar* 22(3), 479–484.

29. Jhooty, J. S., Sokhi, S. S. Bains, H. S., Rewal, H. S. (1977). Evaluation of germplasm of Okra against powdery mildew and Cercospora blight. *News Letter No.2 Vegetable and Hort. for Humid Tropics* pp. 30–32.

30. Johnson, A. W., Burton, G. W., Sumner, D. R., Handoo, Z. (1997). Coastal Bermuda grass rotation and fallow for management of nematodes and soilborne fungi on vegetable crops. *Journal of Nematology* (USA) 29(4), 710–716.

31. Kaswate, N. S., Shinde, S. S. Rathor, R. R. (2003). Effect of biological agents against different isolates of *Rhizoctonia bataticola* in vitro. *Annals of Plant Physiology* 17(2), 167–168.

32. Kumar, C. S. K. V., Rao, A. S. (1976). A report of leaf spot diseases on some vegetable, fodder and ornamental plants. *Current Science* 45(8), 309–310.

33. Mashooda, Begum. Rai, V. R., Lokesh, S. (2003). Effect of plant growth promoting rhizobacteria on seedborne fungal pathogens in okra. *Indian Phytopathology* 56(2), 156–158.

34. Mathivanan, N. Srinivasan, K., Chelliah, S. (2000). Biological control of soil-borne diseases of cotton, eggplant, okra and sunflower by *Trichoderma viride. Zeitschrift fur Pflanzenkrankheiten und Pflanzenschutz* 107(3), 235–244.

35. Muhsin, T. M. A., Zubaidy, S. R., Ali, E. T. (2001). Effect of garlic bulb extract on the growth and enzymatic activities of rhizosphere and rhizoplane fungi. *Mycopathologia* 152(3), 143–146.

36. Naik, K. S., Nagaraja, A. (2000). Chemical control of powdery mildew of okra. *Indian Journal of Plant Protection* 28(1), 41–42.

37. Neeraja, G. Vijaya, M. Chiranjeevi, C., Gautham, B. (2004). Screening of okra hybrids against pest and diseases. *Indian Journal of Plant Protection* 32(1), 129–131.

38. Ndzoumba, B. Conca, G., Portapuglia, A. (1990). Observations on the mycoflora of seeds produced in Gabon. *FAO Plant Protection Bulletin* 38(4), 203–212.

39. Omer, M. E. H. (1972). Chemical control of powdery mildews of cucumber and okra in the Sudan. *Wad Medani, Sudan Experimental Agriculture* 8 (3), 265–270.

40. Patel, N. N., Gohel, V. P., Vala, D. G. (2004). Control of okra wilt with chemical seed treatment. *Plant Disease Research* 19(1), 47–48.

41. Patel, N. N. Vala, D. G. (2004). Evaluation of phyto extracts against the growth of Fusarium solani. *Plant Disease Research* 19(2), 204.

42. Prabhu, A. S., Phatak, K. D., Singh, R. P. (1971). Powdery mildew of 'bhindi' (*Abelmoschus esculentus, L.*) in Delhi State. *Indian Journal of Horticulture* 28(4), 310–312.

43. Ragupathy, N., Thiruvudainambi, S., Thamburaj, S. (1998). Chemical control of powdery mildew disease of bhendi. *South Indian Horticulture* 46(1/2), 102–103.

44. Rahman, M. A., Ali, M. Mian, I. H., Begum, M. M., Kalimuddin, M. (2000). Pesticidal control of Pseudocercospora leaf spot and shoot and fruit borer of okra seed crop. *Bangladesh Journal of Plant Pathology* 16(1–2), 31–34.

45. Ribeiro, R. L. D., Robbs, C. F., Akiba, F., Kimura, O., Sudo, S. (1971). Arquivos da Universidade Federal Rural do Rio de Janeiro. *Inst. Biol.,* UFRRJ, Brazil. 1, 1, 9–13.

46. Shahid, Ahamad, Khan, Anis and Chauhan, S. S. (2001). Studies on Seed Borne Nature of *Macrophomina phaseolina* in okra. *Annals of Plant Protection Sciences* 9(1), 152–154.

47. Shahnaz, Dawar, Perveen, F., Atif, Dawar (2005). Effect of different strains of Rhizobium spp. in the control of root infecting fungi and growth of crop plants. *International Journal of Biology and Biotechnology* 2 (2), 415–418.

48. Siddiqui, I. A. Ehteshamul Haque, S., Zaki, M. J., Ghaffar (2000). Greenhouse evaluation of rhizobia as biocontrol agent of root infecting fungi in okra. *Acta Agrobotanica* 53(1), 13–22.

49. Singh, Akhilesh and Malhotra, S. K. (1994). Host range studies of *Rhizoctonia solani* causing web blight in winged bean. *Bhartiya Krishi Anusandhan Patrika* 9(2), 113–116.

50. Singh, N. B., Sharma, H. K., Srivastava, K. J. (1998). Chemical control of powdery mildew in okra seed crop. News Letter National Horticultural Research and Development Foundation. 18(4), 1–4.

51. Sinha, Asha, Rai, J. P., Singh, H. K. (2001). Seed mycoflora of Okra and its control by fungicides. *Progressive Horticulture* 33(1), 84–89.

52. Sohi, H. S., Sokhi, S. S. (1972). Disease of bhendi (Okra) and their control. *The Lal Baugh* 14(3), 1–6.
53. Sohi, H. S., Sokhi, S. S. (1974). Behavior of okra varieties to damping of powdery mildew and cercospora blight. *Indian Phytopath.* 27, 90–91.
54. Souza, V. L. D., CafeFilho, A. C. (2003). Resistance to Leveillula taurica in the genus Capsicum. *Plant Pathology* 52(5), 613–619.
55. Soylu, S., Yigit, A. (2002). Feeding of mycophagous ladybird, *Psyllobora bisoctonotata* (Muls.), on powdery mildew infested plants. *OILB/SROP Bulletin* 25(10), 183–186.
56. Yadav, S. K., Dhankhar, B. S. (2001). Correlation Studies Between Various Field Parameters and Seed Quality Traits in Okra cv. Varsha Uphar. *Seed Res* 29, 84–88.

CHAPTER 10

RECENT ADVANCES IN THE DIAGNOSIS AND MANAGEMENT OF MAJOR DISEASES OF ONION AND GARLIC IN INDIA

R. K. MISHRA,[1] M. D. OJHA,[2] V. S. PANDEY,[3] and R. B. VERMA[4]

[1]*The Energy and Resources Institute (TERI), India Habitat Centre, Lodi Road, New Delhi, India*

[2]*College of Horticulture, Bihar Agriculture University, Sabour, Bhagalpur, Bihar, India; E-mail: drmdojha@gmail.com*

[3]*National Seed Corporation, Beej Bhawan, Pusa Complex, New Delhi, India*

[4]*Bihar Agricultural University, Sabour–813210, Bhagalpur, Bihar, India*

CONTENTS

10.1 INTRODUCTION

Onion (*Allium cepa* L.) and Garlic (*Allium sativum* L) is the most important *Allium* species cultivated in India and used as vegetable, salad and spice in the daily diet by large population. In India, onion and garlic crops are grown almost all over the country, especially in the states of Haryana, Punjab, Madhya Pradesh, Gujarat, Rajasthan, Uttaranchal, Jammu & Kashmir, Bihar, Uttar Pradesh, Maharashtra, Andhra Pradesh, and Karnataka. Madhya Pradesh is the leading state in garlic production while in onion, Maharashtra is the leading state accounting for more than 22.83% of area and 28.42% of production with an average yield of 12.37 t/ha. In India per hectare yields are highest in Gujarat (22.65 t/ha) followed by Punjab (22.63 t/ha). The area under garlic during 2009–2010 is 0.14 million ha and production is 0.75 million tones with average productivity of 5.38 t/ha. In India per hectare yield are higher in Punjab (16.67 t/ha) followed by H.P. (13.14 t/ha) and Haryana (12.38 t/ha). The crop is attacked by many diseases, which vary from region to region, season to season and variety to variety. Various biotic factors like fungi; bacteria, viruses, phytoplasmas and nematodes are associated with garlic at different stages of growth and cause considerable damage/losses in yield as well as quality of garlic. There are many diseases affecting the garlic production throughout the country. These diseases can also affect at production, harvesting, processing and marketing stages, which lower the quality, reduce the yield there by increase the cost of production and export potential also. The diseases alter the cropping pattern and affect local and export markets. In 1993, in Maharashtra state 60–80% losses were reported due to Purple blotch. Consistent use of fungicides and other chemicals to manage the diseases in crop plants not only poses a serious threat to the environment and mankind but also slowly build up resistance in the pathogens.

Most of the new generation pesticides are systemic in their mode of action leads to certain level of toxicity in the plant system and thus resulting health hazards. Further, it disturbs complete microbial diversity of whole ecosystem. All these factors have led to new dimension in research for biological control and integrated disease management. The effect of disease can be lessened through various means including cultural, chemical and biological. In recent IDM approach has been found to be one of the most important tools to minimize the incidence of diseases of onion and garlic. Important diseases affecting the onion and garlic and their integrated management are reviewed here.

10.2   INTEGRATED DISEASE MANAGEMENT: BASIC CONCEPT

Plant disease control means absolute check of a disease, which is neither possible nor feasible and replaced by more appropriate term management. No pathogen is completely eradicated from natural ecosystem, but its population can be reduced below economic injury level. Disease management does not employ against the only pathogen but also considers the other three components of disease triangle, for example, host, environment and time factor which gives the concept of integrated disease management (IDM). The main objective of IDM is to maintain the loss below an economic injury level and minimize the recurrence of disease by interrupting disease cycle, survival period and inoculums source. Pathogen management involves reduction in inoculum, eradication and prevention of inoculum while management of host involves improving plant vigor, induced resistance through nutrition and genetic manipulation and protection by chemical means. The time factor could be managed by adjustment of planting date so that favorable time of disease should not coincide without any adverse effect on yield. The principles of IDM are based on avoidance, exclusion, eradication, protection, host resistance and therapy. Avoiding disease by planting at times when or in areas where inoculum is ineffective due to environmental conditions. It could be also achieved by selecting geographic area for disease free seed production, using disease escaping varieties, selection of disease free seeds and modification cultural practices. Exclusion means preventing the inoculum from entering or establishing in the field or area where it does not exist. Eradication is applied when pathogen has already been entered in area or crop in spite of the above-mentioned precautions. At these conditions, biological control methods, crop rotation, rouging, sanitation, soil treatments, heat and chemical treatments of diseased plants can be applied. In present scenario integrated disease management (IDM) strategies is the only sustainable approach. IDM should be effective against more one disease, long lasting, economically viable and eco-friendly. There is an urgent need to reinitialize the use of pesticides and develop IDM packages for onion and garlic diseases that are capable to managing plant pathogens more effectively and have minimum impact on humans, wild life and environment.

10.3   INTEGRATED DISEASE MANAGEMENT: KEY STRATEGIES

The success of integrated disease management approach in onion and garlic is only possible when we apply all the following practices and precautions every year on community basis in large area.

- Avoid indiscriminate use of fungicides and do not mix incompatible pesticides, phytohormones and micronutrients at a time.
- Soil as a reservoir of harmful as well beneficial microorganisms, therefore soil health should be properly managed by timely tillage, summer plowing, green manuring, optimum C:N ratio, balanced macro and micronutrients, aeration, etc.
- Summer plowing followed by irrigation in summer and then subsequent plowing is very effective.
- Seed health must be maintained up to prescribed standard and ensure that seed should be free from internal and external pathogen, contamination of any infected crop debris, sclerotial bodies, etc.
- Field sanitation as essential practices where removal and burning of infected crop debris, alternate and collateral weed hosts should be carried out periodically.
- Crop rotation is very important for soil borne diseases where non-host crop preferably cereals should be selected for particular pathogen.
- Avoid off season vegetables and intensive cropping because it prolongs the perpetuation period of a pathogen and do not break the life cycle of the pathogen.
- Knowledge of correct diagnosis of the diseases and disease cycle of the pathogen is very important to apply different methods of critical phase of life cycle for maximum efficacy.
- Application of biocontrol agent's particularly resident antagonist accompanied with green manuring, FYM or any organic matter should be maximized.
- Always prefer to grow tolerant varieties. However, it is difficult to get resistance against diseases with all desirable traits in hybrid.
- Proper post-harvest management practices should be followed to avoid any rotting of onion and garlic during storage and transportation.

Knowledge of the following aspects of disease development is essential for effective and economic management:

- Cause of the diseases.
- Mode of perennation and dissemination of the infectious causes.
- Host-parasite relationship and means of secondary spread.
- Effect of environment on pathogenesis in the plant and spread of the disease in the plant population.

10.4   IMPORTANT DISEASES

10.4.1   PRE-HARVEST DISEASES

(a) Damping-off:

Pathogen:

Pythium sp.
Fusarium sp.
Rhizoctonia sp.

Symptoms:

Damping-off of onion seedlings occurs in two stages: (i) pre-emergence; and (ii) post-emergence. In pre-emergence stage, the younger seedlings are killed before they reach the soil surface. They may, infect, be killed even before the hypocotyls has broken the seed coat. The radicle and plumule, when come out of the seed, undergo complete rotting. Since this happen below the soil surface and the disease often not seen, the failure in emergence of seedlings is attributed to the poor quality of seeds. The post-emergence damping off is characterized by the toppling over of infected of infected seedlings anytime after they emerge from the soil. It usually occurs at or before the ground level and infected tissues soft and water soaked. As the disease advances, the stem becomes constricted at the base and plant collapses. It is observed that most of the loss is due to pre-emergence damping-off. It is more common during Kharif season when temperature and humidity are very high.

Management:

- Proper drainage is essential.
- Sowing of clean uninfected seed on raised beds.
- Crop rotation.
- Soil solarization of nursery beds with transparent polythene for 30 days before sowing gave good control of damping-off.
- Seed treatment with Thiram, Captan @ 2.5 g/kg seed or Carbendazim @ 0.1% or *Trichoderma viride* @ 4–5 g/kg is effective.
- Drenching the nursery by Thiram @ 2.5% or Carbendazim @ 0.1% or *Trichoderma viride* @ 4–5 g/L.

(b) Purple Blotch:

Pathogen: *Alternaria porri.*

Symptoms:

This is a most important disease prevalent in all the onion and garlic growing areas in the country. The disease is reported to occur in hot and humid climate. First it appears on the leaves as small, whitish, sunken lesions. These spots later enlarge and eventually encircle the leaf. Later, darkened zones appear on the surface of the leaves, retaining the characteristic purple color. The leaves and stems fall over gradually. Infection can cause a semi-watery rot of neck of the bulbs, which turn yellow red in color, bulb tissues eventually become papery. Infected plant debris is the main source of inoculums.

Management:

- Use of healthy seeds.

- Crop rotation of 2–3 years with non-related crops should be followed.
- Spraying of Dithane M-45 (2.5gram/liter water), Kavach (2.0gram/liter water) should be sprayed at 15 days intervals just after 45 days after transplanting.
- Spraying of *Trichoderma viride* (5gram/liter) and *Baeveria bassiana* (2–3 g/L) at weekly intervals would be effective against this disease.
- Seed treatment with Captan @ 2.5 g/kg seed, Thiram @2.5 g/kg or *Trichoderma viride* @ 4–5 g/kg seed before nursery sowing.
- Summer plowing is reduced the incidence.
- Grow resistant/tolerant cultivars like, G-1, G-50, and G-323.

(c) Stemphylium Blight:

Pathogen: *Stemphylium vesicarium.*

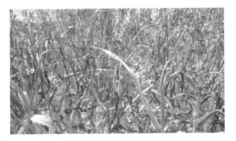

Symptoms:

This is a serious problem in Northern parts of the country especially in bulb crops. The disease is also prevalent in some parts of the southern parts of the country. Symptoms appears as small, yellow to orange flecks or streaks on the leaf. These soon develop into elongated, spindle shaped to ovate elongate, diffused spots, often reaching the leaf lips. They usually turn gray at the center, brown to dark olive brown. The spots frequently coalesce into extended patches blight of the leaves and gradually the entire foliage. The pathogen survives on the dead host tissue and serves as the source of inoculums.

Management:

- Spraying of Mancozeb (2.5gram/liter water) along with Dhanuvit is effective to control the disease. Spraying is started just after appearance of disease and repeated fortnightly.
- Spraying of Sixer (carbendazim + D. M-45) @ 2.5gram/liter of water with Dhanuvit (Sticker) at 15 days intervals.

- Crop rotation with other vegetables and cereals to reduced the disease severity.

(d) Cercospora Leaf Spot:

Pathogen: *Cercospora duddiae.*

Symptoms:

The disease appears on the leaves as small ash colored and irregular shaped on leaf lamina. The spots coalesce gradually and result in blighting of the foliage. High temperature and prolonged wet conditions favors the disease development.

Management:

- Spraying of Dithane M-45 (2.4gram/liter water) and Copper oxychloride (3.0gram/liter water) at 15 days intervals give good results.
- Spraying of Kavach (Chlorothalonil) @ 2.0 gram/liter water and sixer (Carbendazim + D. M-45) @ 2.5 gram/liter water at 15 days intervals.
- Deep summer plowing.

(e) Powdery Mildew:

Pathogen: *Leveillula taurica.*

Symptoms:

Whitish, circular to oblong lesions of variable size occurs on abaxial surface of leaves. Older leaves infected first. The lesions are covered with white powdery mass of fine hypae and conidia.

Management:

- Use of sulphur fungicides @ 0.2% at fortnightly intervals.
- Proper drainage is essential.

(f) Rust:

Pathogen: *Puccinia allii* or *Puccinia porrii*

Symptoms:

This fungus was considered to be minor importance in onion and garlic production. Initial symptoms occur on the foliage and stem as small white flecks that develop into orange spots (spores) or pustules. The bulbs become shrunken and deformed. Heavily infected plants may turn yellow and die. Disease incidence is highest in stressed plants.

Management:

- Use of healthy seeds
- Spraying of hexaconazole @ 0.1% and propiconazole (Tilt) @ 0.1% at 15 days intervals.
- Rotate with non-allium crops.

(g) Anthracnose Disease (Twister):

Pathogen: *Colletotrichum gloeosporioides*

Perfect Stage: *Glomerella cingulata.*

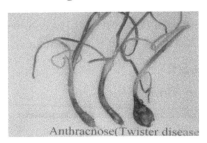

Anthracnose(Twister disease

Symptoms:

Curling and twisting of leaves, chlorosis and abnormal elongation of neck portion are major symptoms. Bulb development is already affected. White oval sunken lesions are developed on the leaves. In advance stage roots become sparse and plant may die. Bulbs are slender and may rot before harvest or during storage.

Management:

- Use of good quality of seeds.
- Seed treatment with Thiram or Capaton (2.0–2.5gram/kg seed) before nursery sowing.
- Seed treatment with *Trichoderma viride* (5.0gram/kg seed).
- Soil treatment with Benomyl or Benlate (@ 2.0gm/m^2) and spraying of D.M-45 (2.5 g/Lit water) were effective.
- Sanitation and destruction of infected plant debris helps in reducing the disease incidence.

(h) Garlic Mosaic:

Pathogen: Virus

Symptoms:

Garlic plants infected with mosaic virus show typical symptoms of chlorotic mottling and strips on the emerging leaf followed by pale yellow broken stripes resulting typical mosaic pattern on matured leaves. Yellowish dots on leaves, whitish leaf margin, or twisting of leaves are also recorded on some cultivars. Generally symptoms are more pronounced in young leaves.

Management:

- Spraying of Monocrotophas @ 0.05%, Endosulphan @ 0.25% or methyl dematon @ 0.075% at 10–15 days intervals.
- Cloves treatment with Bavistin @ 0.1% before planting.
- Use of HNPV and some neem based botanicals at weekly intervals.

(i) Onion Yellow Dwarf:

Pathogen: Virus (Poty virus)

Symptoms:

Onion Yellow Dwarf Virus (OYDV) is an important pathogen of garlic and onion also, causing severe losses in garlic clones. It is an aphid-borne poty virus. It produces symptoms of mild chlorotic strips to bright yellow strips depending on virus isolate and cultivars. Reduction in growth and bulb size also occurs. Infection by other viruses such as Leek yellow stripe, Garlic common latent virus and Shallot latent virus also occurs and may aggravate the symptoms further. However, OYDV is recognized as a major element of the virus disease complex. Enzyme Linked Immunosorbant Assay (ELISA) is the main diagnostic method for large-scale routine detection of

OYDV in garlic. The other method of OYDV detection is based on reverse transcriptase polymerase chain reaction (RT-PCR), which is 10–100 times more sensitive than ELISA.

Management:

- Removal of virus infected plants.
- Alternate spray of systemic insecticides with neem based botanicals.
- Collect healthy seed from disease free plants.

(j) Iris Yellow Spot Virus (IYSV):

Pathogen: Virus (Tospovirus)

Symptoms:

The disease symptoms vary among onion bulb and seed crops, but often appear as straw-colored, diamond shaped lesions on leaves and scapes with twisting or banding flower bearing stalks were observed in onion plants. Some lesions have distinct green center with yellow or tan borders, other lesions appears as concentric rings of alternating green and yellow/ tan tissue. Infected plants can be scattered or generalized through out a field. Large necrotic region may develop on scapes and cause a collapse of the escape. Diseased plants may be scattered or wide spread across a field, but the highest incidence of disease is often found on the field edges.

Management:

- Iris yellow spot virus (IYSV) is a top virus, similar to tomato spotted wilt virus, which is currently through to be vectored solely by onion thrips (*Thrips tabaci*), so alternate spray of softer insecticides formulations (spinosad, neem extract) and organic mulches (straw) is effective against IYSV.
- An integrated approach is essential for management of IYSV.

10.4.2 POST-HARVEST DISEASES

(a) White Rot:

Pathogen: *Selerotium cepivorum.*

Symptoms:

Pathogen is soil inhabiting and invades the roots and the basal part of the bulb scales. The first symptoms of the disease is yellowing and dying back of leaf tips. The roots are generally destroyed and there is semi watery decay of the scale with abundance of superficial white truly mycelium. Brown to black sclerotia is developed on surface or within tissue. The organism is most active when the temperature is cool. In Northern climates it usually attacks in the spring.

Management:

- Removal of infected plants during season reduces the sclerotial population and also avoids incorporation of the same to the soil.
- Solarization of soil at high temperature, for example, 35°c for 18hours or 45°c for 6 hours reduces the incidence.
- Iprodione (Rovral) @ 0.25% was effective and Benomyl @ 0.1% thiophemate methyl @ 0.1% have reduced disease incidence.
- The organisms, which have given promising result, are *Trichoderma viride, Gliocladium Zeae, Penicillium nigricans, Bacillus subtilus* and *Trichoderma harzianum* were highly effective against pest harvest diseases.

(b) Basal or Bottom Rot:

Pathogen: *Fusarium oxysporum f. sp. Cepae*

Symptoms:

In garlic, pre-emergence decay of cloves, stem plate and post-harvest decay of cloves in stored bulbs are main symptoms. Infected garlic show reduced emergence, yellowing and/or browning (necrosis) of leaves beginning at tips. On stem and bulb early in the season with some discoloration on bulb sheath at harvest. This will eventually wither and die.

Management:

- Use of *Trichoderma viride* @ 4–5 kg/ha before transplanting is effecting against this disease.
- Proper drainage is essential during entire cropping period.
- Proper drying and curing.
- Use of *Trichoderma viride*, *Pseudomonas* spp., *Baeveria bassiana* before transplanting is effecting against this disease.
- Deep plowing and avoiding injury during cultural practices.

(c) Brown Rot:

Pathogen: *Pseudomonas aeruginosa*

Symptoms:

Dark brown discoloration in bulb scale is the characteristics feature of this disease. Browning of inner scale along with rotting is the main symptoms of disease. The rotting starts from the inner scale and spread to outer scales. Apparently, the bulb seems to be healthy, but when pressed, the white oozing is noticed from the neck. In several cases the whole lot of bulbs gets rotten giving the bad odor in storage.

Management:

- Proper curing is required before storage.

- Use of maleic hydrazide (20 ml/liter) before one month of harvest.
- Neck cutting is about 2.5–3.0 cm.long above the bulb is reduced the bacterial infection.
- Light irrigation is required during entire cropping period.

(d) Soft Rot:

Pathogen: *Erwinia caratovora*

Symptoms:

Severe discolorations with soft rotting of onion bulbs are the main symptoms of this disease. Severe infection occurs at high temperature. The affected fleshy scale tissues are water soaked and pale yellow to light brown and become soft as the rot progresses. The whole bulb may break down and a watery liquid may ooze from the having foul odor as the disease advanced and, if squeezed.

Management:

- Proper curing is required after harvest.
- Proper drainage is required during cropping period.
- Reduced the doses of nitrogenous fertilizers.

(e) Botrytis rot:

Pathogen: *Botrytis allii.*

Symptoms:

Infection usually takes place through neck tissue and occasionally else where. The first sign is the softening of the effected tissue, which takes a sunken cooked appearance. Later stage, decaying starts, which is separated from healthy tissue by a definite margin. A dense grayish mycelial mat after develops upon the decaying tissue of the scope bearing short conidiophores with conidia. The disease progresses rapidly down the scales of neck tissue. Black sclerotial bodies are developed later. When diseased bulbs are cut open, water soaked brown tissues are seen near the neck region.

Management:

- Seed treatment with benomyl @1 g/kg seed or benomyl + thiram @ 1 g/kg seed reduced the disease incidence.
- Pre harvest spray of benomyl @ 0.1% reduced the fungal infection.
- Proper curing is required before storage.
- Neck cutting is 2.5–3.0 cm above the bulb to reduce the infection.

(f) Pink Root:

Pathogen: *Pyrenochaeta terrestris*

Symptoms:

The characteristic symptom is the pink coloration of the roots. The affected roots initially turn yellow but later on become soft and ultimately take distinctly deeper pink color. The new roots, which grow from the infected plants, get immediately infected and become functionless. The affected plants are not killed but development is retarded as leaf number and size are reduced. Bulbing starts earlier, but the size is reduced at maturity. The disease is confined to the roots only.

Management:

- 5–6 years crop rotation gives good control of pink root.
- Soil solarization also helps in reducing the disease.
- Use of dichloropropene @ 450 L/ha increases the yield.

(g) Black Mold:

Pathogen: *Aspergillus niger*

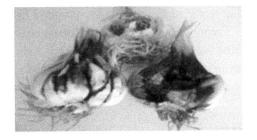

Symptoms:

Fungus infects the neck of bulbs and occasionally penetrates from side and balsa end of bulb, where damage to the dry skins has exposed the bulb scales. In advance stages of the symptoms, the entire surface of bulbs turns black and all scales are infected. It reduces the market value of the bulb.

Management:

- Proper drying and curing of the bulbs after harvest.
- Bulbs are dusted with calcium carbonate and godowns will fumigate with nitrogen trichloride @ 430 mg/m^2.

(h) Blue Mold:

Pathogen: *Penicillium corymbiferum.*

Symptoms:

The first symptoms of the disease are pale yellowish blemishes, watery soft spots, or occasionally a purplish red stain on the scales. A green to blue green mold may develop on the surface of lesions when bulbs are cut longitudinally, one or more of the fleshy scales may appear water soaked and exhibit a light tan or gray color.

Management:

- Proper drying and curing after harvest and before storage.
- Injury should be avoided during post-harvest handling.
- Mercurial dip of bulb/cloves was practiced before drying.
- Bulbs stored at temperature of 5°C or less with RH as low as possible may present infection.

(i) Garlic Bulb Canker:

Pathogen: *Embellissia allii.*

Symptoms:

This is new disease recorded in garlic bulbs (variety: G-313) produced in Himachal Pradesh. The initial symptoms of the disease are small blackish dark on the outer scales of the bulbs, which later enlarges and cover whole bulbs with black powdery mass. Under advance conditions, decay of the cloves has been observed. It was recorded by NHRDF for the first time in India.

Management:

- Spray of Carbendazim (1.0 gram/L) should be given before 15–20 days before harvest of the crops.

- Proper curing of bulbs should be taken before storage.
- Seeds (cloves) will be treated with bavistine (1.0 g/kg seed) before planting.
- The exotic material of garlic not is used for planting.

10.5 FUTURE STRATEGIES FOR DIAGNOSE THE DISEASES

- A series of cultural and chemical control methods for garlic pests and diseases with targets varying with crop growth stage.
- Monitoring of new diseases and appropriate management strategies should be developed.
- Exploitation of disease tolerance observed in the screened improved lines.
- Large-scale use of fungal and bacterial bio-agents in a consortium to achieve better disease management.
- Studies on the use of plant growth promoting rhizobacteria (PGPR) in improving plant growth and offering protection from diseases.
- Develop techniques for quick diagnosis of viral diseases.
- Conversion of onion and garlic waste in to an antimicrobial agent.
- Develop IPM modules for garlic diseases.
- Since many diseases of garlic reported in other parts of the world and not present in our country, strict embargo on import of seed, planting material and the produce for domestic consumption should be followed.
- Quarantine and legislation should be mandatory in these crops.

KEYWORDS

- **blight**
- **bulb**
- **damping off**
- **mildew**
- **onion**
- **spot**

REFERENCES

1. FAO/IPGRI, *Allium* spp. Technical guidelines for the safe movement of germplasm (ed. Diekmann, M.), 1997, No. 18.
2. FAOSTAT data. http://apps.fao.org/default.htm (Access June 12th, 2011).
3. http://extension.usu.edu
4. http://faostat.fao.org/faostat/
5. http://vegetablemdonline.ppath.cornell.edu
6. http://www.apsnet.org
7. http://www.dogr.res.in
8. http://www.garlicaustralia.asn.au
9. http://www.nhrdf.res.in
10. Khar, A., Lawande, K. E., Asha Devi, A. (2008). Analysis of genetic diversity among Indian garlic (*Allium sativum* L.) cultivars and breeding lines using RAPD markers. *Ind. J. of Genet. Pl. Breed*, 68, 52–57.
11. Malathi, S., Mohan, S. (2012). Analysis of genetic variability of *Fusarium oxysporum* f. sp. *cepae* the causal agent of basal rot on onion using RAPD markers Archives of Phytopathology and Plant Protection 45(13), 1519–1526.
12. Mishra, R. K., Gupta, R. P. (2012). *In vitro* evaluation of plant extract, bioagents and fungicides against purple blotch and stemphylium blight of onion. J. Medicinal Plant Research 6(48), 5840–5843.
13. Mishra, R. K., Sharma, P., Gupta, R. P. (2012). First report of *Phoma terrestris* causing Pink root of onion in India. *Vegetos* 25(2), 124–125.
14. Mishra, R. K., Sharma, P., Singh, S., Gupta, R. P. (2009). First report of *Embellissia allii* causing Skin blotch or bulb Canker in onion from India. *Plant Pathology.* 59(4), 807.
15. Mishra, R. K., Singh, S., Pandey, S., Sharma, P., Gupta, R. P. (2010). First report of root knot nematode *Meloidogyne graminicola* on onion in India. International Journal of Nematology. 20(2), 236–237.
16. Mishra, R. K., Srivastava, K. J., Gupta, R. P. (2008). Evaluations of botanical extracts in the management of Stemphylium blight of onion (*Allium cepa* L.) Vegetable Science. 35(2), 169–171.
17. Mishra, R. K., Verma, A., Gupta, R. P. (2009). Screening of garlic promising lines against Purple blotch and Stemphylium blight of Garlic. *Pest Management in Hort. Ecosystem.* 15(2), 22–24.

CHAPTER 11

DAMPING-OFF DISEASE OF SEEDLING IN SOLANACEOUS VEGETABLE: CURRENT STATUS AND MANAGEMENT STRATEGIES

KANIKA PAGOCH,[1] J. N. SRIVASTAVA,[2] and
ASHOK KUMAR SINGH[3]

[1,3]*Sher-e-Kashmir University of Agriculture Science and Technology, Jammu (J&K), India*

[2]*Department of Plant Pathology, Bihar Agricultural University, Sabour–813210, Bhagalpur, Bihar, India*

CONTENTS

11.1 INTRODUCTION

Vegetables play an importance role in balance diet by providing not only energy but also supplying vital protective nutrients like mineral and vitamins. In addition to their role in nutrition, vegetables increase attractiveness and palatability of a diet by providing sensory appeal through their test and flavors. Vegetables are major and very important constitute of human diet (Thamburaj and Singh, 2005).

The vegetable crops propagated by seeds, like cucurbits, beans, radish, turnip, leafy vegetable and okra required to be sown directly in the field whereas some crops like, tomato, brinjal, chili, etc., are first sown in nurseries for raising seedling and then transplanting. The disease is common in nursery bed and young seedlings by several fungi *Pythium, Phytopathora, Fusarium, Rhizoctonia, Sclerotium Colletotrichum.*

Damping off disease of seedlings is widely distributed all over the world. It was first studied by Hens (1874) in Germany. Damping off is a seedling disease common to most of solanaceous vegetable, *viz.*, tomato, brinjal and chili. The disease is of common in nursery bed and young seedlings. Several seed and soil borne fungi can kill before the tender radical and plumule established in the nursery bed (Fageria et al., 2003).

The pathogen attracts to the seed and seedling roots during germination either before or after emergence. Within days, more number of seedlings destroyed is by pathogens, and also later several weeks damping-off seedling, may develop root rot or stem canker (Atkinson, 1895). The pathogen attracts under ground, soil line or crown roots of seedlings. Some damping-off fungi, foliar blight may also occur.

Depending upon host variety and environmental factor, 25–75% losses are caused due to this disease (Gupta and Paul, 2001). Damping off of chili (*Pythium aphanidermatum* (Edson) Fitz) was responsible for 90% mortality of seedlings both in nursery and main field

(Sowmini, 1961). The fungus has wide host range and attracts the plants belonging to families Crucifereae, Leguminoceae and Chenopodiaceae (Alexander, 1931).

The amount of damage the disease causes to seedlings depends on the fungus, soil moisture and soil temperature and other factors rather than upon the particular species of plant concerned. Normally, however, cool wet soils favor the development of the disease (Alexander, 1931).

The disease is responsible for poor germination and stand of seedling in nursery bed and often the infected seedling carry the pathogen to the main field where transplanting is done. Older plants are seldom killed by damping off fungi mainly because the development of secondary stem tissue forms roots and stems still can be attacked, resulting in poor growth and reduced yields (Singh, 1995).

Therefore, vegetables require more attention by the farmers and scientists' at field level, correct diagnosis of malady with suitable control measures need to be explored for better production of vegetables (Fageria et al., 2003).

11.2 SYMPTOMS

Damping off disease in vegetables occurs in two phases based on the time of infection.

11.2.1 PRE-EMERGENCE DAMPING OFF

In this phase, the infection take place before the hypocotyl has broken the seed coat or as soon as the radicle and plumule emerge out of seeds, the seedling disintegrate before they come out of soil surface (Singh, 1995). In fact, the seeds may rot or seedlings are killed before they emerge through the soil surface. This referred to as pre-emergence damping off, which results poor field emergence/poor seed germination. If germination has occurred the hypocotyl (emerging shoot) is showed water-soaked lesion (Singh, 2000). The disease is often not recognized by the farmer who attributes the failure of emergence to poor quality of the seed.

11.2.2 POST-EMERGENCE DAMPING OFF

Post-emergence damping off is characterized by development of disease after the seedlings have emerged out of soil surface but before the stems are lignified (Atkinson, 1895). The infection results as lesion formation on the collar region giving a pinched appearance. The infection point at stem become hard and thin and such symptoms are commonly called "wire stem" appearance at the base of the stem. Infection usually occurs at the ground level or through roots. The infected tissues appear soft and water-soaked. As the disease advances the stem becomes constricted at the base and plants collapse. Seedlings that appear healthy one day may have collapsed by the next morning (Singh, 1995). The top of the plant may appear healthy when it falls over but quickly wilts and dies. The roots may or may not be decayed. Generally, the cotyledons and leaves wilt slightly before the seedlings are prostrated, although sometimes they remain green and turgid until collapse of the seedlings occurs (Brien and Chamberlain, 1937). Transplanting of seedlings on infested soil, escaping damping off and soon dies. In fields and nurseries the disease often occurs in a roughly

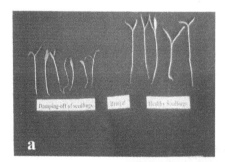 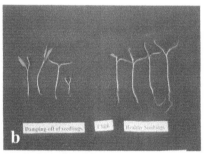

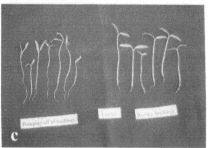

FIGURE 11.1   Damping off symptoms in (a) chilli, (b) brinjal, and (c) tomato seedlings.

circular pattern, or radiates from initial infection points, causing large spots or areas in which nearly all the seedlings are killed. This is because of the tendency of fungi to grow radially from the point of origin thereby causing large spots or areas in which nearly all the seedlings are killed (Wick, 1998). Other above ground symptoms of root rot include stunting, low vigor, or wilting on a warm day. Foliage of such plants may yellow and fall prematurely starting with the oldest leaves.

Besides this, damping off is confused repeatedly with plant injury caused by insect feeding (severe mite, aphid, scale infestation), excessive fertilization, high levels of soluble salts, excessive heat or cold injury, excessive or insufficient soil moisture, insufficient light or nitrogen, root feeding by nematodes or insect larvae or chemical toxicity in air or soil (Brien and Chamberlain, 1937).

11.3   CAUSAL ORGANISMS

The numbers of fungi responsible for damping off include species of *Pythium, Phytopathora, Fusarium, Rhizoctonia, Sclerotium Colletotrichum*. Other fungi that occasionally cause this disease include *Glomerella, Alternaria, Phoma*, and *Botrytis*. These fungi are not host-specific and more or less associated with vegetable crops. But *Pythium* spp. are generally known as damping off fungi (Chupp and Sherf, 1960). These include *P. aphanidermatum* (Eds.) Fitz., *P. debaryanum* (Hesse), *P. butleri* (Subram), *P. ultimum* (Trow) and *P. arrhenomanes* (Drechsler). In which *P. aphanidermatum* is most common fungus responsible for damping off disease.

11.4   DISEASE CYCLE AND ENVIRONMENTAL RELATIONS

Pythium species is soil-borne pathogen and also weak saprophytes and poor parasites. The pathogen perennates in soil through oospores present in plant debris, or most commonly through its mycelium (Rangaswami, 2002). The mycelial stage of the fungus is capable of infecting the host plant and multiplying very rapidly. The species of *Pythium* enter to soil through the pre-colonized host residue carrying oospores and sporangia, which are the survival structures. When proper host and proper growing condition become available, the pathogen infects the seed and seedlings causing damping off disease (Singh, 1995). Germination of these leads to primary infection of seedlings and the asexual spores later formed carry on the secondary infection and rapid spread of disease.

High soil moisture and relatively high soil temperatures favor rapid development of damping off. High soil moisture makes soil nutrients available to the oospores, which germinate and produce zoospores. In the presence of high soil moisture there is rapid dissemination of the zoospores, which attack the seeds and germinating seedlings (Gattani, and Kaul, 1995). With the help of hydrolytic enzymes the fungus causes rapid breakdown of the host tissues prior to actual colonization of the tissues (killing in advance).

After invading the host tissues the fungus rapidly forms oospores. The oospores persist in soil, resisting adverse condition. When there is

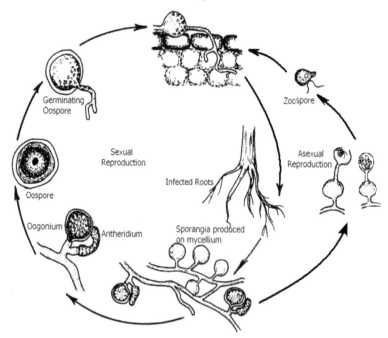

FIGURE 11.2   The disease cycle of *Pythium* damping-off in solanaceous vegetable.

sufficient moisture, they germinate and produce the mycelium, which later forms the asexual stage of reproduction (Chamount, 1979). The zoospores formed in the vesicle of sporangia are commonly released in soil water and spread from place to place. Often, when the climatic conditions are unfavorable for the asexual reproduction of the fungus, sexual reproduction starts, resulting in formation of oospores, which help to survive adverse environmental condition. Thus the fungus capable of living for many years in soil, completing its life cycle both saprophytically and as a facultative parasite (Clinton, 1920).

11.5 EPIDEMIOLOGY

High soil moisture (90–100%) with temperatures between 24° and 30°C favor the development the disease. But the disease is most severe in ill-aerated, ill-drained soil. Such conditions are common in compact, heavy soil. While loose soil, having a good proportion of sand, exhibit less loss from the disease. These types of soils are not suitable for the pathogen (Gupta and Paul, 2001). The infection being favored by poor aeration, narrow spacing, prongation of juvenile stage of the seedlings and general weakening of the plants under such condition. On the whole, the condition which pre-dispose the seedlings to damping- off are over crowding of seeds of seedlings, growth under too damp condition, excess of water/moisture in soil, and the occurrence of too much decaying organic matter (Linderman, 1989).

Nitrogen applications made too early promote damping-off. Germinating seed and new seedlings do not need much supplemental nutrition, the endosperm contains sufficient food required by the seedlings to grow nitrogenous fertilizer (Rajan and Singh, 1974). An ideal nursery should be located on light, well-drained soil conditions (Singh, 2000).

Pythium debaryanum. P. ultimum prefer cool (20°C) condition while *P. aphanidermatum* is more severe at temperatures above 20°C. Presence of seed or root exudates is known to encourage sporangial germination in *P. aphanidermatum* and *P. irregulare* (Baker and Cook, 1974).

11.6 MANAGEMENT STRATEGIES FOR DISEASE

11.6.1 PREVENTION PRACTICES FOR DISEASE

The best way to control damping off disease is to prevent it. There are many prevention techniques, and a combination of them is most effective. Certain cultural and sanitation practices can help in reducing inoculum load of the disease causing pathogens. It is essential to adopt timely control measures to avoid losses due to pathological problems during emergence (Hewson et al., 1998). The following steps are necessary for successful integrated management to avoid, exclude or eradicate the pathogen from the place of activity.

(a) Avoidance of the pathogen can be done by the adopting following practices:

- Selected area for planting a crop and raising a nursery should be free from pathogenic fungi responsible for pre-and post-emergence damping off disease.
- During summer season it is necessary to plow the soil deep in open sun to expose and check the growth of propagules of pathogens existing in the soil at high temperatures (Palti, 1981).
- One of the most important aspects in the control of damping-off is to ensure high quality seed. Avoid sowing of cracked or injured seeds (Suryanarayana, 1978).
- Planting should be done at right time to discourage the pathogen activity and enhance seed germination and emergence (Cook et al., 1978).
- Persistence of the pathogen can be checked by 2–3 years crop rotation involving non-host crops. Long crop rotations are helpful in managing the pathogens. Inter and mixed cropping check the severity of pathogens in soil, which attack the crop even at early stages of emergence (Curl, 1963).

(b) Exclusion of the pathogen can be done by seed treatment, soil treatment and rouging of diseased plants:

- *Seed treatment:* Systematic fungicides like Vitavax, Carbendazim, Benomyl may be used as seed treatment when the infection is within the seed (Evans, 1971). Contact/non-systemic fungicides like Thiram, Captan, Mancozeb, Organomercurials, etc. are also used as seed treatment for damping off pathogen (Tripathi and Grover, 1978; Baylis, 1941; Chupp and Sherf, 1960; Harman, 1991). The pathogens eradicate by physical treatments like hot water treatment at 52°C, for 30 min. Solar energy treatments are also useful pathogens eradication (Pandey and Pandey, 2002).

- The diseased seed and seedlings should be removed and destroyed as soon as observed (Cook et al., 1978).
- Proper cleaning of the seed is necessary before planting.

(c) *Eradication of the pathogen:* The pathogens can be eradicated by several methods.

- *Soil treatment:* Suitable soil treatment can be easily done in limited areas in nurseries, which may control some specific pathogens infesting a particular crop. Such treatment will also eradicate inoculum carried on the seed coat (Chaube and Vashney, 1960).

(i) *Soil Fumigation:* Application of formalin (formalin + charcoal ash @15:85) @30g/sq.ft about 3″deepin soil is effective.

(ii) *Soil Solarization:* Soil borne pathogens are eradicated by solarization (Chen and Katan, 1980). Soil disinfestations by use of white transparent polythene in hot summer are one of most effective approach for management of soil borne diseases particularly in nursery bed. Nursery bed soil can be mulching by white transparent polythene for 14–30 days. The main objective of soil solarization is to eliminate pathogen, insect, biotic agent and also weeds. Dominant soil pathogens and also damping-off causing pathogen *Pythium, Phytopathora, Fusarium, Rhizoctonia, Sclerotium Colletotrichum* spp. are effectively reduced in nursery beds (Pandey and Pandey, 2004).

Solanaceous vegetable nursery bed soil requires mulching only 14 days by white transparent polythene. (Manomohan and Sivaprakasam, 1994). The percentage germination of seed was also enhanced (Katan, et al., 1990). Soil solarization of nursery beds has been found very effective in reducing the damping off in tomato, chili, and brinjal (Pandey and Pandey, 2005).

11.6.2 MANAGEMENT PRACTICES

(a) Cultural practices

Cultural practices, such as thin sowing to avoid overcrowding, use of light soil in the nursery beds, light but frequent watering of the nursery, use of well-decomposed farm yard manure, avoidance of excessive use of nitrate forms of nitrogenous fertilizers, proper drainage, and no repetition of the same crop in the same field are do not favor pathogen development and thus reduce the chances of damping off (Mehrotra and Aneja, 2001; Palti, 1981).

Avoidance of nursery sowing in same bed year after year and also apply crop rotation for 2–3 years with non host crop (Mukhopadhyay, 1994).

Burning of 12 inches thick stack of farm trash over the nursery bed provides partial sterilization of soil and is most common and effective in reducing pathogen population.

(b) Chemical control

Seed treatment: The incidence of damping-off can be reduced by treating the seed with a fungicide prior to sowing. Fungicides will protect seed from the soil-borne pathogens, but have little effect against the seed-borne fungi. Treating of seed with suitable seed protectants is one of most effective control measure against pre-emergence stage of damping off. The protectant are applied in dry or wet form to the seed and a protectants layer around the seed coat which keeps the pathogen away until the seed have emerge (Person and Chilton, 1942; Pandey et al., 2002).

Seed protectants *viz.*, Agrosan GN (An organo mercurial), Captan, Difolatan, Thiram, Bavistin, etc. recommended as dry seed treatment @2.5–3g/kg seed to check damping off (Tripathi and Grover, 1978; Leach, et al., 1945).

Copper sulfate (10%) solution as seed soaking treatment also reduced disease. (Shyam, 1991).

Ceresan, Semesan recommended as dry seed treatment @2.5/kg seed to check damping off (Clinton, 1920).

Thiram and captan and organomercurial fungicides have proved to be highly effective in controlling seed-borne pathogens (seed rots, damping off of seedlings due to *Pythium, Fusarium* spp., etc.) in different vegetables resulting in improved seed germination and crop stand (McCallans, 1948).

The fungicide metalaxyl has systemic properties and may be used prior to sowing to reduce populations of *Pythium* and *Phytophthora* in the soil (Sahni et al., 1967).

Soil treatment:

(i) Soil fumigation:

Periodical soil treatment of the nursery can be done by using formalin dust (15 parts of formalin and 85 parts of charcoal ash) @ 30 g/sq. foot (3-inch deep) check damping off (Gupta and Paul, 2001).

(ii) Drenching of soil:

To control post-emergence stage of damping off disease, soil drenching with fungicides is very effective.

Soil drench with Captan, Difolatan, Captafol, Blitox-50 and Thiram @0.2–0.5% gave better control of disease, but these may prove costly (Tripathi and Grover, 1978).

Provide protection from post-emergence damping off nursery should be drenched with Captan @ 0.2% or Carbendazim 50WP @ 0.1% or Mancozeb75WP @ 0.25% (Joseph, 1997).

A good all-purpose preventive treatment for damping off is a 50–50 mixture of captan and benomyl may be applied as a drench. Drenching of nursery soil with 0.3% captan @ 5 L/m^2 before sowing treated seeds provide excellent control of damping off.

(c) Biological control

(i) Bio-agents against damping off

Effective of fungicides are available to control damping off disease, but chemical control by seed treatment or soil drenching has several limitations besides environmentally hazardous (Richardson, 1991). Therefore, biological control as a mechanism to reduce soil borne plant pathogens is gaining importance in recent years due to chemicals that are used to control the diseases are expensive but also contribute to soil pollutants and adversely affect non-target species (Richardson, 1991). Several microorganisms antagonistic to pathogens are available which can be used as seed or seed bed treatment. *Trichoderma viride*, *T. harzianum*, *T. hamatum*, *T. reesee*, *T. koningii*, have been reported to control species of *Pythium*,

Rhizoctonia, Fusarium, Sclerotium rolfsii, (Bagyaraj and Govindan, 1996; Papavizas and Lumsden, 1980). Damping off due to *Pythium indicum* in tomato was controlled by the application of *T. viride* (Krishnamoorthy and Bhaskaran, 1994).

Chilli damping off due to *P. aphanidermatum* was controlled by seed treatment with conidia of *T. viride* and *T. hamatum* in nursery (Ramanathan and Sivaprakasan, 1994).

Seed and soil treatment of chili and nursery soil with *T. harzianum* and *T. reesee* effectively controlled damping off and enhanced germination and emergence (Krishnamoorthy and Bhaskaran, 1994). Vescicular arbuscular mycorrhizae also control soil borne pathogens and check seed rot and seedling damage problem (Jalali and Thaereja, 1981).

Biological seed treatment can be don by priming on seed, coating the seeds, seedling dipping and dry powder treatment depending upon the nature of biocontrol agents. Generally 6–10 gram *Trichoderma* for one kg of seed is used for seed treatment but spore concentration should be in between 10^6 to 10^9 ml. Similarly 10–25g powder should be applied in per m^2 area depending upon the soil type and organic matter. Seed treatment by *Trichoderma* @1% along with soil application @10g/m^2 for nursery diseases. Some *Trichoderma* spp. Is insensitive to fungicides at lower doses of pencycuron, copper hydroxide, captan that can be incorporated while application (Pandey et al., 2002).

(ii) Leaf extracts against damping off

Plant extracts have also been successfully used to control emergence problems due to pathogens including *P. aphanidermatum* (Jacob et al., 1989).

FIGURE 11.3　Seed and soil treatment with *Trichoderma harzianum* (a), and seed–soil treatment with *Trichoderma viride* (b) in brinjal.

Due to the presence of phenolic substances and resins, gummy and non-volatile substances the plant extracts are effective against *Pythium* spp.

Narayana and Shukla (2001) evaluated the antifungal activity of 37 plants against *P. aphanidermatum* and reported that maximum inhibition (94.4%) and least post-emergence damping off was recorded by *O. paniculata* extracts among all the plants. Tomato seeds soaked in 20% leaf extract of *Bougainvillea glabra* or *Piper betle* for 6 hours before sowing increased germination by 75% and damping off due to *P. aphanidermatum* and *S. rolfsii* was also inhibited by the leaf extract of the two plants (Muthuswamy, 1972). Drenching of soil after sowing, with extracts of *Tamarindus indica* and *Leucaena leucocephala* also found very effective against damping off due to *P. indicum* (Mukhopadhyay, 1992).

11.6.3   INTEGRATED DISEASE MANAGEMENT

Integrated disease management provides a combination of cultural, biological and chemical tools to control and/or manage crop diseases effectively (Stija and Hooda, 1987). Cultural controls keep *Pythium* spp. from reaching the roots while biological and chemical controls inhibit or suppress *Pythium* spp. in the root zone.

(a) Integration of biocontrol agents and fumigants

Fumigation with dazomet, methyl isothiocyanate/1, 3-dichloropropene, and mixtures of methyl bromide and chloropicrin effectively decreases the populations of *Pythium*. Moreover, several biocontrol agents have also been used successfully for the control of damping off (Urech et al., 1977). Strashnow et al(1985) reported that under green house condition the combined treatment of *T. harzianum* (equivalent to 200 kg/ha) with lower dose of methyl bromide completely controlled disease incidence of *R. solani* in bean seedlings. Under field condition the combination of *T. harzianum* (200 kg/ha) and methyl bromide gave significant synergistic effect on damping off of carrot seedlings caused by *R. solani*.

(b) Integration of biocontrol agents and fungicides

It has been observed that in certain cases the bio-control agents or the chemicals alone could not provide satisfactory result for the management of a particular soil borne disease (Sokhi and Thind, 1996). However, the integration of biocontrol agents with certain compatible chemicals may give synergistic effect and provide better disease control than either treatments alone (Sokhi and Thind, 1996). In radish, the conidial suspension of *T. harzianum* and benodanil was found effective to minimize pre-emergence damping off caused by *R. solani*. The control of damping off by both seed treatment with *T. harzianum* and soil mix of benodanil was additive but not interactive (Lifshitz et al., 1985). Bacterial species viz. *Pseudomonas cepacia, P. fluorescens* and *Corynebacterium* species as seed dressing in combination with captan provided effective control of the damping off and root rot of peas caused by *Pythium* and Aphanomyces (Parke et al., 1991). The control of damping off of Tomato seedlings by both seed treatment with *T. harzianum* and drenching soil with fungicides is suitable control measure (Rakesh and Hooda, 2007).

(c) AM fungi and *Azospirillum* in suppression of damping off

AM fungi are known to colonize a number of tropical plants including vegetables. AM association are known to help in the growth of various crops like carrot, tomato, etc. (Sasal, 1991). Reddy et al., reported that G. *fasciculatum* proved as the benefactor and enhanced the plant growth, nutrition status, yield and reduction in disease severity. The dual inoculations (mycorrhiza with fungal pathogen) showed significant suppression in the progression of the pathogen and consequently reduction in the severity of the damping off. Mosse, (1973) indicated that mycorrhizal colonization induce chemical, physiological and morphological alterations in the host plant, which may result in an increase in, host resistance. The potential in biological suppression of soil borne pathogens gives a wider vision of AM fungi, in that they act as an alternative strategy for the host plant in conditions that are deleterious to root growth. Therefore, the introduction and consequent management of such symbiotic colonization could be employed for the advantage of the crop.

Due to the imposition of competition with the pathogen for space, nutrition and host photosynthates (Harley and Smith, 1983) or the alterations of the physiology of the host which induces host defense mechanisms (Schenck and Kellam, 1978) the arbuscular mycorrhizal fungi (AMF) are able to suppress the damping off disease. Suppression of damping off by AMF has already been reported in cucumber and ginger (Joseph, 1997).

Chilli seedlings pre-inoculated with native AMF recorded the least percent disease incidence of 22.3 and this was significantly superior as against the control, which recorded 65.1% disease incidence (Kavitha et al., 2003). Dual inoculation of AMF along with *Azospirillum* also reduced damping off by 72.7% over control. This may be due to the effect of *Azospirillum* and AMF interaction, which makes the plant healthier by way of enhanced uptake of nutrients and trigger the host defense mechanism.

11.6.4 CHEMICAL CONTROL

Although fungicides are slowly taking back seat in our fight against plant pathogens in view of some associated adverse effects on the environment, their role in managing several devastating plant diseases can not be overlooked. Prevent *Pythium* diseases by practicing integrated disease management strategies based on cultural and biological controls. Use fungicides as a last resort at the onset of disease (Thind, 2007). Fungicides have been commonly used for the control of plant diseases the world over since 19th century and in many cases have become an integral component of our crop production system (Gupta and Bilgrami, 1970). Among different methods of disease control, host resistance is still the most preferred choice although lack of its durability has been a persistent drawback. Fungicides, despite certain drawbacks, are considered to play a significant role in containing losses due to plant diseases in the coming years (Mukhopadayay, 1994).

Effectiveness of fungicides to control damping-off is highly variable (Govindappa and Grewal, 1965). Several fungicides are registered for use in vegetable nurseries to control soil borne diseases. Thiram and captan and organomercurial fungicides have proved to be highly effective in controlling seed-borne pathogens (seed rots, damping off of seedlings due to *Pythium, Fusarium* spp., etc.) in different vegetables resulting in improved seed germination and crop stand. Phenylamide fungicides have a unique potential of curbing plant pathogens belonging to the class Oomycetes such as *Pythium, Phytopathora, etc.* Metalaxyl came as a landmark discovery and a breakthrough was achieved in the effective control of diseases caused by Oomyceteous fungi. The fungicide metalaxyl has systemic properties and may be used prior to sowing to reduce populations of *Pythium* and *Phytophthora* in the soil (Urech et al., 1977). Other fungicides having different chemistry but similar anti-oomycete activity spectrum such as prothiocarb and propamocarb (carbamates), hymexazol (isoxazoles), etc. Hymexazole is used as seed dressing or soil drench to control soil-borne *Pythium* spp. and *Aphanomyces* spp. in various vegetables. Pre-emergence damping off of cabbage and cauliflower can be controlled by seed treatment with apron-70 (White et al., 1984) or thiram (Sandu and Gill, 1983).

The first post-plant fungicide application should be made when most seedlings have emerged and the seeds begin to drop from cotyledon leaves. If frequent applications of fungicides are planned, alternation of the captan-benomyl mix with other fungicides is advised to minimize the buildup of resistant pathogens. But in view of the resistance risk associated with most of the systemic, site-specific compounds, there is need to develop and employ an ideal IDM module ensuring integration of cultural, chemical and biological means of disease management options in a holistic manner.

KEYWORDS

- **biocontrol**
- **cultural**
- **damping off**
- **fumigation**
- **fungicide**
- **vegetable**

REFERENCES

1. Alexander, L. J. (1931). The causes and control of damping-off of tomato, brinjal and chili seedlings. Ohio Agr. Exp. Sta. Bull. 496.

2. Atkinson, G. F. (1895). Damping-off. N. Y. (Cornell) Agr. Exp. Sta. Bull. 94, 233–272.

3. Bagyaraj, D. J., M. Govindan. (1996). Microbial control of fungal root pathogens. In Advances in Botany. (Ed.) K. G. Mukerji. APH Publishing Corporation, New Delhi. pp. 293–321.

4. Baker, K. F., R. J. Cook. (1974). Biological control of plant pathogens. W. H. Freeman, San Francisco. pp. 433–437.

5. Baylis, G. T. S. (1941). Fungi which control pre-emergence injury to garden peas. *Ann. Appl. Biol.* 28, 210–218.

6. Brien, R. M., Chamberlain, E. E. (1937). Seedling Damping-Off. *New Zeal. Agr. Jour.* 54, 321–327.

7. Chamount, J. P. (1979). *Bulletunde la Societe Botaque de France.* 126, 537–541.

8. Chaube, H. S., S. Varshney. (2003). Management of seed rot and damping-off diseases. In: Compendium Training Program of CAS in Plant Pathology, GBPUAT, Pantnagar. pp. 281–288.

9. Chen, Y., J. Katan. (1980). Effect of solar heating of soils by transparent polythene mulching on their chemical properties. *Soil Sci.* 130, 271–277.

10. Chupp, C., Sherf, A. F. (1960). Vegetable diseases and their control. Ronald Press, New York.

11. Clinton, G. P. (1920). Damping-off. N.Y. (Cornell) Agr. Exp. Sta. Bull. 222, 475–476.

12. Cook, R. J., Boosalis, M. G., Doupnik, B. (1978). Influence of crop residues on plant diseases. In: Crop Residue Management System. *Ann. Soci. Agron.* 31, 147–163.

13. Curl, E. A. (1963). Control of plant diseases by crop rotation. *Bot. Rev.* 29, 413–479.

14. Evans, E (1971). Systemic fungicides in practices. *Pesticide sci.,* 2, 192–196.

15. Fageria, M. S., Choudhary, B. R., Dhaka, R. S. (2003). Vegetable Crops, Production Technology. Kalyani Publishers, Ludhiana. pp. 25–74.

16. Gattani, M. L., T. N. Kaul. (1951). Damping off of tomato seedlings, its cause and control. *Indian Phytopath.* 4, 156–161.

17. Govindappa, M. H., J. S. Grewal. (1965). Efficiency of different fungicides in controlling damping off of tomato. *Indian J. Agric. Sci.* 35, 210–215.

18. Grewal, J. S., R. P. Singh. (1965). Chemical treatment of seed and nursery bed to control damping off of cabbage and Chilli. *Indian Phytopath.* 18, 225–228.

19. Gupta, S. C., Bilgrami, R. S. (1970). *Proc. Natl. Acad. Science. India.* 40, 6–8.

20. Gupta, V. K., Paul, Y. S. (2001). Fungal diseases of Tomato, Chilli and Brinjal. In: Diseases of Vegetable crops. (Ed.) Gupta and Paul, Pp. 87–101.

21. Harley, J. L., Smith, S. E. (1983). *Mycorrhizal Symbiosis.* Academic Press, London.

22. Harman, G. E. (1991). Seed treatment for biological control of plant diseases. *Crop Protection.* 10, 166–171.

23. Hazarika, D. K., Sharma, P. Paramanick, T., Hazarika, K. Phookan, A. K. (2000). Biological management of Tomato damping–off caused by *Pythium aphanidermatum. Indian J. Plant Pathol.* 18, 36–39.

24. Hewson, R. T., Sagenmuller, A., Scholz-Tonga, E. M., Fabretti, J. P., Lobo, Satin. D. A. G., Blanquat, A. D. E., Schreuder, R. (1998). Global integrated crop management success stories. In: Brighton crop protection conference, pests and diseases. *Proceeding of International Conference, Brighton,* U. K., 1, pp. 161–168.

25. Jacob, C. K., Sivaprakasam, K., Jayarajan, R. (1989). Evaluation of some plant extracts and fungal antagonists for the biological control of pre-and post-emergence damping off of brinjal. *Journal of Biological Control 23,18–23.*

26. Jalali, B. L., Thaereja, M. L. (1981). Management of soil borne plant pathogens by vesicular-arbuscular mycorhizal model system. *Indian Phytopathology.* 34, 115–116.

27. Johenson, James (1914). The control of damping–off disease in plant bed. Wis. Agr. Exp. Sta. Bull. 31, 29–61.

28. Joseph, P. J. (1997). Studies on damping-off of Vegetable. PhD Thesis, Kerala Agricultural University, Thrissur, p. 192.

29. Kavitha, A., Meenakumari, K. S., Sivaprasad, P. (2003). Effect of dual inoculation of native arbuscular mycorrhizal fungi and Azospirillum on suppression of damping off in chili. *Indian Phytopathol.* 56, 112–113.

30. Krishnamurthy, A. S., R. Bhaskaran. (1994). Biological control of damping off disease of tomato caused by *Pthyium indicum* Balakrishan. *J. Bilogical control,* 4, 52–54.

31. Krishnamurthy, A. S., R. Bhaskaran. 1994. *Trichoderma viride* in the control of damping off disease of tomato caused by *Pythium indicum.* In: Crop Diseases-Innovative Techniques & Management. (Ed.) K. Sivaprakasam and K. Seetharaman. Kalyani Publishers, Ludhiana. pp. 199–203.

32. Leach, L. D., Smith, P. G. (1945). Effect of seed treatment on protection, rate of emergence, and growth of garden peas. *Phytopathology* 35,191–206.

33. Lifshitz, R., Lifshitz, S., Baker, R. (1985). Decrease in incidence of *Rhizoctonia* pre-emergence damping off by the use of integrated chemical and biological controls. *Plant Dis.* 69, 431–434.

34. Linderman, R. G. (1989). Organic amendments and soil-borne disease. *Can. J. Pl. Pathol.* 11, 180–183.

35. Manomohan Das, K. Sivaprakasam. (1994). Biological control of damping off disease in chili nursery. In: Crop Diseases. Innovative Techniques and Management. (Ed.) K. Sivaprakasam and K. Seethraman. Kalyani Publishers, Ludhiana pp. 215–243.

36. McCallans, S. E. A. (1948). Evaluation of chemical as seed protectants by greenhouse tests with peas and other seeds. Contrib. Boyee Thompson Inst. Bull., 15, 91–117.

37. Mehrotra, R. S., Aneja, K. R. (2001). An Introduction to Mycology. New Age Int. (P) Ltd., Pub. New Delhi. pp. 107–194.

38. Mosse, B. (1973). Plant growth responses of vesicular-arbuscular mycorrhizae. *New Phytol.* 72, 127–136.

39. Mukhopadhyay, A. N. (1994). Bio-control of soil borne fungal plant pathogens: current status, future prospect and potential limitations. *Indian Phytopathology.* 47, 119–126.

40. Mukhopadhyay, A. N., Shreshta, S. M., Mukherge, P. K. (1992). Biological seed treatments for control of soil borne plant pathogen. *FAO Plant. Prot. Bull.* 40, 110–115.

41. Muthuswami, M. (1972). Studies on damping off of Tomato incited by *Pythium aphanidermatum* (Edson) Fitz. M. Sc. (Ag) Thesis TNAU Coimbatore. Pp. 110.

42. Narayana Bhat, M., Shukla, B. K. (2001). Evaluation of some leaf extracts against *Pythium aphanidermatum* in vitro and pot culture. *Indian Phytopathol.* 54 (3), 395–397.

43. Palti, J. (1981). Cultural practices and infections crop diseases. Springer-Verlag, Berlin.
44. Pandey, K. K., Pandey, P. K. (2001). *In: Sovenior AICRP Vegetable Crops*. Pp. 51–52.
45. Pandey, K. K., Pandey, P. K. (2004). Effect of soil solarization on disease management in vegetable nursery. *Journal of Mycology & Plant Pathology*. 34, 398–401.
46. Pandey, K. K., Pandey, P. K. (2005). Differential response of biocontrol agents against soil pathogens on tomato, chili, and brinjal. *Indian Phytopath*. 58 (3), 329–331.
47. Pandey, K. K., Pandey, P. K., and Satpathy, S. (2002). *I. I.V. R.* Varanasi. *Technical Bulletin* No. 9. pp.22.
48. Papavizas, G. C. Lumsden, R. D. (1980). Biological Control of Soil–borne fungal propagules. *Annu. Rev. Phytopathol*. 18, 389–412.
49. Parke, J. L., Ramd, R. E., Joy, A. E., King, E. B. (1991). Biological control of *Pythium* damping off and *Aphanomyces* root rot of peas by application of *Pseudomonas cepacia and P. fluorescens* to seed. *Plant Dis*. 75, 987–992.
50. Person, L. H., Chilton, J. P. (1942). Seed and soil treatment for the control of damping off. *Jour. Agr. Res*. 75, 161–179.
51. Rajan, K. M., R. S. Singh. (1974). Effect of fertilizers on population of *Pythium aphanidermatum*, associated microflora and seedling stand of tomato. *Indian Phytopath*. 27, 62–69.
52. Rakesh, Kumar and Hooda Indra (2007). Integrated management of damping off of Tomato caused by *Pythium aphanidermatum*. *J. Mycol. Pl. Pathol*. 37(2), 259–264.
53. Ramanathan, A., Sivaprakasam, K. (1994). Effect of seed treatment with plant extract, antagonist and fungicide against damping off disease in chili. In: Crop Diseases Innovative Techniques and Management. (Ed.) K. Sivaprakasam and K. Seetharaman. Kalyani Publishers, Ludhiana. pp. 221–226.
54. Rangaswami, G. (2002). Diseases of Crop Plants in India. Prentice Hall of India Pvt. Ltd., New Delhi. pp. 286–334.
55. Reddy, B. N., Raghavender, C. R., Sreevani, A. (2006). Approach for enhancing mycorrhiza-mediated disease resistance of tomato damping-off. *Indian Phytopath*. 59(3), 299–304.
56. Richardson, M. L. (1991). Chemistry Agriculture and Environment. Cambridge: *R Soc. Chem*. pp. 546.
57. Sahani, M. L., Singh, R. P., R. Singh. (1967). Efficacy of some new fungicides in controlling damping off of chilies. *Indian Phytopath*. 20, 114–117.
58. Sandhu, K. S., Gill, S. P. S. (1983). Effect of seed and soil treatments on damping off of cauliflower. *Pesticides* 16 (11), 17–18.
59. Sasal, K. (1991). Effect of phosphate application on infection of vesicular-arbuscular mycorrhizal fungi on some horticultural crops. *Scientific reports of the Miyagi Agricultural College* 39, 1–9.
60. Satija, D. V., Hooda, Indra (1987). Greenhouse evaluation of seed treatment with fungicides for the control of tomato and chili damping off. *Indian Phytopath*. 40, 222–225.
61. Schenck, N. C., Kellam, M. K. (1978). The influence of vesicular-arbuscular mycorrhizae on disease development. *Florida Agric. Exp. Stn. Tech. Bull*. 16, 798.

62. Shyam, K. R. (1991). Important Disease of summer and winter vegetable crops and their management. In: Horticultural Techniques and Post-Harvest Management. (Ed.) Y. C. Jain et al., pp. 62–69.

63. Singh, R. P. (1995). Plant Pathology. Central book depot, Allahabad. Pp. 187–193.

64. Singh, R. S. (2000). Plant disease. Oxford & IBH publishing Co., New Delhi. pp. 173–177.

65. Sohi, H. P. S. (1982). Studies on fungal diseases of Solanaceous vegetables and their control. In: *Advances in mycology and plant pathology*. Oxford & IBH, Calcutta. pp. 19–30.

66. Sokhi, S. S., Thind, T. S. (1996). Application of fungicides in integrated disease management in vegetables. In: Rational use of fungicides in plant disease control. NATIC, Ludhiana. 133–142.

67. Sokhi, S. S., Thind, T. S. (1996). Integration of fungicides with biocontrol agents in plant disease control. In: Rational use of fungicides in plant disease control. NATIC, Ludhiana. 183–194.

68. Sowmini, R. (1961). Damping-off of seedlings and its management. M.Sc. (Ag.) Thesis, University of Madras.

69. Strashnow, Y., Elad, Y., Sivan, A., Chet, I. (1985). Integrated control of *Rhizoctonia solani* by methyl bromide and *Trichoderma harzianum*. *Plant Pathology*. 34, 146–151.

70. Suryanarayana, D. (1978). Seed pathology. Vikas Pub., New Delhi.

71. Thamburaj, S., Singh, N. (2005). Vegetable, Tubercrops and Spices. Indian council of Agriculture research, New Delhi. Pp. 10–75.

72. Thind, T. S. (2007). Changing cover of fungicide umbrella in crop protection. *Indian Phytopath*. 60 (4), 421–433.

73. Tripathi, N. N., Grover, R. K. (1978). Comparison of fungicides for the control of *Pythium butleri* causing damping off of tomato in nursery beds. *Indian Phytopath*. 30, 489–496.

74. Urech, P. A., Schwinn, F., Staub, T. (1977). A novel fungicide for the control of late blight, downy mildews and related soil-borne diseases. *Proc. Br. Crop Prot. Conf. – Pests and Diseases*, 42, 623–631.

75. White, J. G., Crute, I. R., Wynn, E. C. (1984). A seed treatment for the control of *Pythium* damping off diseases and *Peronospora parasitica* on brassicas. *Ann. Appl. Biol.* 104, 241–247.

76. Wick, R. L. (1998). Damping-off of bedding plants and vegetables. Department of Plant Pathology, Fernald Hall, University of Massachusetts, Amherst, MA.

CHAPTER 12

DOWNY MILDEW OF CUCURBITS AND THEIR MANAGEMENT

C. P. KHARE,[1] J. N. SRIVASTAVA,[2] P. K. TIWARI,[1]
A. KOTESTHANE,[1] and V. S. THRIMURTHI[1]

[1]*Division of Plant pathology Indira Gandhi Agricultural University, Raipur, Chhattisgarh, India*

[2]*Department of Plant Pathology, Bihar Agricultural University, Sabour–813210, Bhagalpur, Bihar, India*

CONTENTS

12.1 INTRODUCTION

Downy mildew (*Pseudoperenospora cubensis*) is an extremely destructive disease of cucurbits, and it was first reported from Cuba in 1868. During 1985–1988, epidemics of downy mildew (*Pseudoperenospora cubensis*) were recorded on cucumbers in Poland grown under plastic and in the field condition. It has now been reported from Japan, England, Brazil, New Jersey, Africa, etc. In India it occurs on all the cultivated cucurbits (Rondomanski and Wozniak, 1989).

It occurs practically on member of Cucurbitaceae, and mostly those, which are cultivated, although it has been observed on the wild cucumbers and few other weed hosts (Doran, 1932). In India, it is present all over the country causing heavy damage on muskmelon, watermelon, cucumber, sponge gourd, and ridge gourd and less destructive on bottle gourd, pumpkin, vegetable marrow, etc. (Gangopadhyay, 1984).

The disease is prevalent in warm temperate and tropical regions of the world with abundant moisture. About 61% reduction in crop yield has been recorded in cucurbits due to early infection of downy mildew (Figure 12.1), late infection being less harmful. The disease is confined mostly to the leaves. The loss of foliage result of early infection precludes normal flower set and fruit development. The fruit of infected plants resulting from the loss of foliage may be poor quality *viz.*, fail to get proper color, testless and look sun burnt (Gupta et al., 2001).

12.2 SYMPTOMS

The firstly, the infected leaves as green to yellowish and dark green areas reflecting a mosaic pattern, and then symptoms characterized as

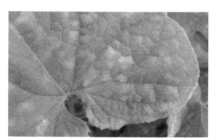

FIGURE 12.1 Downy Mildew Symptoms.

development of irregularly shaped yellowish spots on the upper leaf surface. The spots quickly turn distinctly angular bounded by veins, become yellow and then necrotic and also increase spots in to number and size (Gupta et al., 2001). If the leaf is examined on the opposite side when dew or rain is present, the brown lesion will be covered, or at least bordered, by a pale gray to purple fungus growth (sporangia and sporangiophore). Later, the severely infected leaves become chlorotic, turn brown and shrivel. In rainy humid weather entire vein is killed and showed wilt symptoms and injury to plants a whole may be great enough to cause severe stunting and death. The infection results in the reduction number and size of fruits and prevents fruit maturation and have poor flavor. Pathogens are overwintering as active mycelium on either cultivated or wild cucurbits (Brain and Jhooty, 1976).

Symptoms appear on upper surface of the leaves as angular yellow spots. During favorable condition, the under side of these spots are covered with a grayish moldy growth. The yellow spots on the upper portion of leaves, it appear just like in definite mosaic pattern lesion on the upper surface, As the spots enlarge, a general yellowing of the leaves occurs followed by the death of the tissue. The leaves subsequently wither and die (Thamburaj and Singh, 2005).

12.3   CAUSAL ORGANISM

Pseudoperonospora cubensis (Berk. & Curt.) Rostow is an obligate parasite causing downy mildew of cucurbits. The mycelium of *Pseudoperenospora cubensis* is hyaline, coenocytic and intercellular with small, ovate haustoria (Charles, 1998). The mycelium produces long, branched sporangiophores which come out through stomata on the lower surface of the leaf. Sporangiophore is dichotomously or monochotomously branched with acute angle at the tip and also arise in group of one to five. Sporangia are lemon shaped, grayish to purple in color, avoid to ellipsoidal thin walled and with a apical papilla at distal end. They measure 21–39 × 14–23 micron in size (Singh and Thind, 2005).

Babadoast (2001) stated that the sporangia are lemon-shaped, colored, borne on the gracefully on curved and pointed tips of branched sporangiophore. These are windborne and can successfully disperse to long distances if the air is moist. Sporangia give rise to biflagellate zoospores, which swim, and germinate producing with germ tubes to penetrate the host leaf. Zoospores are biflagellate and 12–13 microns in diameter and Oospores are globose, yellow or hyaline and 23–25 mm in size but production of oospores is extremely rare.

12.4   EPIDEMIOLOGY

Perpetuation of pathogens by mycelium or sporangia from one season to next season, while oospores are not common. In India, agroclimatic conditions are not favorable throughout the year. But the fungus is able to survive on wild cucurbits. The fungus survives between cropping periods and sporangia survive cold weather (Sherfand and Macnab, 1986). Sporangia have been survive below freezing temperature (–18°C) for 3–4 months and may serve as resting structure in the absence of oospores (Lange et al., 1989). Sporangia surviving on the greenhouse crops may form the primary source of infection to the later sown crops in the field (Brain and Jhooty, 1978). Sporangia can be disseminated by water, splashed by rain or carried by cucumber beetles.

The pathogen produce a germ tube in germination process and that germ tube penetrates through stomata of host surface and cause infection. The pathogen can infect the plants at temperatures between 10 and 27°C, with optimum day temperature of 25–30°C and night temperature of 15–21°C (Ullasa and Amin, 1988). Sporulation and infection is arrested above 35°C, but the fungus can survive for several days at that temperature. Relative humidity of more than 75% is conducive for disease development (Cohen and Rotem, 1970).

Initiation and further progress of the disease depend mainly in moisture, temperature having a second negative effect on infectivity of airborne sporangia (Mahrishi and Siradhana, 1988). A film of moisture is necessary on the leaf surface for the infection to occur. The environmental conditions triggering epidemics had been determined to be leaf wetness from 22.00 until 10.00 h and a temperature of 15°C for at least 6 h (Lehmann, 1991). Disease severity was positively correlated with rainfall at 7 and 8–14 d before disease occurrence, but negatively correlated with average RH (Tsai, 1992).

Disease progress was the highest between mid august until September when the maximum temperature was 32–35°C, and the minimum temperature was 21–25°C and RH 75–93% (Gandhi et al., 1996).

Major changes induced in the resistant cultivars include heavy deposition enrichment with lignin like material, which encases the penetrating haustoria. Containment of the host cells and haustoria by such materials interrupt the flow of nutrients from and into the invaded cells (Cohen et al., 1989).

Ma, S. Q. et al. (1990) reported that the pathogen, *Pseudoperonospora cubensis*, is inhibited by high temperature. The damage caused by downy mildew in crops grown in plastic houses can be mitigated by closing doors and ventilation openings to allow the temperature to rise to 40–47°C for 2 h every other day.

According Palti and Rotem (1973) downy mildew epidemics have been reported under semi – arid condition because of agriculture practices such as irrigation, favorable microclimate and abundant pathogen reproduction. Hot and dry weather had a significant influence on the spread of disease in the fields. Huang et al. (1989) reported that increasing rainfall caused downy mildew as epidemic. Epidemics over a large area can result on account of multiple infections appearing uniformly, all over the fields, under favorable weather condition. Proximity of a given fields to a source of inoculum was an important factor in the out break of disease epidemic (Cohen and Rotem, 1970).

12.5   MANAGEMENT PRACTICE

12.5.1   CHEMICAL CONTROL

Fungicide sprays are recommended for all cucurbits. Spray programs for downy mildew on any cucurbit are most effective when initiated prior to the first sign of disease because once downy mildew occurs in a planting, it becomes increasingly difficult for fungicides to control downy mildew. Both systemic and protectant fungicides are use for control of downy mildew. These can also be easily manage by spray of conventional fungicides as recommended against downy mildew.

The losses caused by *Pseudoperonospora cubensis* depend on growth stage at infection, rate of foliage growth and pathogen development (Palti and Cohen, 1980).

Wu (1994) reported that seed treatment, reduced RH, high temperature treatment and fungicides were used to control downy mildew (*Pseudoperonospora cubensis*), anthracnose (*Colletotrichum orbiculare*) and Sclerotinia rot (*S. sclerotiorum*) of cucumbers in the greenhouse.

Firstly collect seed from disease free fruit. If disease appeared on the crop, spray the crop with Mancozeb @ 0.25% (2.5 g/L water), or Zineb @ 0.25% (2.5 g/L water), Copper fungicide @ 0.3% (3 g/L water) or Chlorathalonil @ 0.25% (2.5 g/L water), or Mayalaxyl + Mancozeb @ 0.25% (2.5 g/L water) repeat at weekly interval keeping in view the wet weather condition. Thorough spacing is needed in ensuring coverage of the under surface of leaves as well (Anonymous, 2006).

Treat seeds with Agrosan GN or Emisan @ 2.5g/kg of seed before sowing. (Saha, 2002). Spray the plants with Dithane M-45 or Indofil M-45 @ 0.25% (2.5 g/L water) or Dithane Z-78 @ 0.3% (3 g/L water) or Diconil @ 0.2% (2 g/L water) at 7–10 days interval starting from the first appearance of the disease (Saha, 2002).

The most effective control of downy mildew (*Pseudoperonospora cubensis*) was achieved with Arcerid [metalaxyl + Polykarbacin (metiram)], zineb and tank mixtures of Ridomil [metalaxyl] with Cuprosan (pyrifenox) and copper oxychloride, applied just appearance of the first symptoms (Chaban et al., 1990). While sprays of Ridomil plus M 45 (copper oxychloride + metalaxyl) @ 0.5%, Ridomil plus 48 (copper oxychloride + metalaxyl) @ 0.3%, Ridomil MZ 72 (mancozeb + metalaxyl) @ 0.25%, Mikal Cu (fosetyl) @ 0.6% and Sandofan (copper oxychloride + oxadixyl) @ 0.25% gave the best control against *Pseudoperonospora cubensis* (Manole et al., 1990).

Mah (1985) were tested five fungicides (metalaxyl-mancozeb, triforine, carbendazim, cyperal and diathionon-copper) were tested for their effectiveness in controlling cucumber downy mildew. Metalaxyl-mancozeb gave very good control of the disease, followed next in order of effectiveness by cyperal.

Welt et al. (1990) found that Ridomil and Zineb (metalaxyl + zeneb) gave best control and recommendation for the use of this fungicide are given in order to minimize the risk of occurrence of resistance of the fungus *P. cubensis* to metalaxyl. Mixtures of mancozeb (0.1%) and metalaxyl (0.05%) spray initiated at conductive weather and repeated at 10 days interval are very effective. Metalaxyl resistant strains exhibit cross-resistance to other acyl-alanine fungicides. Use of metalaxyl has been abandoned in some countries for the above reasons. Metalaxyl sensitive or metalaxyl tolerant strains of *P. cubensis* are controlled by dimethomorph on cucumber and melons.

Thind et al. (1991) reported that in laboratory pot house and field studies mancozeb at 0.3% provided good control of *Pseudoperonospora cubensis* when used as a protectant but failed to check established infections even when applied only 24 h after inoculation with a sporangial suspension. Formulations of acylalanines, for example, Ridomil MZ (metalaxyl + mancozeb), Galben M8–65 (benalaxyl + mancozeb) and fosetyl aluminum, showed good protectant and eradicant activity under artificial (laboratory) and natural (field) conditions. Acylon, Pulsan and Caltan were similarly effective. Even after a gap of 15 d between treatments, these fungicides checked the disease. Ridomil MZ at 0.25% had the longest persistence and best eradicant action, no disease developing even when application was delayed for 48 h after inoculation.

Golyshin et al. (1994) tested, Akrobat 50% (dimethomorph) singly or in combination with contact fungicides against *Pseudoperonospora cubensis* on cucumber. Dimethomorph + mancozeb or a tank mixture of dimethomorph + Daconil (chlorothalonil) (3–5 applications) recommended for disease control and preventive sprays recommended for control of primary infection.

Two or three spray of Ridomil MZ-72 (metalaxyl) resulted in less downy mildew disease intensity whereas Spray of folpet, Bordeaux mixture and Aliette (fosetyl) + mencozeb were also effective. Treatments also gave higher monetary result and increase yield (Gaikwad, 1994).

High percentage control of *P. cubensis* on cucurbitaceous crops was achieved by spray applications of fosetyl-aluminum + folpet and using a tank mixture of fosetyl-aluminum with mancozeb, propineb and zineb (Yucel and Gncu, 1994).

Iikweon, et al. (1996) found that when fosetyl was applied 4 times from the start of the disease (downy mildew) at 10- days' interval, a yield index of 161% was obtained.

Aliette (fosetyl) provided good control of downy mildew (*Pseudoperonospora cubensis*) in field and greenhouse cucumbers (Merz et al., 1995).

Brunelli and Collina (1996) reported that among copper oxychloride, copper hydroxide, anilazine, chlorothalonil and fosetyl, Chlorothalonil gave the best control against *Pseudoperenospora cubensis*, followed by fosetyl, which was less persistent, while the other products gave mediocre control.

Egan et al. (1998) reported that RH-7281 is a new, high performance fungicide currently under development for foliar use to control downy mildew.

Fugro et al. (1997) found that of 6 fungicide treatments tested, fosetyl (as Aliette) gave the best disease control. Mancozeb, chlorothalonil, copper oxychloride, carbendazim and mancozeb + fosetyl all controlled the disease to different degrees.

Mercer et al. (1998) reported that RPA 407213 combination with fosetyl-Al was highly active against *P. viticola*, *Pseudoperonospora cubensis* (on Cucurbitaceae) and *Peronospora parasitica* (on Brassicaceae).

Santos et al. (2003) observed that downy mildew caused by *Pseudoperonospora cubensis* is the main disease affecting melon fruits. Different intensities of the disease were achieved by spraying the following fungicide mixtures: methyl hyphenate (hyphenate-methyl) + chlorothalonil or metalaxyl + mancozeb. There was a significant reduction in fruit yield when the disease started at 24 and 36 days after planting, but when the disease started at 47 days, no effect in production was observed.

The fungicidal mixture metalaxyl+mancozeb was highly effective in controlling downy mildew in both dry and rainy seasons, while chlorothalonil+methyl hyphenate was effective only during the dry season (Santos, 2004). ICI A5504, a betamethoxyacrylate compound is particularly effective on cucurbitaceae, providing unique control to both *P. cubensis* and *Sphaeroltheca fuliginea* causing powdery mildew.

Ullasa and Amin (1988) observed that the incidence of *P. cubensis* was most effectively reduced by Daconil (chlorothalonil) followed by Dithane M-45 (mancozeb) in field, 3 to 5 sprays of tank mixtures of dimethomorph + mancozeb or dimethomorph + chlorothalonil are recommended for checking primary infection. Weekly sprays of chlorothalonil or mancozeb give good control as protectants but fail to check established infection.

Daconil (chlorothalonil) and copper oxychloride gave effective disease control (Ullasa and Amin, 1988).

When epidemic of downy mildew developed in susceptible cultivar, Chlorothalonil, Mancozeb were most effective in lowering the infection rate and reducing disease severity (Summer et al., 1981).

Motte et al. (1988) found that the control of downy mildew which include cultural measures to prevent outbreaks and chemical control measures using mancozeb 80 or Ridomil (metalaxyl) + zineb.

Rondomanski and Zurek (1988) recorded that the fungicides tested for control of downy mildew, mancozeb + metalaxyl (as Ridomil MZ 58) was the most effective.

Mahrishi and Siradhana (1988) stated that sprays of Dithane M-45 (mancozeb) @ 0.2% decreased disease intensity, with 3 and 4 sprays at 10 days intervals or 5 sprays at 7 days intervals.

Mancozeb, mancozeb + metalaxyl, mancozeb+ oxadixyl and chlorothalonil, fungicides were effective for disease control (Rondomanski and Wozniak, 1989).

Bedlan (1990) reported that Galben M 8–65 (benalaxyl + mancozeb) is specifically approved for cucumber downy mildew, but it is not very effective.

Metalaxyl + mancozeb and chlorothalonil are the most effective fungicides for treating seedlings before inoculation (Tsai, 1992).

Khalil et al. (1992) reported that all 5 fungicides tested gave adequate control of this disease, but the most effective treatments were Trimeltox forte (200 g/100 liters), Vitigran blue [copper oxychloride] (300 g) and Dithane M-45 [mancozeb] (150 g), reducing disease intensity to 8.62, 7.75 and 9.25%, respectively, and compared with 92.5% in the untreated plot and increasing yields to 12.67, 14.87 and 17.79 t/ha.

Rramirezarredondo (1995) conducted an experiment for chemical control of downy mildew in Squash and found that oxadixyl + mancozeb (0.3+1.6 g/L), metalaxyl + mancozeb (0.2+1.4 g/L) and mancozeb (2.4 g/L) gave the best disease control and yield (2223,1997 and 1919 exportation boxes/ha respectively) compared with control treatment (1706 boxes/ha).

A study was conducted to determine the most effective treatment for controlling downy and powdery mildew (caused by Pseudoperonospora cubensis and Erysiphe cichoracearum, respectively) on bitter gourd (Momordica charantia) and among all treatments, eight sprays of 0.3% copper oxychloride + 0.3% wettable sulfur at 10-day intervals from 30 days after crop sowing was the most effective for the control of downy and powdery mildew diseases of bitter gourd during the rainy season, recording the highest yields and economic returns (Memane and Khetmalas, 2003).

12.5.2 CULTURAL PRACTICES

Destroying of wild cucurbits from vegetable growing areas also help to minimized the disease. Wide spacing between plants and conditions, which do not favor high humidity in the microclimate and expose plants to sunlight reduce the disease. Management practices including irrigation should be used to reduce the relative humidity and also the leaf wetness thereby reducing the chances for downy mildew development and its further spread, Potassium enrichment reduces incidence of downy mildew in cucumber. Follow three-year crop rotation to reduce soil borne inoculum and field sanitation by burning the crop debris after harvest (Saha, 2002).

12.5.3 BOWER SYSTEM

Bower system is superior both in terms of additional yield and return on investment. Different training systems, for example, ground, bush (dry bamboo sticks along with thorny branches), kniffin and bower (both prepared with iron angles and galvanized iron wire) use for less disease incidence of Bitter gourd (cucurbits). Except for the ground training system, vines were trained on the support. The bower system had significantly less incidence of diseases, for example, anthracnose (27.72%), powdery mildew (23.80%) and downy mildew (30.80%), as well as the highest number of fruits per plant (30.46%), longer and dark green fruits, and maximum yield (78.25 q/ha). The percentage of increase in yield over the ground training system was 71.83, 200.65 and 243.80% in the bush, kniffin and bower systems, respectively, which was attributed to low incidence of diseases (Bhokare and Ranpise, 2004). Joshi et al. (1994) conducted an experiment to assess the economic feasibility of different training system for Bitter gourd and found that the bower system is superior both in terms of additional yield and return on investment.

Lin Anchio (1995) conducted a trial and compared the yield of the cucumber cv. swallow obtained using traditional PNG methods with those obtained with pruning and the use of stand support poles (SSPs) and thus concluded that there is no need to prune the lateral branches after topping but the crop should be supported with stand poles.

Jaiswal et al. (1997) carried out a staking trial on cucumber and found that Bhakatpur local performed well for off season production in terms of its good fruit yield, good quality fruit bearing (Shape, size, color status), early fruit bearing. The effect of staking System on days to the first harvest was not Significant at the any site Farmer's practice of staking (i.e., use of bamboo sticks or tree Branches) produced 10.7% and 49% more fruits than the use of plastic string or no staking respectively. However fruit yield did not differ between staking system under low to medium management condition. On average (over location) the no staking treatment gave 28% and 52% more unmarkable fruits than staking with plastic string & the Farmers staking with plastic String and the farmers Staking practice, respectively.

12.5.4 USE OF RESISTANT VARIETY

Pusa hybrid-1and Arka chandan of pumpkin, Panjab Chappan kaddu-1 of summer squash, Pusa hybrid-3 and Pusa summer prolific Round & Long of bottle gourd, BL-240, Hybrid BTH-7, BTH-165, Phule Green, RHR BGH 1, and Arka Sujat of bitter gourd, IIHR-8 oh sponge gourd Poinsette and Priya of Cucumber, Arka manik of water melon, Panjab rasila and pusa Madhuras of Musk melon, cultivar are tolerant to downy mildew disease (Thamburaj and Singh, 2005).

Spurling (1973) reported that downy mildew (*P. cubensis*) is a serious disease of introduced cultivars of cucurbits grown by smallholders but local cultivars are rarely damaged.

Reddy et al. (1995) reported that the incidence of downy mildew varied significantly with cultivar and plant growth. Cultiver Karlhatti showed complete resistance to downy mildew during the early stages of growth, while other cultivars were susceptible. The local cultivars Chitradurga and Bellary exhibited less disease incidence at 40 days after sowing (40 DAS) as compared with the other cultivars. At later stages of growth 40 days after sowing (60 DAS), the local cultivars Siddavanahalli and Bellary exhibited the lowest incidence.

12.5.5 BIOLOGICAL CONTROL

Aqueous extracts of horse and cow manure have provided good control in Germany to cucumbers. The extracts increase chlorophyll content and Peroxidase activity of treated plants, also inhibit release of zoospores from sporangia. Winterscheidt et al. (1990) reported that watery extracts of composted cattle manure, composted sea algae, composts grape and composts manure significantly reduced the infection of cucumber leaves (*Cucumis sativus*) by *Pseudoperonospora cubensis*. All efficient extracts inhibited the germination of Sporangia with zoospores. No induced resistance of the host was observed. The extracts had no curative effects.

Ma-LiPing et al. (1996) also reported that compost extracts from horse and cow manures gave good control of *Pseudoperonospora cubensis* under greenhouse conditions, with relative efficiencies of 67.33 and 66.1% compared with untreated plants. Sheep and pig manures were less effective (46.5 and 57.3%, respectively).

Abou-Hadid et al. (2003) reported that *Trichoderma harzianum* and *Trichoderma hamatum* were the most effective antagonists against the pathogens powdery or Downy mildew disease.

KEYWORDS

- biocontol
- cucurbits
- fungicide
- mildew
- resistant
- variety

REFERENCES

1. Abou-Hadid, A. F., Abed, E. L., Moneim, M. L., Tia, M. M. M., Aly, A. Z., Tohamy, M. R. A. (2003). Biological control of some cucumber diseases under organic agriculture. *Acta Horti.*, 608, 227–236.

2. Anonymous (2006). Package and Practices for vegetable crops. Directorate of Extension Education, Sher-e-Kashmir University of Agricultural Science and Technology Jammu. pp. 107.

3. Badadoost, M. (2001). Downy mildew of cucurbits. *Bull. Univ. of Illinois.* 345.

4. Bains, S. S., Jhooty, J. S. (1976). Overwintering of *Pseudopaeronospora cubensis* causing downy mildew of Musk melon. *Indian Phyto Path.* 29, 213–214.

5. Bains, S. S., Jhooty, J. S. (1978). Epidemiological studies on downy mildew of Musk melon caused by *Pseudopaeronospora cubensis. Indian Phyto Path.* 31, 42–46.

6. Bedlan, G. (1990). Problems in treatment of cucumber downy mildew under glass. *Pflanzenschutz-Wien.* 3, 5–7.

7. Bhokare, S. S., Ranpise, S. A. (2004). Effect of different training systems on incidence of pest, diseases and yield of bitter gourd (*Momordica charantia L.*) cv. Konkan tara under Konkan conditions of Maharashtra. *Haryana J. Hort. Sci.* 33, 139–141.

8. Brunelli, A., Collina, M. (1996). The protection of melon from *Pseudoperonospora cubensis. Colture-Protette.* 25(12), 107–108.

9. Chaban, V. V., Kitsno, L. V., Nedobytkyn, V. A. (1990). Testing fungicides against downy mildew on cucumber. *Ukraine. Zashchita-RastenMoskva*, 9, 27–28.

10. Charles Chupp. (1998). Manual of vegetable plant diseases. Discovery Publishing House, New Delhi. pp186.189.

11. Cohen, Y., Eyal, H., Hanania, J., Malik, J. (1989). Ultra structure of *Pseudopaeronospora cubensis* in Musk melon genotypes susceptible and resistant to downy mildew. *Physiol Mol. Plant Pathol.*, 34, 27–40.

12. Cohen, Y., Rotem, J. (1970). The relationship of sporulation to photosynthesis in some obligatory and facaltative parasites. Phytopathology, 60, 1600–1604.

13. Doran, W. L. (1932). Downy mildew of cucurbits. Mass. Agri. Expt. Sta. Bull. 283.

14. Egan, A. R., Michelotti, E. L., Young, D. H., Wilson, W. J., Mattioda, H. (1998). RH-7281, a novel fungicide for control of downy mildew and late blight. *Brighton Crop Protection Conference: Pests and Diseases. Volume 2, Proceedings of an International Conference, Brighton, UK,* 335–342.

15. Fugro, P. A., Rajput, J. C., Mandokhot, A. M. (1997). Sources of resistance to downy mildew in ridge gourd and chemical control. *Indian Phytopath.*, 50(1), 125–126.

16. Gaikwad, A. P., Karkeli, M. S. (1994). Control of downey mildew of cucumber with new fungicide. *J. of Maharashtra Agril. Univ.*, 19 (3), 445–446.

17. Gandhi, S. K., Maheshwari, S. K., Mehta, N. (1996). Epidemiological relationship between downy mildew of ridge gourd and metrological factors. *Pl. Dis. Res.*, 11(1), 62–66.

18. Gangopadhyay, S.1984. Advance in vegetable disease. Associated publishing company, New Delhi: 151–161.

19. Golyshin, N. M., Maslova, A.A., Goncharova, T. F. 1994. Fungicides against peronosporacae. *Zashchita Rastenii Moskva.* 4, 18.

20. Gonzalez, M., Barrios, F., and Rodriguez, F. 1992. Evaluation of the effectiveness of different fungicides against the mildew of cucumber caused by *Pseudoperonospora cubensis. Proteccion de Plantas.* 2(4), 75–86.

21. Gupta, V. K. And Pal, Y. S. (2001) Fungal disease of cucurbits. In: Diseases of vegetable crops (Ed.) Gupta, V. K. and Pal, Y. S., 71–75pp.

22. Huang, X. M., Yang, Z., Lu, Y. H. (1989). Effects of climatical factors on the occurrence of cucumber downy mildew in the suburbs of Shanghai. *Acta Agriculture Shanghai.* 5(1), 83–88.

23. Ilkweon, Y., Hanwoo, D, Yongseub, S., Sugan, B., Sungkuk, C., Boosull, C. (1996). Timely application of fosetyl-Al for the control of downy mildew on oriental melon (*Cucumis melo* L.). *J. Agril. Sci. Crop Protection.* 38(2), 378–381.

24. Jaiswal, J. P., Bhattarai, S. P., Subedi, P. P. (1997). Effect of different staking system in cucumber var Bhaktapur local for off-season production. *Working paper Lumle Re. Alternaria solani* in hydroponically grown tomato. *Phytopath.* 89, 722–727.

25. Jhooty, J. S., Bains, S. S., Parkash, V. (1989). Epidemiology and control of cucurbit downy mildew. *Perspectives in phytopath.* 301–314.

26. Joshi, V. R., Lawande, K. E., and Pol, P. S. (1994). Studies on the economic feasibility of different training system in Bitter gourd. *J. Mah. Agril. Univ.*19 (2), 238–240.

27. Khalil, M. R., Khan, N. U., Younas, M., Shah, N. H., Hassan, G. (1992). Control of downy mildew (*Pseudoperonospora cubensis* (Berk. & Curt.) Rostow) of melon (*Cucumis melo*) with different fungicides. *Pak. J. Phytopath.* 4(1–2), 50–53.

28. Lange, L., Eden, U., Olison, L. W. (1989). Zoosoprogensis in *Pseudopaeronospora cubensis* the causal agent of cucurbit downy mildew. *Nordic J. Bot.* 8, 497–504.

29. Lehmann, M. 1991. Study on forecasting the beginning of epidemics of downy mildew (*Pseudoperonospora cubensis*) in field cucumbers. *Pflanzenschutz Wien.* 2, 4–6.

30. Lin Anchio. (1995). Effect of pruning and use of stand support poles on cucumber yield in Papua. *Harvest (Port Moresby).* 17(1–2), 9–16.

31. Ma, S. Q., Liang, H. H., Ma, J. X. (1990). Study on the ecological way of preventing cucumber downy mildew: a report of a control experiment on the environmental temperature. *Chinese Journal of Applied Ecology.* 1(2), 136–141.

32. Mah, S. Y. (1988). A comparison of five fungicides and two-spore germination suppressing agents for control of cucumber downy mildew (*Pseudoperonospora cubensis* Berk. and Curt.). *Teknologi Sayur Sayuran* (Malaysia). 61–65pp.

33. Mahrishi, R. P., Siradhana, B. S. (1988). Effect of nutrition on downy mildew disease caused by *Pseudoperonospora cubensis* (Berk. & Curt.) Rostow. on muskmelon. *Annals of Arid Zone.* 27(2), 153–155.

34. Mahrishi, R. P., Siradhana, B. S. (1988). Studies on downy mildew of cucurbits in Rajasthan: incidence, distribution, host range and yield losses in muskmelon. *Annals of Arid Zone.* 27 (1), 67–70.

35. Ma-Liping., Gao-Fen., Wu-Yingpeng and Qiao-Xiong Wu. (1996). The inhibitory effects of compost extracts on cucumber downy mildew and the possible mechanism. *Acta Phytophylacica Sinica.* 23(1), 56–60.

36. Manole, N., Costache, M., Varadie, P., Paraschiv, G., Gogoci, I., Szabo, A., Mirghis, R. (1990). Epidemiology and integrated control of *Pseudoperonospora cubensis* (Berk. et Curt) Rostov in cucumber. *Academia de Stiinte Agricole si Silvice.* 23.

37. Memane, S. A., Khetmalas, M. B. (2003). Chemical control of downy and powdery mildew diseases of bitter gourd during rainy season. *J. Maha. Agril. Univ.* 28(3), 283–284.

38. Mercer, R. T., Lacroix, G., Gouot, J. M., Latorse, M. P. (1998). RPA 407213, a novel fungicide for the control of downy mildews, late blight and other diseases on a range of crops. *Brighton Crop Protection Conference: Pests & Diseases. Volume 2, Proceedings of an International Conference, Brighton, UK,* 319–326pp.

39. Merz, F., Schrameyer, K., Sell, P. (1995). Establishment of a standard treatment against downy mildew in cucumbers. *Gemuse-Munchen.* 31(7), 438–439.

40. Motte, G., Muller, R., Auerswald, H., Beer, M. (1988). Monitoring and control of downy mildew of cucumbers. *Gartenbau.* 35(1), 15–16.

41. Palti, J., Cohen, Y. (1980). Downy mildew of cucurbits (*Pseudoperonospora cubensis*) the fungus and its hosts, distribution, epidemiology and control. *Phytoparasitica.* 8(2), 109–147.

42. Palti, J., Rotem, J. (1973). Epidemiological limitations of the forecasting of downy mildew and late blight in Israel. Phytoparasitica 1,119–126.

43. Ramirezarredondo, J. A. (1995). Chemical control of downy mildew (*P.cubensis* (Bert. & Curt.) Rostow.) In sqush in the Mayo valley. *Mexcio Revista Mexicana de Fitopatologia.* 13(2), 126–130.

44. Reddy, B. S., Thammaiah, N., Patil, R. V., Nandihalli, B. S. (1995). Studies on the performance of bitter gourd genotypes. *Advances in Agril Res. In India.* 4, 103–108.

45. Rondomanski, W., Wozniak, J. (1989). Distribution and chemical control of downy mildew on cucumber. *Biuletyn-Warzywniczy.* 145–149.

46. Rondomanski, W., Zurek, B. (1988). Downy mildew of cucurbits-a new threat to cucumber in Poland. *Roczniki Nauk Rolniczych, E Ochrona Roslin.* 17 (1), 199–209.

47. Saha, L. R. (2002). A hand book of Plant diseases. Kalyani Publisher Ludhiana. 361–362.

48. Santos, A. A. Dos., Cardoso, J. E., Vidal, J. C., Silva, M. C. L. (2004). Evaluation of chemical products on the control of downy mildew and stem canker of melon. *Revista Ciencia Agronomica.* 35(2), 390–393.

49. Santos, A. A. Dos., Cardoso, J. E., Vidal, J. C., Vianaf, M. P., Rossetti, A. G. (2003). Effect of the downy mildew initiation on the melon fruit production. *Fitopatologia Brasileira.* 28(5), 548–551.

50. Sherif, A. F., Macnab, A. A. (1986). *Vegetable disease and Their control. John Willey and Sons New York, USA.* 728 pp.

51. Singh, P. P., Thind, T. S. (2005). Diseases of cucurbits and their management. In: Diseases of fruit and vegetable (Ed.) T. S. Thind. Kalyani Pub. Ludhina.290–305.

52. Spurling, AT.1973. A review of mildew problems on cucurbit vegetable crops in Malawi. *PANS.* 19, 1, 42–45.

53. Summer, D. R., Phathak, S. C., Smitte, D., Johnson, A., Glaze, N. C. (1981). Control of cucumber foliar diseases, fruit rot and nematode by chemicals applied through overhead sprinkler irrigation. *Pl. Dis.* 65 (5), 401–404.

54. Thamburaj, S., Singh, N. (2005). Vegetable, Tubercrops and Spices. Indian council of Agriculture research, New Delhi. 10–75.

55. Thind, T. S., Singh, P. P., Sokhi, S. S., Grewal, R. K. (1991). Application timing and choice of fungicides for the control of downy mildew of muskmelon. *Pl. Dis. Res.* 6 (1), 49–53.

56. Tsai, W. H., Tu, C. C., Lo, C. T. (1992). Ecology and control of downy mildew on cucurbits. *Plant Protection Bulletin-Taipei.* 34 (2), 149–161.

57. Ullasa, B. A., Amin, K. S. (1988). Ridgegourd downy mildew epidemics in relation to fungicidal sprays and yield loss. *Mysore J. of Agril. Sci.* 22(1), 62–67.

58. Weit, B., Neuhaus, W. (1990). Biology and control of cucumber downy mildew (*Pseudoperonospora cubensis*). *Nachrichtenblatt fur den Pflanzenschutz in der DDR.* 44(1), 5–8.

59. Winterscheidt, H., Minassian, V., Weltzien, H. C. (1990). Studies on biological control of cucumber downy mildew (*Pseudoperonospora cubensis*) with compost extracts. *Gesunde Pflanzen (Germany, F. R.).* 42(7), 235–238.

60. Wu, S. Q. (1994). Integrated management of cucumber diseases in greenhouse. *Bulletin of Agricultural Science and Technology.* 2, 24.

61. Yucel, S., Gncu, M. (1994). Studies on chemical control of downy mildew (*Pseudoperonospora cubensis* Berk. and Curt.) on cucurbits in the Mediterranean region. *Bitki Koruma Bulteni.* 31(1–4), 109–118.

CHAPTER 13

DISEASES OF COLOCASIA CROP AND THEIR MANAGEMENT

R. C. SHAKYWAR, M. PATHAK and K. M. SINGH

College of Horticulture and Forestry, Central Agricultural University, Pasighat – 791102, Arunachal Pradesh, India, E-mail: rcshakywar@gmail.com

CONTENTS

13.1 INTRODUCTION

Colocasia (*Colocasia esculenta*) is tuber/rhizome crop and belongs to the family Araceae. It is also locally known as *Arvi* or *Ghuiyan* in north India and *Kachchu* in northeastern parts of India. This crop grows wild in sub-Himalayan tract, peninsular region and northeastern region, mostly in waterlogged humid tropical areas. It shows extensive genetic variability in eastern region. Like other tuber crops, polyploidy as well as diploidization have occurred in this crop also during the course of evolution. The variable diploid (2n) chromosome numbers 24, 28, 42 and 48 and triploid (3n) 42 have been recorded in various cultivars. Most of the important characters such as, plant height, size of leaves, tuber weight and number and yield show polygenic inheritance. This crop is mainly cultivated in Bangladesh, Brazil, Egypt, India, Indonesia, Malaysia, Nigeria, Pacific region, Philippines, West Indies and few other countries. In India, it is grown in Andhra Pradesh, Bengal, Bihar, Gujarat, Karnataka, Kerala, Madhya Pradesh, Maharashtra (Konkan region), Tamil Nadu, Uttar Pradesh and West Bengal. It grows well in lowland and upland areas (Swarup, 2006). All the above ground (leaves) and underground (cormel) parts of this crop are edible. However, it is mainly grown for corms. The cormels and leaves are eaten fried and cooked vegetable. Delicious dish is prepared frying the rolled leaves dipped in gram paste (*besan*). The corms are used for culinary purposes and in preparation of chips. Cormels are rich source of starch, protein, vitamin C, calcium and phosphorus (Fageria et al., 2006). Besides, this crop is of great medicinal value and is included in many Ayurvedic preparations. The juice from petioles or whole leaves is used for styptics, poultices and pulmonary congestion. Taro lactic is now a commonly prescribed infant food in Hawaii and is fed to a newborn baby. Patients suffering from ulcers and other alimentary disorders or convalescents derive great relief from this easily digestible, nutritive food. It is also strongly recommended in prenatal diets as well as to nursing mothers (Chadha, 2003). Planting of *Colocasia esculenta* var. *antiquorum* is normally done during the rainy season (June–July) and summer season (February–March) in northern states. September–October is the best time of planting in southern parts of Gujarat. However, it can be planted at any time where irrigation facilities are available. The results obtained at Central Tuber Crops Research Institute indicated that April–June was the ideal time of planting under rainfed conditions. The crop is harvested after 6–8 months of planting. It grows in all kinds of soils but thrives best in deep, well-drained, well-manured, friable loam. Where rainfall is sufficient, the fields are frequently irrigated. It is often grown in Kitchen gardens under intensive cultivation and irrigation (Shakywar et al., 2012).

Understanding the diseases and their behavior is basic to successful and economic cultivation of colocasia. In this chapter, attempt has been made to present the current knowledge about the colocasia diseases including bacterial diseases, fungal disease, viral and phytoplasmal diseases, nematodes diseases, phanerogamic plants, non parasitic diseases and physiological disorders and their management practices.

13.2 BACTERIAL DISEASES

13.2.1 SOFT ROT

13.2.1.1 Symptoms

Bacterial soft rot is a strong smelling watery soft rot ranging in color from white to dark blue. Wounds and bruises caused by the feeding of insects and other animals and those inflicted at harvest are the most common infection courts for this disease.

13.2.1.2 Causal organism

Erwinia carotovora or *E. chrysanthemi*

13.2.1.3 Etiology

The bacterium is a Gram-negative, rod-shaped that lives alone or aggregates into pair's chains, non spore forming and peritrichously flagellated.

It is a facultative anaerobe that catalase negative and oxidase positive. Bacterium produces a number of extracellular plant cell wall degrading enzymes such as pectic enzymes that degrade pectin, cellulase that degrades cellulose, hemicellulases, arabanases, cyanoses and protease. As a mesophilic bacterium, *Erwinia carotovora* thrives the most in the temperature range 27–30°C (Bell et al., 2004).

13.2.1.4　Mode of spread

The bacterium is survived in infected plant debris in the soil. The disease spread through contaminated soil, rhizomes and maggot flies.

13.2.1.5　Favorable Conditions for Disease Development

The fungus is favor high relative humidity 94–100% and temperature 21–29°C. Abundant moisture is required for invasion of the bacteria.

13.2.1.6　Disease cycle

The bacterium is ubiquitous in the environment, meaning that it is always present. Soft rot can occur at anytime as long as the right conditions have been met. Infection occurs due to human plant interactions with harvesting, planting, irrigation, insecticide and fungicide applications under pressure, pruning and propagating. Once a plant has been infected the bacterium can live on old foliage, tubers, soil on colocasia, field equipment and in water.

13.2.1.7　Management

1. Careful handling of corms to minimize injury at harvest air-drying of corms and storage at low temperatures of only the sound corms.
2. Sowing healthy seeds.
3. Keeping the crops weed free.

13.3 FUNGAL DISEASES

13.3.1 INTRODUCTION

Leaf blight of colocasia is caused by *Phytophthora colocasiae* (Raciborksi, 1900). It is the most destructive amongst all the diseases, which infect crop and appear in epidemic form in congenial environmental conditions, like late blight of potato (Butler and Kulkarni, 1913; Mendiola and Espino, 1916). In India, leaf blight of colocasia has been reported to be serious disease in different states like as Punjab (Luthra, 1938), Himachal Pradesh (Paharia and Mathur, 1961), Orissa (Thankappan, 1985), Arunachal Pradesh (Chaudhary and Rai, 1988) and Uttar Pradesh (Singh, 2000).

13.3.2 LEAF BLIGHT

13.3.2.1 Symptoms

The early stages of the disease are characterized by small circular water-soaked lesions 1–2 cm in diameter, generally dark brown or purple. A clear amber fluid exudes from the center of the lesion. This liquid turns bright yellow or dark purple when it dries. The lesions rapidly enlarge and take on a zonate appearance. The zonation is the result of the temperature-related growth response of the fungus with rapid growth during the warm days followed by slow growth during the cooler nights. The sporangia appear

FIGURE 13.1 Blight symptoms on colocasia leaf.

as white fuzz on both sides of the leaf. The ring of sporangia is particularly prominent in the morning before the leaves dry. After initial establishment lesion development is rapid until the leaf is entirely colonized and collapses (Walker, 1952).

13.3.2.2 Causal Organism

Phytophthora colocasiae Raciborski.

13.3.2.3 Etiology

The fungus is reproduce by sexually or asexually methods. In many species, sexual structures have never been observed or have only been observed in laboratory matings. In homothallic species, sexual structures occur in single culture. Heterothallic species have mating strains designated as A1 and A2. When mated, antheridia introduce gametes into oogonia, either by the oogonium passing through the antheridium (amphigyny) or by the antheridium attaching to the proximal (lower) half of the oogonium (paragyny) and the union producing oospores. Like animals, but not like most true Fungi, meiosis is gametic and somatic nuclei are diploid. Asexual (mitotic) spore types are chlamydospores and sporangia, which

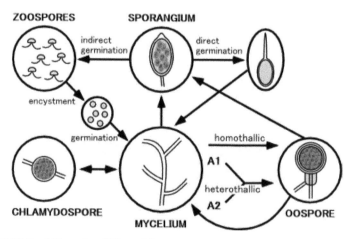

FIGURE 13.2 Life cycle of *Phytophthora.*

produce zoospores. Chlamydospores are usually spherical, pigmented and may have a thickened cell wall to aid in its role as a survival structure. Sporangia may be retained by the subtending hyphae (non-caducous) or be shed readily by wind or water tension (caducous) acting as dispersal structures. Moreover, sporangia may release zoospores, which have two unlike flagella, which they use to swim towards a host plant.

13.3.2.3   Mode of Spread

The fungus is soil borne and spores of the fungus are moved in wind-driven rain and dew to new areas of the same leaf to nearby plants or new plantings.

13.3.2.4   Favorable Conditions for Disease Development

Rapid spread of the disease is favored by temperatures 25–28°C and relative humidity 65% during the day, cooler temperatures 20–22°C and humidity 100% at night, when the spores are produced and light rains or heavy dew in the morning to scatter the spores and allow germination and infection.

13.3.2.5   Disease Cycle

The disease cycle of colocasia leaf blight is given in Figure 13.3 (Macpherson, 2000). Sporangia with apical papilla – a small rounded process at the top are produced on slender sporangiophores which branch with a swelling at the point of branching. Sporangia are ovoid to ellipsoid, mostly 45–50 × 23µm. Chlamydospores are thick-walled, usually 26–30µm diameter. Oospores require the opposite mating type of *P. colocasiae* or a different species of *Phytophthora* (Cho et al., 2004 and CAB International, 2002). The spores are very delicate and on sunny days they shrivel and die within 2–3 hours as humidity falls. During hot, dry weather it is common for lesions in the field to stop expanding and for the necrotic centers to drop out. Many of these 'shot holes' expand no further; others will resume development (often from

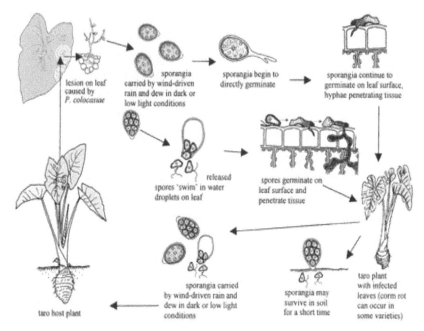

FIGURE 13.3   Disease cycle of leaf blight of colocasia.

one point at the margin) under conditions of heavy rain. The most rapid expansion of lesions occurs when cool, showery weather allows fungal growth in tissues both night and day (Fullerton and Tyson, 2003). Corms are infected from spores washed from leaf infections into the soil. At harvest stage, the spores invade the corms where the suckers are removed (Jackson, 1999).

13.3.2.6 Integrated Disease Management

1. Sanitation by pruning and removing infected leaves biweekly appears to help reduce disease incidence.
2. Exploitation of date of sowing to manage the leaf blight of taro (Shakywar et al., 2013a).
3. Exclusion through quarantine will protect areas still free of the pathogen.
4. Use of soil amendment like mustard, mahua and neem cake before sowing (Shakywar and Pathak, 2012).

5. Use of tolerant varieties of taro crop (Shakywar et al., 2013b, 2013c).
6. Copper fungicides applied with low volume spraying equipment are effective against the disease.
7. Spraying should begin when the colocasia is four months old with application every week during the rainy weather and every two weeks during the dry weather.
8. Fungicide application should continue until the plants are 9 months.
9. Foliar application of Ridomil MZ 72 WP @ 2 g/L of water at weekly intervals.

13.3.3   CORM SOFT ROT

Corm soft rot of colocasia is widespread and occurs in irrigated (wetlands) crops as well as those grown under rainfed (dry land) conditions. Many different *Pythium* species have been found in diseased plants.

13.3.3.1   Symptoms

The first sign of the disease, on plants grown in both dry and wetland situations is a slowing of leaf production. It is due to restriction of water movement to the leaves as roots are attacked. On young plants root decays is followed by rot of the corm piece of the planting sett. Leaves collapse and plant dies. On older, established plants are outer leaves wilt and dies prematurely. The leaf blade of the two or three remaining leaves are crinkled slightly rolled or curled inwards, their colored in unhealthy grayish blue green and the margin are pale yellow. Plants remain stunted new leaf production is slow and show the corm are small.

13.3.3.2   Causal Organism

Pythium aphanidermatum, P. carolinianum, P. graminicola, P. irregular, P. myriotylum, P. splendens and *P. vexan.*

13.3.3.3   Etiology

The hyphae are hyaline and mycelium aseptate (coenocytic). *Pythium* species requires examination under a microscope of the sporangia, oogonia and antheridia. Sporangia are the asexual spores and in the case of *P. aphanidermatum*. They are lobate (inflated). The apluerotic oogonium (oospores does not fill the oogonium) and intercalary (rarely terminal) attachment of the antheridia further distinguishes *P. aphanidermatum* from other *Pythium* species.

13.3.3.4   Mode of Spread

The disease spread is rapid in wetland situations and also in dry land areas where rainfall is high. Water is required for zoospores movement.

13.3.3.5   Favorable Conditions

Temperature above 25°C is required for more *Pythium* species to grow in the soil and infected plants.

13.3.3.6   Disease Cycle

Pythium can be introduced into a field in soil growing media, plant refuse and irrigation water. *Pythium* spreads by forming sporangia, sack-like structures, each releasing hundreds of swimming zoospores. Zoospores that reach the plant root surface encyst, germinate and colonize the root tissue by producing fine thread like structures of hyphae, collectively called mycelium. These hyphae release hydrolytic enzymes to destroy the root tissue and absorb nutrients as a food source. *Pythium* forms oospores and chlamydospores on decaying plant roots, which can survive, prolonged adverse conditions in soil, and water leading to subsequent infection (Zamir et al., 2003).

13.3.3.7   Integrated Disease Management

1. Diseased plants are easily removed from the soil by hand.
2. Infected land should be carefully inspected for symptoms of *Pythium* infection before planting a new site.

3. If rot are found on the corm pieces. They should be cut out.
4. Field soil, debris, pond and stream water, and roots and plant refuse of previous crops can contain *Pythium*.
5. Use of bio-agent for corm treatment @ 5g/kg before planting.
6. The severity of soft rot may be reduced in soil by incorporating 112 kg Captan 50 WP/ha into the acid soils before planting.
7. Corms should be dipped in Ridomil solution @ 2 g/kg for ½ to 1 h before planting.

13.3.4 PHYLLOSTICTA LEAF SPOT

Phyllosticta leaf spot can often be seen on dry land colocasia in Hawaii, especially in the high rainfall areas of the islands. It is also known in American Samoa.

13.3.4.1 Symptoms

The spots on the leaves vary from 8–25 mm or more and are oval or irregular in shape. The young spots are buff to reddish brown. Older spots are dark brown with a chlorotic region surrounding the lesion. The centers of the infected area frequently rot out to produce a shot-hole type lesion. *Phyllosticta* spots generally resemble those caused by *Phytophthora colocasiae* except for the absence of sporangia produced on *Phytophthora colocasiae* lesions.

13.3.4.2 Causal Organism

Phyllosticta colocasiophila Weedon

13.3.4.3 Etiology

The fungus is produces single celled, hyaline conidia within pycnidia (possibly on leaves or stems). The spores are produced in pycnidia (flask-shaped structures, containing the spores conidia), which are buried in the disease leaf tissue. The conidia are cylindrical to oval.

They are extruded from the pycnidia and remain at the opening (ostiole) as small pink clusters (Gerlach, 1988). They also have a mucilage sheath around the conidium and a mucilaginous appendage at one end. The pycnidial fungi on colocasia have relatively small conidia without a sheath or appendages, for example, they are *Phoma* rather than *Phyllosticta*.

13.3.4.4   Mode of Spread

The disease is spread through spores.

13.3.4.5   Favorable Conditions

The disease is favor by cloudy, rainy weather for a prolonged time (2–3 weeks) accompanied by cool winds is conducive to infection and disease development.

13.3.4.6   Disease Cycle

The fungus produce two types of spores (conidia and ascospores) germinate when moisture is present. Conidia can quickly be carried from diseased plants to healthy ones by splashing rainwater, sprinklers or watering. In addition, the ascospores are discharged into the air and can travel between plants on a breeze or current. If they land on a moist leaf, ascospores germinate, infect the colocasia host and begin the cycle one more.

13.3.4.7   Integrated Disease Management

1. Remove all dead plant material and allow for adequate air circulation between and around plants.
2. Collecting and burning the diseased leaves seems to be of some value.
3. Keep the growing environment clean.

4. Applications of fungicides such as Dithane M-45, Captan, Ferbam and Mancozeb or Thiophanate Methyl @ 3 g/L of water will help control infection levels and can prevent new infections in healthy plants.

13.3.5 SCLEROTIUM OR SOUTHERN BLIGHT

Sclerotium blight of colocasia is a generally problem of dry although wetland is frequently infected. This disease has been reported in Fiji (Dumbleton, 1954), the Philippines (Fajardo and Mendoza, 1935), Hawaii (Parris, 1941) and India (Goyal et al., 1974).

13.3.5.1 Symptoms

This disease appears on over mature corms and plant stress. Affected plants are usually stunted and corms rotted at the base where abundant sclerotia of the pathogen develop. Sclerotia abundantly produced on infected corms persist in the soil causing serious outbreaks of the disease in warm, wet weather following a significant dry spell. They also float on the water of paddy, infecting the dead petioles of the colocasia when the opportunity presents itself and subsequently invading the corm and producing a rot in the field and in storage under some conditions.

13.3.5.2 Causal Organism

Sclerotium rolfsii Sacc. Perfect stage: *Pellicularia rolfsii* (Curze) West (syn. *Corticium rolfsii* Curzi).

13.3.5.3 Etiology

Sclerotia are usually relatively large, hard and composed of very compact hyphae. Observe sclerotia of fungus under the dissecting microscope. Mount and crush some of the sclerotia and observe under the compound microscope. The sclerotia are small, almost spherical lemon yellow to dark brown bodies 53 resembling cabbage seeds. The rotted tissue is ocherous to brown and soft with a tendency to stringiness. A dense white mycelium may cover the tissue. In the wetland culture the rot frequently starts at the waterline on the corm rather than at its base.

13.3.5.4 Mode of Spread

Sclerotium rolfsii may survive saprophytically on plant debris or as sclerotia in the soil. *The fungus is spread by the sclerotia, which also serve as overwintering structures.*

13.3.5.5 Favorable Conditions

When sufficient moisture is present sclerotia germinate and infect young or old roots, dead leaf petioles and over mature corms. The disease is usually serious during warm wet periods.

13.3.5.6 Disease Cycle

Sclerotium rolfsii affects the lower stems, roots and leaf of plant. The disease is characterized by the presence of a white, web-like mycelium, which often forms at the bases and on the lower stems of affected trees. Tree death usually occurs rapidly. Light brown to yellow, round sclerotia form in the mycelial mat.

13.3.5.7 Integrated Disease Management

1. Hooding of paddy fields in early stages of disease development is an excellent cultural control method in Hawaii.
2. For dry land colocasia, harvesting the colocasia before it becomes over mature will reduce losses to this disease.
3. Burying plant debris after harvest by deep plowing is suggested for controlling this disease in other crops (Graham et al., 1972 and Brandes et al., 1959).
4. *Avoid planting sites where the disease has been severe on previous crops.*

13.4   MINOR FUNGAL DISEASES

13.4.1   *CLADOSPORIUM LEAF SPOT* (CLADOSPORIUM COLOCASIAE *SAWADA*)

Cladosporium colocasiae causes a relatively harmless disease common on dry land colocasia in Hawaii (Parris, 1941). Bugnicourt (1958) reports that *C. colocasiae* is frequently present in the planting of colocasia in irrigated terraces of New Caledonia. According to Trujillo (1967), it is present in the New Hebrides, Western and American Samoa the Carolines and the Marianas.

13.4.1.1   Symptoms

The disease attacks both wetland and upland colocasia and occurs mainly on the older leaves. On the upper surface the spot appears as a diffuse light yellow to copper area. On the lower leaf surface the spots are dark brown due to superficial hyphae, sporophores and conidia of the fungus. The lesions are generally 5–10 mm in diameter.

13.4.2   *SPONGY BLACK ROT* (BOTRYODIPLODIA THEOBROMAE *PAT.*)

Botryodiplodia theobromae causes a spongy rot, occasionally becoming dry and powdery ranging in color from cream to grayish brown and frequently becoming dark blue to black with an indistinct margin between healthy and diseased tissue. The fungus is capable of invading undamaged corms under conditions of high relative humidity.

13.4.3   *BLACK ROT* (CERATOCYSTIS FIMBRIATA *ELL. AND HAIST.*)

Ceratocystis fimbriata causes a soft dark to charcoal black rot with a fragrant banana odor, starting from natural or mechanical wounds in corms.

13.4.4   *RHIZOPUS ROT* (RHIZOPUS STOLONIFER *SACC.*)

Rhizopus stolonifer has caused serious losses in corms stored at moderate temperatures and high humidity.

13.4.4.1   Symptoms

Rhizopus rot is a white to cream-colored soft rot ranging in consistency from cheesy to watery with a slight yeasty odor. The skin of the corm generally remains intact until the rot is very advanced. External development of mycelium is sparse. However, sporulation at breaks in the skin and wounds resulting from the removal of cormels are extensive covering these areas with a black powdery layer.

13.4.4.2   Management

1. The disease can be minimized through removal of the roots and soil from the corm.
2. The corms rinsing well with clean water and dipping them into a 0.5% solution of NaOCI for approximately one minute air drying and storing the corms in a cool, clean area of approximately 50% relative humidity (Ooka, 1981).

13.4.5   *FUSARIUM DRY ROT* [FUSARIUM SOLANI *(MARS.) SYN. AND HANS.]*

Fusarium dry rot is a brown rot, mostly dry and powdery but sometimes becoming wet and soft in later stages, with a distinct margin between healthy and diseased tissues.

13.5   VIRAL DISEASES

13.5.1   *COLOCASIA BOBONE DISEASE VIRUS (CBDV)*

CBDV is a rhabdo virus that has been identified only in *Colocasia esculenta* (Brunt et al., 1996). CBDV causes bobone disease and probably also causes the more severe alomae disease (James et al., 1973).

13.5.1.1 Symptoms

Plants with bobone disease are stunted often severely have thickened, malformed, brittle leaves, galls on their petioles and larger veins. Usually, only a few leaves are affected by bobone disease and healthy leaves are produced after several weeks in an apparent recovery (Cook, 1978; Carmichael et al., 2008). In the beginning, alomae disease may be indistinguishable from bobone disease, but the plants with alomae develop chlorosis or progressive necrosis. Some plants collapse, all finally rot and die (Cook, 1978; QUT, 2003). After plants recover from bobone disease the symptoms may return (Carmichael et al., 2008) indicating that the plants still harbor the virus. Some plants infected with CBDV do not develop bobone or alomae disease, but instead have milder symptoms or may be nearly symptomless (Shaw et al., 1979; Revill et al., 2005a).

13.5.1.2 Etiology

The etiology of Alomae requires additional studies. A purification technique is to get virus preparations suitable for production of virus-specific antisera as well as for use in biochemically and physically characterizing the particles needs to be developed. Vectors and host ranges, especially of the small bacilliform particles, need to be clarified. There is strong evidence that CBDV causes bobone disease and considerable evidence that it is required for alomae disease. However, tests have not been done to confirm the etiology and four other viruses have been detected in plants with bobone and alomae diseases: *Dasheen mosaic virus* (DsMV), *Taro bacilliform virus* (TaBV), *Taro vein chlorosis virus* (TaVCV) and taro reovirus (TaRV) (James et al., 1973; Shaw et al., 1979; Revill et al., 2005a). It is likely that one or both diseases result from synergistic interactions between two or more of the viruses when they co-infect taro plants (Bos, 1999; Revill et al., 2005a). It has been proposed that co-infections of CBDV and TaBV produce alomae disease but the evidence is weak at present (James et al., 1973; Revill et al., 2005a). The possibility that DsMV, TaRV or TaVCV are involved in alomae disease, probably when co-infecting with CBDV cannot be discounted (Revill et al., 2005a). Cultivar susceptibilities may also be significant (Cook, 1978; Carmichael et al., 2008).

13.5.1.3   Mode of Transmission

The virus is transmitted by the plant hoppers *Tarophagus proserpina*, *Tarophagus colocasiae* and *Tarophagus persephone* (QUT, 2003; CAB International, 2011).

13.5.1.4   Management

1. Rouging plants infected with Bobone and Alomae to reduce the reservoir of pathogens.
2. The use of resistant varieties appears to be the most practical approach to managing the diseases.
3. Use of suitable insecticides for insect vectors.

13.5.2   *DASHEEN MOSAIC*

Dasheen mosaic virus (DsMV) is a potyvirus that infects a wide range of commercially important Araceae both edible and ornamental has a worldwide distribution (Zettler and Hartman, 1987; Brunt et al., 1996; Elliott et al., 1997; Simone and Zettler, 2009). The virus is present in most colocasia growing regions (Zettler and Hartman, 1987; Zettler et al., 1989).

13.5.2.1   Symptoms

The foliar symptoms include a dispersed and veinal mosaic pattern on the leaves. Leaf distortion is generally mild to moderate. Plants generally become asymptomatic three to four months after initial symptom expression. Symptom expression seems to be more pronounced during the cooler months of the year in Hawaii. Apparently this virus does not cause appreciable yield reduction in the varieties grown commercially, and the quality of the corm is not affected.

13.5.2.2   Causal agents

Dasheen mosaic virus

FIGURE 13.4   Colocasia leaf infected by Dasheen mosaic virus.

13.5.2.3   Mode of Spread

The virus is transmitted in a non-persistent manner by the aphids *Myzus persicae*, *Aphis craccivora* and *Aphis gossypii*.

13.5.2.4   Detection and Identification of Virus

Dasheen mosaic virus a flexuous rod 750 nm was initially described in 1970 as a poly virus infecting members of the Araceae (Zettler et al., 1970). The virus is well characterized (Hartman, 1974; Zettler et al., 1970). Purification techniques for the virus and production of virus specific antisera have been developed (Abo EI-Nil et al., 1975). A strain of DsMV known as FP-DsMV (French Polynesia Dasheen Mosaic Virus) has been reported in colocasia in French Polynesia.

13.5.2.5   Management

1. Varietal resistance appears to be a good method for reducing the incidence of this disease in taro.
2. To manage the aphids by systemic insecticides.

13.6   NEMATODE DISEASES

Several nematode species are commonly reported on colocasia crop. A little work has been done on the effect of these invertebrates on colocasia yield. The following nematodes have been reported on colocasia or taro: *Pratylenchus* spp. (Rabbe, Connors, Martinez, 1981); *Helicotylenchus* spp., *H. dihystera* (Cobb) Sher, *Rotylenchulus reniformis* and *Meloidogyne* spp. (Parris, 1940; Rabbe, et al., 1981), *M. incognita*, *M. javanica, Longidarus sylphus* and *Tylenchorhynchus* spp. *Meloidogyne* spp. (Byars, 1917; Nirula, 1959), *Pratylenchus* spp. (Kumar and Souza, 1969) and *Aphelechoides* spp. (Tandon and Singh, 1974) have been reported on colocasia or taro crop in a different place.

13.6.1   ROOT-KNOT NEMATODES (MELOIDOGYNE SPP.)

The nematode (*Meloidogyne* spp.) damage dry land colocasia when the crop is planted in infested soils. Galls on the root and swelling and malformations on the corm are characteristic of attack by this nematode. Severe attacks will stunt the plants and render it chlorotic.

13.6.2   MANAGEMENT

1. Always use clean/healthy planting materials.
2. Treatment of colocasia corms with hot water at 50°C for 40 minutes kills the nematodes in the corms (Byas, 1917).
3. Fumigation with dichloropropene, fenamiphos is desirable for manage of root knot nematodes in heavily infected soils.
4. Further root and corm feeding nematodes may also be controlled by soil fumigation.

13.7   NON-PARASITIC DISEASES/PHYSIOLOGICAL DISORDERS

Starch, present in normal corms is deficient or absent in those with 'loliloli, a term used in Hawaii to describe a physiological disorder of colocasia. Although, the normal corm is firm, crisp and resilient to the touch, loliloli colocasia is soft, spongy and water exudes when affected parts are squeezed. Loliloli colocasia is the result of withdrawal of starch from the corm. This starch is converted into sugar, which is used by the plant to develop new leaves and other parts. Any action that encourages resumption of vegetative growth in mature taro is likely to result in loliloli colocasia.

13.7.1 MANAGEMENT

Use of nitrogenous fertilizers after the corm has formed or the natural growth-decadence of the plant has started should be avoided to reduce chances of loliloli colocasia occurring.

KEYWORDS

- **bacteria**
- **blight**
- **fungi**
- **rhizome**
- **rot**
- **tuber**

REFERENCES

1. Abo El-Nil, M. M., F. W. Zettler, Hiebert, E. (1975). Purification of dasheen mosaic virus. *Proc. Amer. Phytopathol. Soc.*, 2, 73.
2. Bell, K. S., M. Sebaihia, L. Pritchard Holden, Toth, I. K. (2004). Genome sequence of the enterobacterial phytopathogens *Erwinia carotovora* subsp. *atroseptica* and characterization of virulence factors, *Proceedings of the National Academy of Sciences* of the United States of America 2004 Jul 27,101(30), 11105–11110.
3. Bos, L. (1999). Plant viruses, unique and intriguing pathogens – *A textbook of plant virology.* Backhuys Publishers, Leiden, The Netherlands.
4. Brandes, G. A., T. M. Cordero, Skiles, R. L. (1959). Compendium of plant diseases. Philadelphia: Rohm, Haas.
5. Brunt, A. A., K. Crabtree, M. J. Dallwitz, A. J. Gibbs, L. Watson, Zurcher, E. J. (1996). Plant viruses online: descriptions and lists from the VIDE database. http://pvo.bio-mirror.cn/refs.htm (accessed January 2011).
6. Bugnicourt, F. (1954). *Phytopathologie. Courr. Cherch.*, 8, 159–186.
7. Butler, E. S., Kulkarni, G. S. (1913). Colocasia blight caused by *Phytophthora colocasiae. Agric. India*, 5, 233–261.

8. Byas, L. P. (1917). A nematode disease of the dasheen and its control by hot water treatment. *Phytopathology*, 7, 66.

9. CAB International (2002). Crop Protection Compendium. Wallingford, UK: CAB International Accessed: 13th April, 2007. http://www.cabicompendium.org/cpc/datasheet.asp?CCODE=PHYTCL.

10. CAB International (2011). Crop protection compendium. CAB International, Wallingford, UK. http://www.cabi.org/cpc/ (accessed May 2011).

11. Carmichael, A., R. Harding, G. Jackson, S. Kumar, S. N. Lal, R. Masamdu, J. Wright, Clarke, A. R. (2008). *Taro Pest: An illustrated guide to pests and diseases of taro in the South Pacific.* ACIAR Monograph No. 132. http://www.aciar.gov.au/publication/MN132 (accessed January 2011).

12. Chadha, K. L. (2003). *Handbook of Horticulture*, Directorate of Information and publications of Agriculture ICAR, Krishi Anusandhan, Bahvan, Pusa, New Delhi 110–012, 507–508.

13. Chaudhary, R. G., Rai, M. (1988). Effect of plastic film mulching on increasing potato yield, *Acta Agric. Zhejiangensis*, 9(2), 83–86.

14. Cho, J. (2004). Breeding Hawaiian taros for the future. In: *Third taro symposium: 21–23 May 2003, Nadi, Fiji Islands*. Secretariat of the Pacific Community, Fiji: 192–196.

15. Cook, S. C. A. (1978). *Taro Diseases*: In *Pest control in tropical root crops* (eds Anonymous): 177–207. PANS Manual No. 4. Centre for Overseas Pest Research, London, UK.

16. Devitt, L. C., G. Hafner, J. L. Dale, Harding, R. M. (2001). Partial characterization of a new dsRNA virus infecting taro. In *Abstracts of the 1ˢᵗ Australian virology group meeting*. Australian Society for Microbiology, Parkville, Australia.

17. Dumbleton, L. J. (1954). A list of plant diseases recorded in South Pacific territories. South Pacific Commission Technical Paper 78.

18. Elliott, M. S., F. W. Zettler, Brown, L. G. (1997). *Dasheen mosaic potyvirus of edible and ornamental aroids*. Florida Department of Agriculture, Consumer Services, Plant Pathology Circular No. 384, Florida, USA.

19. Fageria, M. S., B. R. Chaudhary, Dhaka, R. S. (2006). *Vegetable Crops Production Technology* (Vol. II). Kalyani Publishers. New Delhi, Noida (U. P.) Hyderabad, Chennai, Kolkata: 249–252.

20. Fajardo, T. G., Mendoza, J. M. (1935). Studies on the *Sclerotium rolfsii* Sacc. *Philipp. J. Agriculture*: 387–425.

21. Farreyrol, K., M. N. Pearson, M. Grisoni, D. Cohen, Beck, D. (2006). Vanilla mosaic virus isolates from French Polynesia and the Cook Islands are Dasheen mosaic virus strains that exclusively infect vanilla. *Archives of Virology*, 151, 905–919.

22. Fauquet, C. M., M. A. Mayo, J. Maniloff, U. Desselberger, Ball, L. A. (2005). *Virus taxonomy: classification and nomenclature of viruses: eighth report of the International Committee on the Taxonomy of Viruses.* Elsevier Academic Press, London, UK.

23. Fullerton, R., Tyson, J. (2003). The biology of *Phytophthora colocasiae* and implication for its management. In: *Third Taro Symposium, 2003. Nadi, Fiji Islands*: Secretariat of the Pacific Community, 107–111.

24. Gerlach, W. (1988). *Plant diseases of Western Samoa*. Samoan German Crop Protection Project, Apia, Western Samoa. Deutsche Gesellschaft für Technische Zusammenarbeit (GTZ) GmbH, Eschborn, Germany.

25. Gibbs, A. J., A. M. Mackenzie, K. J. Wei, Gibbs, M. J. (2008a). The potyviruses of Australia. *Archives of Virology*, 153, 1411–1420.
26. Gibbs, A. J., K. Ohshima, M. J. Phillips, Gibbs, M. J. (2008b). The prehistory of potyviruses: their initial radiation was during the dawn of agriculture. *PLoS ONE* 3, e2523. http://www.pubmedcentral.nih.gov/picrender.fcgi?artid=2429970&blobtype=pdf (accessed January 2011).
27. Goyal, J. P., S. M. Naik, V. N. Pathak, H. C. Sharma, Prasad, R. (1974). Evaluation of some modern fungicides for the control of *Sclerotium rolfsii* Sacc. inciting the tuber rot of *Colocasia antiquorum*, L. (*Arvi*). Nova Hedwigia Beiheftezur, 47, 205–210.
28. Graham, J. H., K. W. Kreitlow, Faulkner, L. R. (1972). Diseases In: Alfalfa science, Technology, (eds. C. H. Hanson), *Amer. Soc. Agron.* No. 15, 497–526.
29. Hartman, R. D. (1974). Dasheen mosaic virus and other phytopathogens eliminated from caladium, taro and cocoyam. *Phytopathology*, 64, 237–240.
30. Hu, J. S., S. Meleisea, M. Wang, M. A. Shaarawy, Zettler, F. W. (1995). Dasheen mosaic potyvirus in Hawaiian taro. *Australasian Plant Pathology*, 24, 112–117.
31. Jackson, G. (1999). Taro leaf blight. *Pest Advisory Leaflet* No.3. Plant Protection Service, Secretariat of the Pacific Community, Suva, Fiji Islands. 2 pp.
32. Jackson, G. V. H., F. Volsoni, J. Kumar, M. N. Pearson, Morton, J. R. (2001). Comparison of the growth of in vitro produced pathogen tested *colocasia* taro and field collected planting material. *New Zealand Journal of Crop, Horticultural Science*, 29, 171–176.
33. James, M., R. H. Kenton, and Woods, R. E. (1973). Virus like particles associated with taro diseases of *Colocasia esculenta* (L.) Schott. in the Solomon Islands. *J. Gen. Virology*, 21, 145–153.
34. Kumar, C. R. M., Souza, D. (1969). A note on *Paratylenchus mutabilis* Colbran 1968 (Nematoda: Criconematidae) from India. *Curr. Sci.*, 38, 71–72.
35. Luthra, J. C. (1938). India some new disease observed in the Punjab and mycological experiments in progress during the year 1937. *Int. Bull. Pl. Prot.*, 13(4), 73–74.
36. Macpherson, C. (2000). *Plant protection in the Pacific Islands – A course for senior high school students.* Students Edition. Secretariat of the Pacific Community Regional Media Centre. 128 pp.
37. Mendiola, N., Espino, R. B. (1916). Some Phycomycetous disease of cultivated plants in the Philippines. *Philippines Agriculturist, Forster*, 5, 67–72.
38. Nelson, S. C. (2008). Dasheen mosaic of edible and ornamental aroids. Cooperative Extension Service PD-44. College of Tropical Agriculture, Human Resources, University of Hawaii at Manoa. http://www.ctahr.hawaii.edu/oc/freepubs/pdf/PD-44.pdf (accessed January 2011).
39. Nirula, N. K. (1959). Root knot nematode on colocasia. *Curr. Sci.*, 28, 125–126.
40. Ooka, J. J. (1981). *Rhizopus stolonifer* rot of taro. *Phytopathology*, 71, 246.
41. Paharia, K. D., Mathur, P. N. (1961). New host plant of colocasia blight (*Phytophthora colocasiae* Racib.) *Curr. Sci.*, 30(9), 345.
42. Parris, G. K. (1941). Diseases of taro in Hawaii and their control. *Hawaii Agric. Exp. Sta. Circ.* No. 18. 29 pp.
43. QUT, (2003). *Development and application of virus indexing protocols for the international movement of taro germplasm.* Plant Technology Program, Science Research Centre, Queensland University of Technology, Australia, The Regional Germplasm Centre, Secretariat of the Pacific Community, Suva, Fiji.

44. Rabbe, R. D., L. L. Conners, and Martinez, A. P. (1981). Check list of plant diseases in Hawaii. HITAHR, Information Text Series 022. 313 pp.

45. Raciborksi, M. (1900). Parasitic he Algen, Pilze, Java's (Java's parasitic algae and fungi). I. Batavia. Page 9. (Cited in Waterhouse 1970a, *Phytophthora colocasiae*).

46. Revill, P. A., G. V. H Jackson, G. J. Hafner, I. Yang, M. K. Maino, M. L. Dowling, L. C. Devitt, J. L. Dale, Harding, R. M. (2005a). Incidence and distribution of viruses of taro (*Colocasia esculenta*) in Pacific Island countries. *Australasian Plant Pathology*, 34, 327–331.

47. Shakywar, R. C., Pathak, S. P. (2012). Integrated disease management of leaf blight of taro, *Indian Phytopath*, 65 (3), 294–296.

48. Shakywar, R. C., S. P. Pathak, Kumar Sunil (2013c). Influence of weather factors on leaf blight of taro against different genotypes, *J. Soils, Crops*, 23 (1), 25–29.

49. Shakywar, R. C., S. P. Pathak, Krishna, S. Tomar, Pathak, M. (2013a). Effect of sowing dates on *Phytophthora* blight of taro (*Colocasia esculenta* var. *antiquorum*). *HortFlora Research Spectrum*, 2(2), 166–168.

50. Shakywar, R. C., S. P., Pathak, Mahesh Pathak, K. S. Tomar, Singh Hem. (2013b). Developmental behavior of leaf blight of taro caused by *Phytophthora colocasiae*. *Vegetos*, 26 (1), 167–170. DOI:10.5958/j.2229–4473.26.1.024.

51. Shakywar, R. C., Satya Pathak, Kumar Sunil (2012). Epidemiology and management of leaf blight of taro. LAP, Lambert Academic Publishing GmbH & Co. Germany. pp.140.

52. Shaw, D. E., R. T. Plumb, Jackson, G. V. H. (1979). Virus diseases of taro (*Colocasia esculenta*) and *Xanthosoma* spp. in Papua New Guinea. *Papua New Guinea Agricultural Journal*, 30, 71–97.

53. Simone, G. W., Zettler, F. W. (2009). Dasheen mosaic disease of Araceous foliage plants. Plant Pathology Fact Sheet PP-42. University of Florida. http://plantpath.ifas. ufl.edu/takextpub/Factsheets/pp0042.pdf (accessed January 2011).

54. Singh, H. K. (2000). Studies on *Phytophthora* leaf blight of colocasia [*Colocasia esculenta* (L.) Schott.]. PhD Thesis, NDUA&T. Kumarganj, Faizabad.

55. Swarup, Vishnu (2006). *Vegetable Science, Technology in India.* Kalyani Publishers, Ludhiyana, New Delhi, Noida (U. P.): 623–627.

56. Tandon, R. S., Singh, S. P. (1974). Two new species of the genus *Aphelenchoides fischer* (Nematoda, Aphelenchoididae) from the roots of *Colocasia antiquorum* Schott. *Indian, J. Entomol.*, 36, 44–50.

57. Trujillo, E. E. (1967). Diseases of the genus colocasia in the Pacific area and their control. *Proc. Int. Symposium on Tropical Root Crops*. 2(IV):13–19.

58. Walker, J. C. (1952). *Diseases of Vegetable Crops*. McGraw-Hill, New York.

59. Zamir, P. K., R. Yip, Hartman, R. D. (2003). Biological control of damping off and root rot caused by *Pythium aphanidermatum* on greenhouse cucumbers. *Can. J. Plant Pathol.*, 25, 411–417.

60. Zettler, F. W., Jackson, G. V. H., Frison, E. A. (1989). *FAO/IBPGR technical guidelines for the safe movement of edible aroid germplasm*. International Board for Plant Genetic Resources, Food & Agriculture Organization of the United Nations, Rome, Italy.

61. Zettler, F. W., M. J. Foxe, R. D. Hartman, J. R. Edwardson, Christie, R. C. (1970). Filamentous viruses infecting dasheen and other araceous plants. *Phytopathology*, 60, 893–897.

CHAPTER 14

DISEASES OF ROSE CROP AND THEIR MANAGEMENT

R. C. SHAKYWAR, K. S. TOMAR and R. K. PATIDAR

College of Horticulture and Forestry, Central Agricultural University, Pasighat – 791102, Arunachal Pradesh, India; E-mail: rcshakywar@gmail.com

CONTENTS

14.1 INTRODUCTION

Rose belongs to the family Roasaceae and all species of this flower, with minor exceptions belong to the genus *Rosa*. The genus contains about 120 species and there are more than 30,000 cultivars differing in form, shape, size, color, fragrance and flowering habit in cultivation. It is a mainly a shrub, though some are of creeping habit. The rose has four basic growth types: bush, climbing, ground cover and standard. The rose stems are normally covered with few to numerous thorns, although few cultivars are thorn less. The rose has compound leaf with five or seven leaflets. The inflorescence in the rose is determinant type and in the form of panicle, corymb or solitary. Flower colors range from orange, pink, red, white, yellow and combinations of these colors. The fertilized flowers form attractive fruits, which are termed as 'hip.' It is very prosperous in vitamin A, B and C. It has been growing for millions of years. Thirty million years old fossils of rose have been found in Oregon and Colorado (Mukhopadhyay, 1990).

The Greek and the Roman mythologies describe roses in their ancient civilization. Rose was the symbol of 'Venus' a deity of love and peace. Among all the flowers, rose hypnotized the mankind most and attained a unique status in the human hearts. Rose is the most ancient and popular flower grown the world over. It is a versatile plant adapted to varying climatic conditions. In our ancient Sanskrit literature the rose mentioned as *Atimanjula, Taruni Pushpa* and *Semantika* but it is certain that the Mugulas were responsible for making this flower popular again. The First Mugal emperor Babar introduced the musk and damask roses in our country. The empress Nur Jahan is supposed to have discovered the *attar* of roses (Chadha, 2001).

In India, it is cultivated commercially for cut flowers both for conventional flower market and modern florist shops. Rose flowers without stem and loose flower petals are used in traditional markets for making garlands for offering in temples, while the florist shops sell cut roses with stems mainly for bouquets and floral arrangements. In recent times, about 60 units have been established under joint ventures around Bangalore, Chandigarh, Hyderabad, Gurgaon (Haryana), Mumbai, Nasik, Pune and Saharanpur (Uttar Pradesh) for growing roses in greenhouses for export of flowers to Germany, Holland, Japan and other European countries. Besides, the Damask rose (*R. damascena*) and Edouard rose (*R. bourboniana*) are cultivated for rose *attar* and other products like as *gulkand*, *gulabjal* and *pankhurj*. The rose is grown in about 6,000 ha area. Andhra Pradesh, Bihar, Gujarat, Haryana, Jammu Kashmir, Karnataka, Madhya Pradesh, Maharashtra, Punjab, Tamil Nadu, Uttar Pradesh and West Bengal are major rose growing states of India (Dutta, 2002). A garden follower would like to have healthy plants by following proper cultivation practices including disease management, rather than try to care them after the appearance of diseases and insect-pests. Most of the ornamental shrubs are considered to be hardy plants. However, rose are attacked by various biotic, meso-biotic and abiotic factors. In this chapter, introduction, economic importance, symptoms, causal organism, etiology, mode of spread, favorable conditions, disease cycle and integrated management strategy of rose crop have been described.

14.2 BACTERIAL DISEASES

14.2.1 CROWN GALL

14.2.1.1 Symptoms

Galls are usually round to irregular in appearance and may have a rough exterior. Upon cutting across a gall, a disorganized callus type of tissue is commonly found. Sometimes aerial galls are produced on the stem, leaf petioles and cut ends of stem where flowers have been removed.

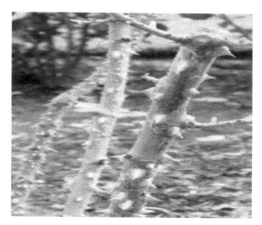

FIGURE 14.1   Crown gall symptoms on rose plant.

14.2.1.2   Causal Organism

Agrobacterium tumefaciens (Smith and Townsend).

14.2.1.3   Etiology

The bacterium is aerobic, gram negative, rod shaped and non-sporulating. They are motile and have 1–5 lateral flagella, but non-motile variants are also in existence. The colonies of bacterium are quite variable, but generally white, convex, circular, glistening and translucent.

14.2.1.4   Mode of Spread/Survival

The bacterium can be spread through infected plants, movements of contaminated soil and water tools, overhead sprinkler irrigation or floodwater and insect.

14.2.1.5   Favorable Condition

The optimum temperature 27°C for growth of bacterium.

14.2.1.6   Disease Cycle

The bacteria survives many years in the soil and can be moved with water or other infected plant parts. Once in the host the tumor inducing principle carried on a small circular portion of DNA is incorporated into the plant cell and overgrowth begins to form. In some plants the bacteria can be systemic and galls may begin to form at many sites on the plant. As the galls develop callus tissue is laid down which is susceptible to other types of breakdown, decay or sloughing. In this way the soil around a plant can become infested with the bacteria (Horst, 1983g).

14.2.1.7   Integrated Disease Management

1. If crown gall is detected the plant may survive many years but could serve as a reservoir for the bacteria.
2. Heating of soil to a temperature of 60°C for 30 minutes (Baker, 1969).
3. Tools used for pruning cutting or cultivation should be thoroughly washed and disinfected at frequent intervals.
4. Cutting out galls and dipping the roots and lower stem for 2 hours in 500 ppm of Streptomycin before planting.

14.3   FUNGAL DISEASES

14.3.1   ANTHRACNOSE

14.3.1.1   Symptoms

The disease initially produced leaf spots are about 0.5 cm diameter and blackish in color, which makes them easily confused with black spot. As the development progresses the spots become purple to brown and finally light brown or tan with a red or purple margin. Stems, hips and pedicles can also be infected as lesions mature small black dots will appear in the papery tan centers. Spotting, yellowing, severe defoliation and shot hole can result under moist spring conditions.

14.3.1.2 Causal Organism

Sphaceloma rosarum.

14.3.1.3 Etiology

These black dots are the spore producing acervuli of the fungus.

14.3.1.4 Mode of Spread/Survival

The disease spread by wind or carried in water droplets.

14.3.1.5 Favorable Conditions

Cool moist conditions are favoring the development of disease.

14.3.1.6 Disease Cycle

The fungus over winters in old lesions on leaves and canes. Warm spring conditions promote the development and release of spores from the acervuli in old lesions. The spores are believed to be carried by water or rain to newly expending leaves and stems. Not much is known about the sexual stage of the fungus or conditions favorable for spore germination (Horst, 1983a).

14.3.1.7 Integrated Disease Management

1. Removal of old leaves from around the base of plants.
2. Pruning out canes that have infections will do much to reduce the inoculums levels in spring season.
3. Generally, the same spray program that is used for Black Spot should work for Anthracnose also.

14.3.2 BLACK SPOT

This disease may also be called leaf spot, leaf blotch, star sooty mold and several other names. The disease was first reported from Sweden in 1815. It is the most important disease of outdoor roses on a worldwide including India.

14.3.2.1 Symptoms

Small black spots of 2 mm diameter can be found on upper leaf surfaces as well as immature canes. These spots are black and sooty enlarging to 12 mm with generally circular appearance but having a feathery edge. Further development involves the appearance of yellow margins around the spot a yellow condition that can extend into the entire leaf. Behind schedule in the growing season defoliation can occur along with purple red, raised irregular blotches on immature wood.

14.3.2.2 Causal Organism

Diplocarpon rosae Wolf.

FIGURE 14.2 Black spot symptoms on rose leaves.

14.3.2.3   Etiology

The sexual stage (*Marssonina rosae*) was first described by Wolf in New York 1912. The apothecia of the fungus are globose to disc shaped, sub-cuticular, radiate and 100–300 μm in diameter. Asci are inoperculate, oblong, cylindrical, short stalk 70–80 × 12–18 μm and contain a pore. Ascus contains eight ascospores, which are oblong, elliptical, hyaline, unequally two celled at the septum and are 20–25 × 5–6 μm in size (Horst, 1983b).

14.3.2.4   Mode of Spread/Survival

The disease dispersed by infected stem tissues and leaves.

14.3.2.5   Favorable Conditions

Cause infection at 6–33°C and optimum temperature 24°C and relative humidity 85% (open conditions) are require for the development of disease.

14.3.2.6   Disease Cycle

Temperatures rise in the spring the fungus produces conidia, which are moved by air currents or splashing water to new developing and expanding leaves which are very susceptible. Conidia germinate under conditions of free moister and need free water conditions for at least seven hours in order for infection to occur. The fungus is considered an obligate parasite producing haustoria in the host cell. As the infection develops, the leaf spot expands taking on the characteristic black appearance and producing conidia as secondary inoculum.

14.3.2.7   Integrated Disease Management

1. Removal and destruction of infected leaves and burn it.
2. Pruning of plants within 2.5–5 cm of the bud union was quite effective in reducing the disease incidence.

3. Foliar application of Mancozeb/Indophil M-45 @ 3 g/L of water.
4. Applying antibiotics @ 0.1% before appearance of the first symptom (Upadhayay and Bhandari, 1985).
5. Protectant sprays can be used on a one-week interval or after periods of rain to prevent infection.
6. Foliar application of systemic fungicides Hexaconazole (Contaf) @ 2 mL/L of water (Kumar et al., 2013).

14.3.3 BOTRYTIS BLIGHT

The disease appears frequently on roses grown in greenhouse or in pen field condition throughout the world. The disease was first described by Martin and Jenkins (1928).

14.3.3.1 Symptoms

The fungus mainly attacks flowers and flowering stems (Nichols and Nelson, 1969). The most common symptoms usually are seen on young flower buds, which droop, turn black at the base and later produce the cottony gray-black mycelium of the fungus. Flowers can also be affected in the same way cut ends will have the black canker like symptoms with presence of mycelium. Any time conditions are cool and wet a gray-black mycelial growth will indicate Botrytis blight.

14.3.3.2 Causal Organism

Botrytis cineria Pers. ex Fr.

14.3.3.3 Etiology

The fungus has branched conidiophores are 2 mm or more in length and 16–30 μm thick on which conidia, 6–8×4–11 μm size are formed. They are single celled, ellipsoidal, hyaline to pale brown. Sclerotia of different shapes and size are also produced by the pathogen.

14.3.3.4   Mode of Spread/Survival

The conidia are air borne highest disease spread during rainy season (Pawsey and Health, 1964).

14.3.3.5   Favorable Conditions

Optimum temperature of 18–25°C (Ellis and Waller, 1974) and relative humidity 94–100% for conidial germination (Snow, 1949).

14.3.3.6   Disease Cycle

Under cool wet conditions profuse sporulation results and spores are moved to roses by air currents or blowing rain. A minor wound in a bud or flower, or perhaps a pruning cut will provide the initial point of entry. The fungus is a low level parasite and will colonize wound sites as well as dead plant materials.

14.3.3.7   Integrated Disease Management

1. Elimination of opportunistic colonization on dead plant material the amount of sporulation can be reduced.
2. Good ventilation is also essential in reducing disease incidence.
3. Use of bio-agents and plant growth regulator (PGR) are effective measures.
4. Uses of boric acid @ 3 g/L of water, benomyl, chlorothalonil @ 0.2% have proven effective as spray treatment.

14.3.4   CANKER

The problems can be especially acute on old established roses that have lost some vigor and on young bare root roses emerging from cold storage.

14.3.4.1 Symptoms

In early spring, pruned stems provide wound sites, which can be colonized by canker causing fungi. The stems will yellow, often have red spots and later become brown or black. Black erumpent spots can often be found in the discolored tissue, which is the fruiting structure of the fungus containing spores.

14.3.4.2 Causal Organism

Coniothyrium wernsdorffiae and *C. fuckelii.*

14.3.4.3 Disease Cycle

The canker fungus is most active during the cold time of the year when roses are not actively growing. Pruning cuts or wounds on stems provide sites of entry for germinating spores. The fungi are not high-level pathogens and cannot produce the disease when conditions are favorable for plant growth. During the dormant months, fungi colonize the tissues, sporulate and are spread to other pruning or wound sites. The disease can be extensive and severe under the ideal conditions for development (Horst, 1983c).

14.3.4.4 Integrated Disease Management

1. Removal of infected canes (in spring season) and a general spray program for fungal diseases should reduce canker problem and protect the plants until they can become vigorous growers once again.
2. Promoting vigorous growth and removal of dead canes and stubs will help to reduce the primary source of inoculum.
3. Pruning before winter always makes an angular cut close to an active bud so the callus can form a protective layer before winter.
4. A dormant spray could be used to protect pruning cuts and wounds during the cold and wet winter.

14.3.5   *DOWNY MILDEW*

14.3.5.1   Symptoms

Under cool and moist spring conditions, young leaves stems and flowers may manifest purple to red or brown irregular spots. As the disease advances, lesions on leaves become angular and black with the possible appearance of white mycelium on the underside of the leaf. Advanced infections will have yellowing of leaves with brown necrotic areas and noticeable leaf abscission.

14.3.5.2   Causal Organism

Peronospora sparsa.

14.3.5.3   Etiology

The sporangia can germinate directly or form zoospores, which swim in free water on the plant surfaces. Encysted zoospores or sporangia germinate and penetrate the leaf surface producing intercellular mycelium. As development progresses sporangiophores are pushed through stomata on the underside of leaves and sporangia or oospores are formed to complete the life cycle (Horst, 1983d).

14.3.5.4   Mode of Spread/Survival

Wind or water borne.

14.3.5.5   Favorable Conditions

Optimum temperature of 21–25°C and relative humidity 96–100% for sporangial germination.

14.3.5.6   Integrated Disease Management

1. Sanitation in the garden will reduce the primary source of inoculums.
2. Ventilation and reducing humidity below 85% will reduce disease development.
3. Foliar sprays of Ridomil MZ-72 WP @ 2 g/L of water and Mancozeb @ 3 g/L of water are effective.

14.3.6   POWDERY MILDEW

14.3.6.1   Symptoms

The first symptoms appear as slightly raised blister like areas on the upper leaf surfaces. Later, the young increasing leaves become twisted, distorted and covered with a white powdery mass of mycelium and spores. Young peduncles, sepals, petals and stems may also show distortion while growing tips and buds may be killed. Infected older leaves and stems may remain symptomless.

14.3.6.2   Causal Organism

Sphaerotheca pannosa (Wallr. Ex Fr.) var. *rosae.*

FIGURE 14.3   Powdery mildew symptoms on rose leaves.

14.3.6.3 Etiology

The appendages of fungus are vestigial or lacking and its ascocarps are usually embedded in felt like mycelium.

14.3.6.4 Mode of Spread/Survival

Air movement is important in spreading of primary and secondary spores.

14.3.6.5 Favorable Conditions

Sporulation of fungus on detached leaves in 4 days at 20°C, 7 days at 15°C, 11 days at 10°C and 28 days at 3°C (Horst, 1983e).

14.3.6.6 Disease Cycle

The fungus can over winter as dormant mycelium or resting spores (cleistothecia) on infected stems or leaves. As conditions warm in spring, dormant mycelia becomes active producing asexual spores (conidia) while the cleistothecia germinate forming ascospores. Conidia and ascospores are then carried by the wind to susceptible young plant parts. After spores germinate they form a short mycelium and directly penetrate the epidermis forming haustoria inside the plant cell. The haustoria are a fungus structure that takes the nutrients from the host plant. Successful infection will result in further development of mycelium, colonization of more plant tissue and production of secondary spores (conidia).

14.3.6.7 Integrated Disease Management

1. Dormant pruning and cleaning up old leaves can remove substantial amounts of primary inoculum.
2. Sanitation should always be the initial means of manage.
3. Fungicides in a wettable powder formulation may provide better coverage if used with a spreader sticker.

4. Use of sulfur dust @ 3 g/L of water.
5. Several new fungicide, for example, Karathane @ 0.2% are more effective against the disease.

14.3.7 RUST

14.3.7.1 Symptoms

Minute orange pycnia appear on upper surface. Yellow to brown pustules impart rusty appearance to the shoots.

14.3.7.2 Causal Organism

Phragmidium rosae.

14.3.7.3 Etiology

The fungus produces aecial stage in spring season. They are produced as yellowish lesions on the lower surface.

14.3.7.4 Mode of Spread/Survival

Air borne.

14.3.7.5 Favorable Conditions

The optimum temperature range 15–21°C for aeciospores and urediospores germination while teliospores are well produced at 18°C.

14.3.7.6 Disease Cycle

Rust fungi are obligate parasites. They cannot be cultured on nutrient media. As the infection precedes the various spore stages develop on rose, there is no alternate host for rose rust. Re infection and spread occurs through aeciospores and urediospores. Spore germination requires cool summer temperatures and continuous moisture for at least two hours so the germ tubes can enter the leaf stomata (Horst, 1983f).

14.3.7.7   Integrated Disease Management

1.   Sanitation should be practiced to reduce inoculum and prevent early season infections.
2.   Pruning very dense bushes will help to reduce the moisture levels inside of plants and prevent some infections.
3.   Preventative fungicidal sprays should be applied every 7–10 days when conditions are favorable for rust development.
4.   Spraying of wettable sulfur or ferbam @ 0.2% or their mixture at the time of appearance.

14.4   VIRAL DISEASES

14.4.1   PRUNUS NECROTIC RING SPOT VIRUS (PNRSV)

Mosaic is probably the most commonly found virus on roses but many other virus diseases also exist. Symptoms of virus are usually dramatic manifestations of coloration, spotting or irregular distorted growth of leaves, flowers or growing points.

14.4.1.1   Symptoms

Rose mosaic usually appears in spring as a distortion of growing tips or expanding leaves. Later, the leaves can appear to be wavy and yellow lightening patterns, oak leaf patterns or simply gold to yellow veins. Plants infected with virus usually are slower to develop in spring than healthy plants and usually produce fewer good quality blooms. During the warm summer typical symptoms can disappear only to come back as fall and cooler temperatures arrive (Horst, 1983h).

14.4.1.2   Name of Virus

Prunus necrotic ring spot virus (PNRSV).

14.4.1.3   Transmission of Virus

The disease is mainly transmitted by pollen, insect feeding or simply by mechanical contact.

14.4.1.4   Management

1. To purchase only quality planting materials, which have no disease symptoms?
2. Some pathologists suspect that mosaic may be pollen transmitted which could prompt removal if other roses in the garden are valuable and not already infected.
3. Some exhibition gardens the disease can actually be very common.
4. Propagation of buds from infected roses will probably result in transmission of the disease if the buds actually take.

14.4.2   ROSE ROSETTE

14.4.2.1   Symptoms

Symptoms of virus are usually dramatic manifestations of coloration, spotting or irregular distorted growth of leaves flowers or growing points. Plants infected with virus usually are slower to develop in spring than healthy plants and usually produce fewer good quality blooms. During the warm summer typical symptoms can disappear only to come back as fall and cooler temperatures arrive.

14.4.2.2   Causal Organism

This disease appears to be vectored by a blister mite but the causal agent has not been identified. Some plant pathologists think that it is a virus while others, including entomologists, think that the disease is the result of the blister mite feeding.

Mosaic is probably the most commonly found virus on roses but many other virus diseases also exist.

14.4.2.3   Mode of Transmission

Mosaic is transmitted by pollen, insect feeding or simply by mechanical contact.

14.4.2.4   Management

1. Purchase only quality planting materials, which have no disease symptoms.
2. Removal and destruction of viral infected plant parts immediately and burn it.
3. To manage the insect vector with a suitable systemic insecticides.

14.5   NEMATODE DISEASES

Nematodes are small soil-inhabiting microscopic, long, thin worms animals and also called roundworm because shape as well as cross section of nematode is round. Several types of nematodes damaged rose crops may be ectoparasite, semiendoparasite and endoparasite. Nematode affected plants may be stunted, weak, lack normal green color and do not flower as profusely and have a shorter life span and unproductive plants. A common way to identify the problem is infected plants will wilt rapidly in hot weather. Many rose growers have observed plants that have failed to respond to good cultural practices and exhibit chlorosis, dwarfing and reduced vigor. These symptoms may be caused by plant parasitic nematodes (Swarup and Dasgupta, 1986). The most important nematode to cause the significant damage to the rose are Root-knot (*Meloidogyne hapla*), Dagger nematode (*Xiphenema* spp.), Lesion (*Pratylenchus penetrans* and *P. vulnus*), Stunt (*Tylenchorhynchus* spp.), Ring (*Criconemella* spp.), Spiral (*Helicotylenchus* spp.), etc.

14.5.1   ROOT KNOT NEMATODE

Root knot nematodes (*Meloidogyne* spp.) cause the disease of plants referred to as 'root knot.' Low plant vigor, small yellow leaves, early leaf shed, stunting and reduced bud formation are the foliar symptoms of nematode damage on roses. These symptoms are easily confused with those of nutrient deficiency. Plant decline usually occurs gradually over a period of several years until poor flower and foliage quality is easily visible. Discoloration and a reduction in the number of fibrous roots are often associated with nematode damage on roses. When feeder roots are under attack, plants cannot receive the proper water and nutrients they need for survival and may die. The nematode laid the egg in the form of egg mass on the root surface in which 300–500 eggs are present. Eggs are hatched within a week and also depend on climatic condition (Sasser et al., 1985). The total life cycle will be completed in around 25–30 days.

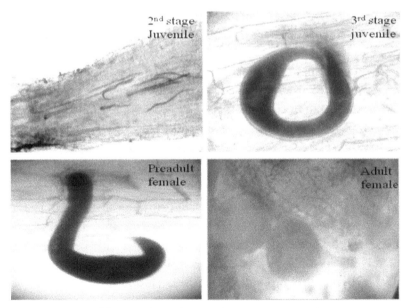

Life cycle of root-knot nematode

FIGURE 14.4   Life cycle of root-knot nematode.

14.5.2   LESION NEMATODE

Lesion nematodes (*Pratylenchus* spp.) are migratory endoparasite and polyphagous in nature. Nematodes can attack a number of hosts including trees, grass, vegetables, fruit crop, flowers, etc. This nematode causes a "lesion disease" on root surfaces of rose. A large number of lesion nematodes in a root frequently cause the root to turn brown and die a common symptom of infection. Lesions are areas of plant tissue that appear dark from a red-brown to black where the tissue is dying. In severe infections of lesion nematodes the entire root system may rot and die particularly under attack by other plant pathogens. All mobile stages of root lesion nematode enter the root and burrow tunnels through the root cortex. Eggs lay inside root tissues or in the soil hatch and emerging juveniles enter or remain in the roots and cause root injury (Chen and Dickson, 2004).

14.5.3   STING NEMATODES

Sting nematode disease is caused by the pathogen *Belonolaimus longicaudatus*. These nematodes attack a number of hosts including trees, grass, vegetables and other crops. Sting nematodes parasitically feed on the tips of roots. Plants may suffer partial destruction of roots that renders the plant incapable of absorbing necessary water the plant above ground may suffer from overall decline the plant may experience stunted growth and deformed plant parts and in severe cases the entire root system may die.

14.5.4   DAGGER NEMATODE

Xiphinema spp. is one of the largest plant parasitic nematode. In rose this nematode is commonly observed in soil and roots. Females lay eggs singly in the soil near plants and they hatch to produce first stage juveniles. Males are rarely observed. Dagger nematodes typically have 3–4 juvenile stages require 6–12 months to complete their life cycle and may live up to 3 years under favorable soil conditions. Dagger nematode feeding causes some necrosis and stunting and swelling of root tips. Several lateral roots may appear above the damaged root tips. Root-tip swelling may be confused with the galls of root-knot nematodes.

14.5.5 SPIRAL NEMATODE

Helicotylenchus spp. feed at or near the root tips causing devitalization the most severe type of root injury and induce stubbiness to primary and lateral roots and coarse root systems lacking feeder roots. They may contribute to stress on plants especially when present in high number in rose Spiral nematodes laid their eggs in soil near the roots.

14.5.6 RING NEMATODE

Criconemoides spp. is small, cigar-shaped, strictly ectoparasite nematodes that feed on the roots and cause browning on tender tissues. Several species of *Criconemoides* and related genera are found to affect crop plants especially fruit and ornamental crops. *Criconemoides* is commonly observed in many different flowers especially those with sandy or light soil types. The life cycle takes 25–35 days. After feeding for several days on roots, females deposit single eggs every 2–4 days. Second stage juveniles (J2) hatch from the egg in 11–15 days, molt to J3 in 3–5 days, molt to J4 in 4–7 days and become adults 5–6 days later.

14.5.7 MANAGEMENT OF NEMATODE DISEASES

Prevention is the best management strategy for nematode pests of rose crop. Before planting bare root or container grown roses, inspect their roots for signs of nematode injury. Collect a soil sample for nematode analysis, particularly if nematode damage was suspected on the previous crop. Nematode management must focus on reducing nematode numbers to levels below the damage threshold rather than eradication. The following points should be operated to manage the nematode of rose.

1. The field should be clean and add organic amendments (like oil cakes) to soil to increase potential for nematode resistance.
2. Marigold (*Tagetes* spp.) and *Brassica* spp. can be used as green manure crops as they contain Glucosinilate or isothiocyanate to manage many plant parasitic nematodes (Zasada et al., 2003).

3. Soil solarization can be useful against nematodes diseases. Then cover the soil with clear 4mm thick plastic. The aim is to raise the temperature between 45°C and 50°C in the top 10 cm of soil for 4–6 week. This is high enough to kill disease pathogens but most beneficial soil organisms will survive. Soil Solarization can be used for Commercial Cut Flower Farms to control the nematodes (Chellemi, 2009).

4. Host resistance: Rootstocks are reported to have resistance to PPNs. In rose rootstock, *Rosa fortuniana* is resistant against rose nematodes. *Rosa manetti* resulted free of galls and harbored incipient *M. hapla* (Voisin et al., 1996). Schneider et al. (1995) found *Rosa multiflora*s a rootstock has resistant against *Pratylenchus vulnus*.

5. Root-knot nematodes (*Meloidogyne hapla*) were eliminated from rose plants when the wash roots were dipped for 30 min in 0.1% solutions of prophos (Mocap) and Bayer 68138 (Nemacur) at 16°C (Dale, 2012).

6. Crop rotation should be used with non-host crops (2–3 years).

7. *Pasteuria penetrans*, soil-inhabiting bacteria can be used for managing of sting nematode populations.

8. A chemical option includes nematicide application; use carbamate or organophosphate to decrease sting nematode infestations.

14.6 NON-PARASITIC DISEASES/PHYSIOLOGICAL DISORDERS

14.6.1 WINTER INJURY

Winter injury results from many environmental factors, which have little in common other that they occur during the winter. Examples include late spring frosts, cool summers followed by warm autumns and sudden drops in temperature, dramatic temperature fluctuations, freeze-thaw cycles, lack of snow cover, unusually warm midwinter temperatures, extended periods of extreme or abnormally cold temperatures, and drying winds. Winter injury is important in and of itself but it also predisposes and weakens plants and subsequently makes them more vulnerable to secondary or opportunistic pests. Another important characteristic of winter injury is that quite often the symptoms are not evident until sometime after the injury has occurred. Symptoms of winter injury are highly variable and are manifest as buds that fail to open in spring or shoot that wilt and collapse shortly after emergence or suddenly collapse during the heat of the summer. In some cases, canes are blackened and dead by spring.

14.6.2 MANAGEMENT

1. Winter injury can be minimized by maintaining plant vigor by following a program of sound cultural care. For example, one of the most effective defenses against winter injury is to stop fertilizing early enough in the season so the plants have a chance to go into natural dormancy.
2. In spring, any dead canes which can serve as sites for secondary invaders or opportunistic pests should be pruned and removed from the planting.
3. Winter protection in the form of winter mulching is also helpful for bud-grafted plants such as hybrid teas. It is usually not necessary for species, shrub, old garden, or climbing roses.
4. Roses can be mulched with loose soil, compost, or leaves mounded around the base after the first hard frost.
5. Winter mulch should be removed in early spring when new growth begins. It is also important to select cultivars or species of rose that are known to be hardy in Connecticut.

KEYWORDS

- bacteria
- disorders
- fungus
- nematode
- rose
- virus

REFERENCES

1. Baker, K. F. (1969). Soil treatments. pp. 40–50. In: Roses (Mastalerz JW and Langhans RW Eds.). Pa. Flowers Growers. NY states Flower Growers Assoc. Inc. and Roses, Inc.
2. Chadha, K. L. (2001). *Handbook of Horticulture*, ICAR, New Delhi pp. 578–582.
3. Chellemi, D. O. (2009). Evaluation of soil solarization on commercial cut flower farms methyl bromide alternatives outreach conference.
4. Chen, Z. X., Dickson, D. W. (2004). Nematology: Advances and Perspectives Vol. 1, Nematode Morphology, Physiology and Ecology. CABI: Walling ford.
5. Dale, P. S. (2012). Elimination of root-knot nematodes from roses by chemical bare root dips. *New Zealand Journal of Experimental Agriculture* 1(2), 1973.
6. Datta, S. K. (2002). *Rose* Manual NBRI, Lucknow.
7. Ellis, M. S., J. M. Waller (1974). *Sclerotinia fuckeliana*. Description of pathogenic fungi and bacteria. CMI No. 431, 2pp.
8. Horst, R. K. (1983a). Compendium of Rose Diseases. The American Phytopathological Society. St. Paul, Minnesota. pp. 20–21.
9. Horst, R. K. (1983b). Compendium of rose diseases. *The American Phytopathological Society*. St. Paul, Minnesota. pp.7–11.
10. Horst, R. K. (1983c). Compendium of Rose Diseases. *The American Phytopathological Society*. St. Paul, Minnesota. pp. 14–17.
11. Horst, R. K. (1983d). Compendium of Rose Diseases. *The American Phytopathological Society*. St. Paul, Minnesota. pp. 13–14.
12. Horst, R. K. (1983e). Compendium of Rose Diseases. *The American Phytopathological Society*. St. Paul, Minnesota. pp. 5–7.
13. Horst, R. K. (1983f). Compendium of Rose Diseases. The American Phytopathological Society, St. Paul, Minnesota. pp. 11–12.
14. Horst, R. K. (1983g). Compendium of Rose Diseases. *The American Phytopathological Society*. St. Paul, Minnesota. pp. 23–26.
15. Horst, R. K. (1983h). Compendium of Rose Diseases. *The American Phytopathological Society*. St. Paul, Minnesota. pp. 26–27.
16. Kumar Sunil, R.C. Shakywar, Krishna S. Tomar, Pathak M. (2013). Varietal screening of rose (*Rosa* x *hybrid*) cultivars and in vitro efficacy of fungicides against black spot disease (*Diplocarpon rosae* Wolf.) in Arunachal Pradesh Conditions. *Global Journal of Environmental Science and Technology*. 1 (1), 15–19.
17. Martin, G. H., Jenkins, A. E. (1928). Preliminary list of fungi and diseases of roses in the U.S. *Plant Dis. Rep. (Suppl.)*. 63, 356.
18. Mukhopadhyay, A. (1990). *Roses* National Book Trust, India, New Delhi.
19. Nicolus, L. P., Nelson, P. E. (1969). Foliage diseases pp.185. In: *Roses: A Manual on the Culture, Management, Diseases, Insects, Economics and Breeding of Greenhouse Roses* (Mastalerz J. W., Langhans R.W. Eds.). NY State Flower Growers Assoc. Inc., and Roses Inc. USA.
20. Pawsey, R. G., Heath, L. A. (1964). An investigation of the spore population of the air of the Nottingham. I. The results of Petri dish trapping over one year. *Trans. Br. Mycol. Soc.* 47, 351–355.

21. Sasser, J. N., Carter, C. C. (1985). An Advanced Treatise on *Meloidogyne*: Volume I, Biology and Control. Department of Plant Pathology and Genetics, North Carolina State University and the United States Agency for International Development, Raleigh, NC.

22. Schneider, J. H. M., J. J. S. Jacob, Van de Pol, P. A. (1995). *Rosa multiflora* 'Ludiek' a rootstock with resistant features to the root lesion nematode *Pratylenchus vulnus.* 63 (1), 37–45.

23. Snow, D. (1949). The germination of mold spores at controlled humidities. *Ann. App. Biol.* 36, 1–17.

24. Swarup, S., Dasgupta, D. R. (1986). Plant Parasitic Nematodes of India: Problem and Progress. Printed by Ravi Sachdeva at allied publisher Pvt. Ltd. New Delhi-110064, 312–327.

25. Upadhyaya, J., Bhandari T. P. S. (1985). Chemical control of black leaf spot of rose. *Prog. Hortic.* 17, 67–68.

26. Voisin, R., J. C. Minot, D. Esmenjaud, Y. Jacob, G. Pelloli, Aloisi, S. (1996). Host suitability of rose rootstocks to the root-knot nematode *Meloidogyne hapla* using a high inoculums pressure test. *Acta Hort.* (ISHS) 424, 237–240.

27. Zasada, I. A., H. Ferris, C. L. Elmore, J. A. Roncoroni, J. D. MacDonald, L. R. Bolkan, Yakabe, L. E. (2003). Field application of brassicaceous amendments for control of soil borne pests and pathogens. *Plant Health Progress.*

CHAPTER 15

USE OF *Stevia rebaudiana* AND ITS DISEASES WITH MANAGEMENT

K. K. CHANDRA

Department of Forestry, Wildlife and Environmental Sciences, Guru Ghasidas Vishwavidhyalaya, Bilaspur–495009, Chhattisgarh, India

CONTENTS

15.1   INTRODUCTION

Nowadays, in many countries maintaining a healthy diet turned into a challenge for the majority of people. On the other hand, refined sugar consumption has increased rapidly in recent years, resulting in inability to weight management, positive caloric balance, obesity and weight gain. Furthermore, this inadequate dietary habit leads to cancer, inflammatory bowel disease, type 2 diabetes and dental caries. Considering the overconsumption of refined sugar, a natural non-caloric sweetener may draw the attention of individuals who are suffering from complications associated with high levels of sugar consumption.

Stevia (*Stevia rebaudiana*) is an important medicinal crop grown in India and in other part of the World. It is estimated that as many as 200 species of Stevia are native to South America; however, no other Stevia plants have exhibited the same intensity of sweetness as *S. rebaudiana*. It is grown commercially in Brazil, Paraguay, Uruguay, Central America, Israel, Thailand, China and India. The leaves has been used as an approved sweetener in Japan and Korea for decades and the turning point to become a mainstream sweetener came in 2008 when steviol glycosides, the sweetening components of the leaf, were deemed to be safe. The US Food and Drug Administration granted GRAS (Generally Recognized as Safe) status to Rebaudioside A one particular steviol glycoside found in Stevia. Since then, approval by legislators across the world has opened the door to new formulations and reformulations of foods and beverages.

The global sweetener market had a calculated value of $58.3 billion in 2010, and is dominated by refined sugar. Stevia is one of the fastest growing products in the sugar substitute market toward the use of natural sweeteners. The demand for Stevia sweeteners is skyrocketing. The biggest drivers behind Stevia demand include the global rise of health problems such as obesity, diabetes and cardiovascular disease. Experts predict that the global Stevia industry could be worth $10 billion by 2015. The World Health Organization (WHO) estimates that Stevia has the potential to replace 20–30% of all dietary sweeteners. Thus it has enormous investment opportunity.

15.2   HISTORY OF STEVIA FOR ITS USE AS MEDICINE

For hundreds of years, indigenous peoples in Brazil and Paraguay have used the leaves of Stevia as a sweetener. The Guarani Indians of Paraguay call it kaa jheé and have used it to sweeten their yerba mate tea for centuries. They have also used stevia to sweeten other teas and foods and have used it medicinally as a cardiotonic, for obesity, hypertension, and heartburn, and to help lower uric acid levels. It was first studied in 1899 by Paraguayan botanist Moises S. Bertoni, who wrote some of the earliest articles on stevia in the early 1900s.

In addition to being a sweetener, stevia is considered in Brazilian herbal medicine to be hypoglycemic, hypotensive, diuretic, cardiotonic, and tonic. The leaf is used for diabetes, obesity, cavities, hypertension, fatigue, depression, sweet cravings, and infections. The leaf is employed in traditional medical systems in Paraguay for the same purposes as in Brazil.

Europeans first learned about Stevia in the sixteenth century, when conquistadores sent word to Spain that the natives of South America were using the plant to sweeten herbal tea. Since then Stevia has been used widely throughout Europe and Asia. In the United States, herbalists use the leaf for diabetes, high blood pressure, infections, and as a sweetening agent. In Japan and Brazil, stevia is approved as a food additive and sugar substitute.

Stevia is a completely safe specific herb for diabetes a hypoglycemia and thereby helps to restore normal pancreatic function. It is used as a flavor enhancer contains a variety of constituents, besides the stevioides and rebaudiosides. Stevia also contains an extremely rich volatile oil comprising rich proportions of sesquiterpenes. The sweet and functional components of stevia are stevioside and rebaudiosides A, B, C and D. Alongside the mentioned compounds flavonoids also can be found in stevia.

Many preclinical and some clinical studies indicate that compounds present in stevia can produce beneficial antihypertensive, anti-hyperglycemic, antioxidant, anti-carcinogenic, chemoprotective, anti-inflammatory and antiviral effects on human health. Some studies showed that stevioside enhances both insulin secretion and insulin sensitivity. In addition, Stevioside also enhances glucose stimulated insulin secretion but does not affect fasting insulinemia. Studies in rats and dogs suggested that stevioside induces vasorelaxation. On the other hand, stevia showed high levels of antioxidant activities due to the scavenging of free radical electrons and superoxides. Immune system support and beneficial effects on treatment of inflammatory bowel disease are among the other health promoting effects of this sweet herb.

Although many studies reported health-promoting effects of this natural product, still long way is ahead for clinical evidences and demonstration of metabolic pathways regarding to such benefits.

15.3   PHYTOCHEMICALS OF STEVIA

Over 100 phytochemicals have been discovered in stevia since. It is rich in terpenes and flavonoids. The constituents responsible for stevia's sweetness were documented in 1931, when eight novel plant chemicals called glycosides were discovered and named. Of these eight glycosides, one called stevioside is considered the sweetest – and has been tested to be approximately 300 times sweeter than sugar. Stevioside, comprising 6–18% of the stevia leaf, is also the most prevalent glycoside in the leaf. Other sweet constituents include steviolbioside, rebausiosides A-E, and dulcoside A.

TABLE 15.1   Stevia Herbal Properties and Actions

Main Actions	Other Actions	Standard Dosage
Naturally sweetens	Kills bacteria	Leaves
Lowers blood sugar	Kills fungi	Ground leaves: 1/4 tsp
Increases urination	Kills viruses	1 tsp of sugar
Lowers blood pressure	Reduces inflammation	Infusion: 1 cup 2–3
Dilates blood vessels		Times daily

The main plant chemicals in stevia include: apigenin, austroinulin, avicularin, beta-sitosterol, caffeic acid, campesterol, caryophyllene, centaureidin, chlorogenic acid, chlorophyll, cosmosiin, cynaroside, daucosterol, diterpene glycosides, dulcosides A-B, foeniculin, formic acid, gibberellic acid, gibberellin, indole-3-acetonitrile, isoquercitrin, isosteviol, jhanol, kaempferol, kaurene, lupeol, luteolin, polystachoside, quercetin, quercitrin, rebaudioside A-F, scopoletin, sterebin A-H, steviol, steviolbioside, steviolmonoside, stevioside, stevioside a-3, stigmasterol, umbelliferone, and xanthophylls.

15.4   CLINICAL RESEARCH

The great interest in stevia as a non-caloric, natural sweetener has fueled many studies on it. The main sweet chemical, stevioside, has been found to be nontoxic in acute toxicity studies with rats, rabbits, guinea pigs, and birds. It also has been shown not to cause cellular changes (mutagenic) or to have any effect on fertility. The natural stevia leaf also has been found to be nontoxic and has no mutagenic activity. Studies conflict as to the effect of stevia leaf on fertility. The majority of clinical studies show stevia leaf to have no effect on fertility in both males and females. In one study, however, a water extract of the leaf was shown to reduce testosterone levels and sperm count in male rats.

Brazilian scientists recorded stevioside's ability to lower systemic blood pressure in rats in 1991. Then in 2000, a double-blind, placebo-controlled study was undertaken with 106 Chinese hypertensive men and women. Sixty subjects were given capsules containing stevioside (250 mg) or placebo thrice daily and followed up at monthly intervals for one year. After three months, the systolic and diastolic blood pressure of the stevioside group decreased significantly and the effect persisted over the whole year. The researchers concluded, "This study shows that oral stevioside is a well tolerated and effective modality that may be considered as an alternative or supplementary therapy for patients with hypertension." Another team of scientists tested the hypoglycemic effects of the individual glycoside chemicals in stevia and attributed the effect on glucose production to the glycosides steviol, isosteviol, and glucosilsteviol. The main sweetening glycoside, stevioside, did not produce this effect.

Researchers in Denmark published a study (2000), which demonstrated that the *in vitro* hypoglycemic actions of stevioside and steviol are a result of their ability to stimulate insulin secretion via a direct action on beta cells. They concluded, "Results indicate that the compounds may have a potential role as antihyperglycemic agents in the treatment of type 2 diabetes mellitus."

Stevia's effects and uses as a heart tonic to normalize blood pressure levels, to regulate heartbeat, and for other cardiopulmonary indications first were reported in rat studies (in 1978). In humans, a hot water extract of the leaf has been shown to lower both systolic and diastolic blood pressure. Several earlier studies on both stevia extracts, as well as its isolated glycosides, demonstrated this hypotensive action (as well as a diuretic action). In hypertensive rats the leaf extract increased renal plasma flow, urinary flow, sodium excretion and filtration rate. In addition to its studied hypotensive effects, a Brazilian research group demonstrated that water extracts of stevia leaves had a hypoglycemic effect and increased glucose tolerance in humans, reporting that it "significantly decreased plasma glucose levels during the test and after overnight fasting in all volunteers." In another human study, blood sugar was reduced by 35% 6–8 h after oral ingestion of a hot water extract of the leaf.

In other research, stevia has demonstrated antimicrobial, antibacterial, antiviral, and antiyeast activity. A water extract was shown to help prevent dental cavities by inhibiting the bacteria Streptococcus mutans that stimulates plaque formation. Additionally, a U.S. patent was filed in 1993 on an extract of stevia that claimed it to have vasodilatory activity and deemed it effective for various skin diseases (acne, heat rash, pruritis) and diseases caused by blood circulation insufficiency.

15.5 CURRENT PRACTICAL USE

For nearly 20 years, millions of consumers in Japan and Brazil, where stevia is approved as a food additive, have been using stevia extracts as safe, natural, non-caloric sweeteners. Japan is the largest consumer of stevia leaves and extracts in the world, and there it is used to sweeten everything from soy sauce to pickles, confections, and soft drinks. Even multinational giants like Coca-Cola and Beatrice Foods use stevia extracts to sweeten foods (as a replacement for NutraSweet and saccharin) for sale in Japan, Brazil, and other countries where it is approved as a food additive. Not so in the United States, however, where stevia is specifically prohibited from use as a sweetener or as a food additive.

Today, stevia leaves and leaf extracts are commonly found in most health food stores, however; they may only be sold in the United States as dietary/herbal supplements, not as food additives or sweeteners.

15.6 PRODUCTION CONSIDERATIONS

Stevia plant grows well in well-drained beds or large containers with fertile loam soil. It favor in warm conditions similar to those preferred by basil. Plants grown in warm climates grows to 60 cm tall and wide whereas in cool summers areas it attained height up to 35–40 cm. Grow three to five plants for a year's supply of dried leaves to a small family.

Seeds are rarely available because of production problems and poor germination, so plants are generally used instead. In garden beds, space plants 25 to 30 cm apart in the row, with two rows per bed.

Foods & Beverages Launched With Stevia

Source: Datamonitor, January – July 2011

FIGURE 15.1 Percentage of Food and beverage products launched with Stevia Worldwide.

15.7 SOIL AND FERTILIZER

Most garden soil is suitable for growing stevia. Loam or sandy loam soil that has been amended with compost is ideal. To grow stevia in heavy clay soil, loosen the clay with organic matter by adding humus or compost. Soil texture that will provide consistent moisture retention is important. Raised beds are useful in areas where the soil may be waterlogged. Stevia occurs naturally on soils of pH 4 to 5, but thrives with soil pH as high as 7.5. However, Stevia does not tolerate saline soils (Shock, 1982).

15.8 LIGHT

Stevia prefers full sun, except in areas where summers are very hot or dry. Partial shade in the afternoon is a good idea for areas with excessive summer heat. Some growers in hot areas use agro shade net to protect their plants from relentless Sun. Shade cloth also helps reduce moisture evaporation from the soil. Stevia plants started indoors needs bright light therefore artificial lights with a timer are a practical way to provide consistent indoor lighting.

15.9 WATER AND HUMIDITY

Stevia prefers consistent soil moisture, with no dry spells periods, avoid getting the leaves wet when watering stevia plant. Wet foliage and waterlogged soil can cause fungus diseases to develop. Remove and destroy affected parts of the plant to remedy a fungus problem.

15.10 TEMPERATURE

Stevia seeds require very warm temperatures of 30–35°C to germinate. Stevia survives winters only in the warmest areas such as southern California, Florida, and Mexico. Research in Japan indicates a critical winter soil temperature of 32°F to 35°F (Sumida, 1980). Stevia is a weak perennial, so plants grown as perennials should be replaced every few years. In colder areas, Stevia is planted after the last frost and treated as an annual. Longer summer days found at higher latitudes favor leaf yield and Stevioside content (Shock, 1982). To over winter stevia plants from the garden, trim them back to a few inches tall and place them in pots. Potted stevia plants will live through the winter indoors under artificial lights in a constant temperature as low as 55 degrees. To maximize leaf production, trimming back the plants several times is required to induce branching.

15.11   PLANT CARE

Stevia should be treated as a vegetable crop. When hot weather sets in, beds should be mulched 6 to 10 cm deep with organic residue such as grass clippings, chopped leaves, straw, hay, or compost. This protects the shallow feeder roots and hold in moisture. A consistent moisture supply is important for Stevia. Irrigate once or twice a week, whenever rain fails to water the plants. Sandy soils require more frequent irrigation. Trickle irrigation is ideal, ensuring consistent moisture levels without wetting leaves.

15.12   DISEASES

A survey of the literature reports the occurrence of only a few fungal diseases on S. rebaudiana. These include Erysiphe cichoracearum, Rhizoctonia solani, Sclerotium dephinii, Septoria steviae, Sclerotinia sclerotiorum, Sclerotium rolfsii and A. steviae (Thomas, 2000; Lovering and Reeleeder, 1996; Kamalakannan et al., 2006; Ishiba et al., 1982), A. alternata (Maiti et al., 2007).

15.13   ROOT ROT CAUSED BY *SCLEROTIUM ROLFSII*

The first symptoms appeared as yellowing and drooping of leaves, with wilting of plants and white cottony mycelial growth at the collar region (Figure 15.2). The mycelial growth spread to the stem and roots, with

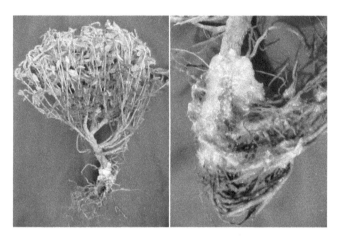

FIGURE 15.2   White mycelia at collar region rotting of the roots

associated tissue rotting (Figure 15.3). On the diseased areas, brown scle-rotia were observed first time in India by Kamalakannan et al. (2006). The mycelium of the fungus in growth medium was hyaline, branched at clamp connections and septate. The abundant sclerotia were round to oblong, initially white and later brown, with an average diameter of 0.5–2.0 mm.

Stevia plants are usually full grown before diseases appear. As har-vest time nears, commercial growers watch plants closely and harvest the entire crop at the first sign of disease.

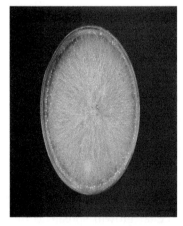

FIGURE 15.3   White mycelia at collar region rotting stem

15.14 STEM ROT CAUSED BY *SCLEROTINIA SCLEROTIORUM*

Chang et al. (1997) observed a stem rot disease of stevia for the first time in India and identified the causal agent as *Sclerotinia sclerotiorum*. The disease was found in 4-month-old plants were growing in loam soil. Diseased stems showed dark brown lesions above and at soil level when plant height reached approximately 30 cm. Under dry conditions, mild stem lesions caused plant stunting with lower leaves turning black and curling downward. Wilted leaf symptoms gradually spread upward in affected plants. Partial wilting symptoms appeared when girdling was restricted to branches. The entire plant collapsed when girdling of the crown and roots occurred. Superficial white mycelium developed over the basal part of affected stems under moist conditions, especially after rainy periods. Black, round to oblong sclerotia, 3.5–10.1 mm in size, formed externally on the crown areas after plant death. This is the first report on stevia of sclerotinia stem rot, a disease that could significantly reduce foliar growth and stevioside production in field plantings. Megeji et al. (2005) recorded a stem rot disease on stevia at Palampur, Himachal Pradesh, India by visual observation without confirming the pathogen.

15.15 LEAF SPOT CAUSED BY *ALTERNARIA ALTERNATE*

Maiti et al. (2007) first time reported the disease in India. The fungus causes leaf spot during February when temperature ranges from 20–25°C. Symptoms initially appeared as small circular spots, light brown in color. Later, many became irregular and dark brown to gray, while others remained circular with concentric rings or zones. On severely infected leaves several spots coalesced to form large necrotic areas. On older leaves concentric spots were more common at the tips. Leaf spots varied from 2–18 mm (Figure 15.4) in diameter. Conidial dimensions varied from 10–40 × 6–12 µm, mid to dark brown or olive-brown in color, short beaked, borne in long chains, oval and bean shaped with 3–5 transverse septa (Figure 15.5).

The pathogen isolated as a pure culture on potato dextrose agar media. The fungus produced abundant branched septate, brownish

FIGURE 15.4   Infected plant with Alternaria alternata

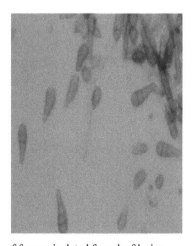

FIGURE 15.5   Conidia of fungus isolated form leaf lesion

mycelia; conidiophores simple, olive-brown, septate, variable in length with terminal conidia, which were solitary or in short chains. Conidial characteristics from culture were similar to the conidia isolated from infected plants.

Septoria steviae leaf spot disease has been reported in Canada by Lovering and Reeleder (1996), Chang et al. (1997) and Brandle et al. (1998). It was characterized by depressed, angular, shiny olive gray lesions, sometimes surrounded by a chlorotic halo, that rapidly coalesce.

15.16 LEAF MOTTLING CAUSED BY VIRUS

Tomato spotted wilt virus (TSWV) (genus Tospovirus, family Bunyaviridae) was first reported in Stevia in northern Greece by Chatzivassiliou et al. (2000); however, only 7% plants were shown symptoms. The diseased plants expressed chlorotic and necrotic rings and line patterns on systemically infected leaves followed by systemic mosaic and mottling of the leaves. Adult thrips (Thrips tabaci) was identified as causing vector for the transmission of TSWV disease in stevia.

15.17 BRONZING OF LEAF AND STUNT PLANT GROWTH CAUSED BY PHYTOPLASMA

This is first reported by Samad et al. (2009) in Stevia. Affected plants in the field expressed a quick decline consisting of growth cessation, bronzing of mature leaves, wilting, and death, resulting in a significant reduction in biomass and quality. Typical phytoplasma-like (pleomorphic) bodies ranging from 450 to 900 nm were observed in the phloem cells of infected plants. Phytoplasma was transmitted in stevia by leafhopper as reported by Samad et al. (2009).

KEYWORDS

- mottling
- phytoplasma
- rot
- spot
- stevia
- virus

REFERENCES

1. Brandle, J. E., Starratt, A. N., Gijzen, M. (1998). "Stevia rebaudiana: Its biological, chemical and agricultural properties." Canadian Journal of Plant Science. 78, pp. 527–536 (1998).
2. Chang, K. F., Howard, R. J., Gaudiel, R. G., (1997). First report on Stevia as a host for Sclerotinia sclerotiorum. Plant Disease 81, 311.
3. Chatzivassiliou, E. K., Peters, D., Lolas, P. (2007). Occurrence of Tomato spotted wilt virus in Stevia rebaudiana and Solanum tuberosum in Northern Greece. Ann. Appl. Biol. 137:127, 2000.
4. Ishiba, C., Yokoyama, T., Tani, T. (1982). Black spot disease of Steviae caused by Alternaria steviae, new species. Annals of the Phytopathological Society of Japan 48, 44–51.
5. Kamalakannan, A., Valluvaparidasan, V., Chitra, K., Rajeshwari, E., Salaheddin, K., Ladhalakshmi, D., Chandrasekaran, A. (2007). First report of root rot of stevia caused by Sclerotium rolfsii in India. Plant Pathology 56(2), 350.
6. Lovering, N. M., Reeleeder, R. D., (1996). First report Septoria steviae on Stevia (Stevia rebaudia) in North America. Plant Disease 80, 959.
7. Maiti, C. K., Sen, S., Acharya, R., Acharya, K. (2007). First report of Alternaria alternata causing leaf spot on Stevia rebaudiana. Plant Pathology 56(4), 723–723.
8. Megeji, N. W., Kumar, J. K., Virendra Singh, Kaul, V. K., Ahuja, P. S., (2005). Introducing Stevia rebaudiana, a natural zero-calorie sweetener. Current Science 88, 801–804.
9. Samad, A., Dharni, S., Singh, M., Yadav, S., Khan, A., Shukla, A. K. (2011). First Report of a Natural Infection of Stevia rebaudiana by a Group 16SrXXIV Phytoplasma in India. Plant Disease, vol. 95(12), 1582.
10. Shock, C. C. (1982). Experimental cultivation of Rebaudi's stevia in California. Univ. California, Davis Agron. Progr. Rep. 122.
11. Sumida, Tetsuya. (1980). Studies on Stevia rebaudiana Bertoni as a New Possible Crop for Sweetening Resource in Japan. Journal of the Central Agricultural Station. 31, 67–71.
12. Thomas, Li, S. C. (2000). Medicinal plants culture, Utilization and Phytopharmacology, Technomic Publishing Co. Inc. Lancaster Basel, 517.

CHAPTER 16

MAJOR DISEASES OF BETELVINE AND THEIR MANAGEMENT

PRABHAT KUMAR,[1] SHIVNATH DAS,[1] AJIT KUMAR PANDEY,[1] and SANTOSH KUMAR[2]

[1]*Betelvine Research Centre, Islampur, Nalanda–801303, India, E-mail: prabhathau@gmail.com*

[2]*Jute Research Station, Katihar, Bihar Agriculture University, Sabour, Bhagalpur, India*

CONTENTS

16.1 INTRODUCTION

Betelvine (*Piper betle* L.) is an important plantation crop of India belonging to the family Piperaceae. It is also known as Pan, Nagaballi, Nagurvel, Saptaseera, Sompatra, Tamalapaku, Tambul, Tambuli, Vaksha Patra, Vettilai, Voojangalata, etc., in different parts of the country (Guha and Jain, 1997). Betelvine is cultivated in many parts of world including India, Bangladesh, Malaysia, Srilanka, Pakistan, Mauritius and Myanmar for its leaves and used for mastication along with areca nut due to its stimulatory aromatic taste (Satyabrata et al., 1995; Mithila et al., 2000). Betel leaves is also known for its medicinal attributes containing some vitamins, enzymes, thiamine, riboflavin, tannin, iodine, iron, calcium, minerals, protein, essential oil and medicine for liver, brain and heart diseases (Chopra et al., 1956; Khanra, 1997). Its leaves also contain anti-oxidant properties due to the presence of phenols; particularly hydroxylchavicol (4-allyl pyrocatechol) and its aromatic volatile oil contain a phenol called chavicol, which has powerful antiseptic properties. The presence of aromatic volatile oil also gives rise to a sensation of warmth and well-being in the mouth and stomach. It is also known to produce a primary stimulation of the central nervous system and the betel leaf is also believed to be a common household remedy for various ailments (Guha, 2006; Ramamurthi and Usha Rani, 2012).

Betelvine is a native crop of tropical south East Asia. The most probable place of origin of betel is Malaysia (Chattopadhyay and Maity, 1967). India is the largest producer of betel leaves in the world (Arulmozhiyan et al., 2005). It is cultivated in India on about 75,000 ha area with an annual production worth about Rs. 1000 million (Dasgupta, 2011; Vijaykumar and Arumugam, 2012). On an average about 66% of such production is contributed by the state of West Bengal, where it is cultivated on about 20,000 ha area (Guha, 2006).

There are about a hundred varieties of betelvine grown across the world, of which about 40 are found in India, and of these, 30 grown in West Bengal alone. It is also cultivated in other states like Assam, Bihar, Madhya Pradesh, Maharashtra, Orissa, Tripura and Uttar Pradesh (Maity, 1989; Samanta, 1994; Guha 1997; Ramamurthi and Usha Rani, 2012).

The betelvine is an evergreen dioecious and perennial creeper, grown in conservatories (*Baroj*) under shady and humid conditions that are necessary for the growth of plant. The shady and moist atmosphere also favor the development of many diseases incited by fungi, bacteria, viruses and nematodes that greatly affect the growth of plants and causes heavy losses to farmers (Mathew et al., 1978; Maiti and Sen, 1979; Singh and Rao, 1988; Chattopadhayay and Maiti, 1990; Goswami et al., 2002; Akhter et al., 2011). Betelvine is subjected to attack of many diseases like phytophthora leaf and foot rot caused by *P. palmivora or P. parasitica*, anthracnose caused by *Colletotrichum capsici* (Syd.) Butler and Bisby and bacterial leaf spot caused by X*anthomonas campestris* pv. *betlicola* (Patel, Kulkarni and Dhande) Dye, basal rot caused by *Sclerotium rolfsii* and root knot caused by nematode (*Meloidogyne incognita*) that are the main yield limiting factors of the betelvine cultivation all over India.

16.2   FOOT ROT AND LEAF ROT

Dastur (1927) first reported foot rot and leaf rot diseases of betelvine (*P. betle* L.) from Durg, caused by *P. parasitica* var. *piperina*. The disease has been also reported from almost all betelvine-growing countries in the world including Indonesia, Myanmar (Su, 1931), Sri Lanka (Paul, 1939) and Bangladesh (Roy, 1948; Turner, 1969) etc. Waterhouse (1963) reported it is caused by *Phytophthora nicotianae* var. *parasitica*. Maiti and Sen (1979) considered it is as *P. palmivora* while Marimuthu (1991) called the pathogen *P. palmivora* MF_4. In recent literature *P. palmivora* MF_4 has been called *P. capcici* (Melhotra and Aggarwal, 2003). The highest intensity of foot and leaf rot has been also recorded in Midnapore and Nadia district of West Bengal (Dasgupta and Sen, 1999) and other states of India. The extent of losses may vary from 30–100% in case of foot rot and leaf rot leading to almost total crop failure (Maiti and Sen, 1979; Maiti and Sen, 1982; Dasgupta et al., 2000).

16.2.1   SYMPTOMATOLOGY

Dastur (1935) gave an accurate description of the symptoms of foot rot or wet rot disease associated with wilting of vines is common. The leaves and shoots turn yellow, wither and finally dry out to a pale brown color (Figure 16.1). In the diseased plants fine young roots are infected first. Gradually the rotting spreads through older roots and ultimately reaches the foot or collar region of the plant. In a diseased plant, the whole underground portion gets more or less completely rotten. The soft tissues of old roots and the inter-nodal portion of the cuttings are completely decomposed by the pathogen, leaving only the fibrous portion.

The disease leaf rot is characterized by the presence of circular black or brownish water soaked spots. These spots rapidly increase in size and coalesce with each other, involving a major area in the leaf blade, which undergoes rotting when the weather is continuously wet. The central rotten portion of the spot drops out, leaving a hole with irregular edges. The symptoms develop on any part of the leaves,

FIGURE 16.1 Symptom of foot rot.

including tips and margins (Figure 16.2). If the conditions still continue to be favorable, the rot proceeds to the petiole and eventually to the stem (Figure 16.3). If drier conditions are present, the infected leaf shows wrinkles and becomes reduced in size. The infection also remains localized and the infected black area is surrounded by a brown zone and presents a dry parched appearance. Infection is mainly confined to the leaves, which are located within a couple of feet from the ground surface.

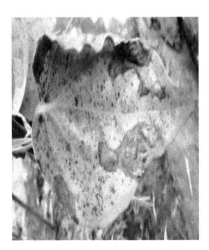

FIGURE 16.2 Symptom of leaf rot.

FIGURE 16.3 Symptom on stem.

16.2.2 ETIOLOGY AND EPIDEMIOLOGY

Foot rot and leaf rot is caused by the fungus *Phytophthora parasitica* var. *piperina* or *Phytophthora palmivora.* The pathogen survives in the soil as a saprophyte during adverse climatic condition by means of oospores and chlamydospores that attacks on roots, stem and leaves. It produces large number of spores at low temperatures under moist conditions. Secondary spread is carried from vine to vine by disseminated sporangia and zoospores through irrigation water and rains accompanied by wind (Rangaswami and Mahadevan, 2006). The foot rot syndrome is also produced by a number of pathogens including *Phytophthora parasitica* var. *piperina*, *Phytophthora nicotianae* var. *parasitica*, species of *Rhizoctonia*, *Pythium* and *Sclerotium rolfsii* Sac. (Dasgupta et al., 2008). Melhotra and Tiwari (1967) demonstrated that the pathogen *Phytophthora* was survived as mycelium in the host tissues, and after five weeks chlamyospores or resting sporangia is formed as well as pathogen can be isolated up to 17 weeks from the host tissues. Chaurasia (1976) reported the oospore of the pathogen in nature first time from India. The disease is primarily carried through plantation of infected cutting (Chawdhury, 1944; Asthana, 1947). The disease appears at the onset of monsoon and remains in high intensity throughout the rainy season. It wanes during the winter and may also occur in summer months when sudden hail storms occur. Whereas, under dry conditions the progress of the disease is slow. During rainy season when very wet conditions prevail for a number of days, leaves of betel vine is infected by the disease known as leaf rot. The incidence of leaf rot is observed in the month from June to August. The appearance and spread of the disease is dependent on external factors. Low temperature (15.6–27.4°C), high humidity (91.0–99.3%) and diffused light prevail inside the baroj that favor the growth of vine and are also congenial for the growth of pathogen (Datta et al., 2010). The peak infection was noted by Huq (2011) during the second week of August when the average temperature, relative humidly and rainfall were 29.6°C, 94.6% and 13.4 mm, respectively.

16.2.3 MANAGEMENT

The phyto-sanitary approach including Collection, removal and destruction of dead and dying plants, burned or buried in the soil is recommended to reduce the incidence of foot rot and leaf rot disease in betel orchard (Melhotra and Aggarwal, 2003). Thyagarajan et al. (1972) recommended the combined application of phosphorus and nitrogen. The mortality of vines was more when high doses of nitrogen were applied but was considerably lowered with phosphorus alone or in combination with a low dosage of nitrogen. Saksena and Melhotra (1970) recommended the removal of collateral hosts such as *Colocasia* species growing around the orchard. Singh and Chand (1973) suggested judicious use of water to prevent dissemination of the pathogen. The pathogen overwinters or over summers in the soil so; judicious rotation of the crop in betelvine orchards may check the spread of disease. Paddy and banana rotation in Salem and Tanjore districts in Tamil Nadu state has been found to considerable safeguard the crop against the incidence of disease (Melhotra and Aggarwal, 2003). A combined effect of flooding followed by hot weather and deep plowing has been found to reduce the incidence of the disease (Tiwari and Mehrotra, 1974). Screenings of betel cultivar led to the identification of a few that are tolerant to this disease such as Halisahar Sanchi, Pachaikodi and Karapaku (Maiti, 1994). Volatile oils, like those from *Luvanga scandans* and *Mentha arvensis* were sown to inhibit growth of *Phytophthora* sp. (Chaurasia and Vyas, 1977). Many workers suggested that biocontrol agents like *Trichoderma harzianum* and *Pseudomonas fluorescens* were applied with oil cakes at quarterly intervals for management of this disease (Dasgupta et al., 2003; Sengupta et al., 2011). Phytophthora rot of betelvine is also reduced by dipping of cuttings in a *T. viride* spore suspension before planting and amendment of *T. viride* in soil multiplied on corn straw and til oil cake (Mehrotra and Tiwari, 1976). Chaurasia (1976) demonstrated the effectiveness of *Aspergillus flavus, A. oryzae* and *Penicillium* sp. in the control of disease. Dasgupta and Maiti, (2008) reported the disease is controlled by soil drenching with 1% Bordeaux mixture. Saksena (1977) obtained best results by dipping the cutting vine in Streptomycin sulfate solution (500 ppm) followed by application of 1% Bordeaux mixture in soil twice a month. The chemical spray near the root zone with fosetyl-Al (%) was also effective in controlling this disease (Mohanty and Dasgupta, 2008). Ayyavoo and Samiyappan (1981) reported the efficacy of Deconil-2787 (0.1%) during winter at 15 days of interval was effective. Johri et al. (1984) reported about 80% mortality of vine checked by application of Aliette and Captaf. The disease is also controlled by spraying of 0.5% Bordeaux mixture, Peronox (0.35%), Fytolan (0.2%), Dithan Z-78 (0.2%) or Blitox-50 (0.25%) three to four times at 8–10 days interval (Balasubrahmanyam et al., 1988; Rangaswami and Mahadevan, 2006). Magdum et al. (2009) emphasized on integrated disease management including sanitation followed by one soil application of 1.0% bordeaux mixture at pre-monsoon and one soil application of *T. horzianum* after one month of bordeaux mixture application as well as one soil application of 1.0% bordeaux mixture at 2 months after its first application was significantly reduced the incidence both leaf rot and foot rot.

16.4   LEAF SPOT/ANTHRACNOSE

Leaf spot caused by *Colletotrichum capsici* is a major disease of the betelvine (Bhale et al., 1987). It is also known as anthracnose of Pan. Leaf spot of betelvine was first identified by Roy (1948) in Bangladesh. It may cause 10–60% yield loss and also reduce market value of the crop (Singh and Joshi, 1971; Maiti and Sen, 1982).

16.4.1 SYMPTOMATOLOGY

The diseased leaves are characterized by the presence of circular to irregular, light to dark brown lesions on the leaves that is surrounded by a yellow hallo (Figure 16.4). The center of such spots are later turned straw yellow in color. The spots often coalesced to form bigger patches. On stems, branches and petiole small black, irregular specks are seen which occasionally ruptured the cortex underneath. Often the spot grow along the length of stem in which case the part of vine above, the diseased internodes is also wilted (Naik and Hiremath, 1986).

16.4.2 ETIOLOGY AND EPIDEMIOLOGY

Leaf spot or anthracnose is caused by *Colletotrichum capsici* (Syd.) Butler and Bisby. The pathogen *C. gloeosporioides* was also reported on this crop from Karnataka (Naik and Hiremath, 1986) with incidence of disease 19%. The fungus colony was grayish black and smooth. Conidia were oblong hyaline, non-septate with rounded ends, having oil globules in the center, formed in culture and measured size of 8.6–19.9 × 3.5–6.5μ (Naik and Hiremath, 1986). The disease usually appears after rains and only the leaves are affected. It is reported that environmental factors like temperature, relative humidity and rainfall have crucial role in development of leaf spot of betelvine (Maiti and Sen, 1982). Roy (1948) recorded severe leaf

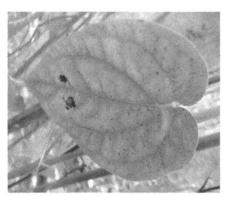

FIGURE 16.4 Symptom of anthracnose on leaf and petiole.

spot of betelvine due to anthracnose when rainfall was high. Dasgupta and Sen (1985) found that 92% relative humidity was critical moisture level for severe leaf spotting and heavy loss of betel vine. Under dry weather the progress of disease severity is slow where as during moist weather leave spot enlarges rapidly causing rotten of whole leaves (Basak et al., 1992). Goswami et al. (2002) reported maximum disease severity observed in the months of June and July when all the three weather factors were higher as compare to other months of year where as the incidence of leaf spot was also observed during March to May. The infection was gradually attained maximum when average temperature, relative humidly and rainfall parameters were 26.7°C, 88.3% and 19.4mm, respectively (Huq, 2011).

16.4.3   MANAGEMENT

The fungicides- Benomyl and Thiophanate-methyl were found to be effective in the control of anthracnose of betelvine (Maiti et al., 1978; Saleem, 2000). Acharya and Das (1995) revealed significant anthracnose disease control of betelvine by foliar sprays of 0.05% bitertanol, followed by 0.2% mancozeb, 0.1% ziram (as Cuman-L) and 0.5% Bordeaux mixture (0.5%). Balasubrahmanyam et al. (1988) reported Copper oxychloride 50 WP (0.25%) was found effective against this disease.

16.5   POWDERY MILDEW

The disease is caused by a species of *Odium*, which was first reported from Ceylon by Stevenson (1926) and later by Mitra (1930) from Burma and by Narasimhan (1933) from Mysore as well as from Bombay by Uppal and Kamat (1938).

16.5.1   SYMPTOMATOLOGY

The disease is easily recognized by the appearance of yellow spots, which are slightly raised and irregular in outline, and correspond in extent to white powdery patches of mildews on upper surface of leaves (Figure 16.5).

FIGURE 16.5   Symptom of powdery mildew.

The patches of mildew are also sometimes found on the lower surface of leaves. These patches are first small but increase in extent as they grow together. The disease appears on the undersurface of the leaves as white to brown powdery patches. These patches gradually increase in size and often coalesce with each other. They vary in size from a few to 40 mm in diameter and are covered by dusty growth which is fairly thick in cases of sever attack. Areas on the upper surface corresponding to patches on the under surface appear yellowish, raised and irregular in outline. Young leaves when attacked fail to grow and become deformed, the surface being cracked and the margin turned inwards. Such leaves present a pale appearance and drop with slight disturbance. The disease is more prevalent in old plantations (Uppal and Kamat, 1938).

16.5.2 ETIOLOGY AND EPIDEMIOLOGY

The causative fungus *Odium piperis* of betelvine is an ectophytes, which feeds by sending globular haustoria into epidermal cells of the leaves. The disease has been reported to be in the leaves only and it has been found to disappear during the hot season and disease is spread by secondary spread of conidia.

16.5.3   MANAGEMENT

The disease was managed by dusting of wettable sulfur on leaves and plucking of infected leaves as soon as they are ready for harvest. Maiti (1994) reported betelvine cultivars- male clones of Kapoori, Tellaku, Vellairettala, Ambadi Badam and Kulgedu were found resistant to powdery mildew.

16.6   SCLEROTIAL WILT/STEM ROT/COLLAR ROT/BASAL ROT

Singh and Chand (1972) reported in a survey of betel gardens in different localities *Sclerotium rolfsii* was found to be responsible for plant losses of 42–62%.

16.6.1   SYMPTOMATOLOGY

Betelvine of all ages are vulnerable to infection particularly at the collar region. The first visible symptom of collar rot disease was observed as yellowing or wilting of lower leaves (Figure 16.6). Later on, the yellowing progressed to upper leaves. The fungal mycelium first appeared at the base of the vines near the soil line. The pathogen then grew upwards covering

FIGURE 16.6   Symptom of sclerotium wilt.

the stem with a cottony-white mass of mycelia. Later on, water-soaked and gray lesion appeared on the vines, which turned brown, resulting in the rotting of whole plant system. A large number of small, light brown, mustard-seed like sclerotia developed in the collar zone. After the pathogen established itself, its subsequent advancement and production of mycelia and sclerotia were quite rapid. The infected vines, which were 3–4 weeks old ultimately toppled down and died (Singh, 2002).

16.6.2   ETIOLOGY AND EPIDEMIOLOGY

Basal rot is caused by *Sclerotium rolfsii Dastur.* Mridha and Alamgir (1987) studied the growth and sclerotia production of *S. rolfsii* isolated from betelvine on different media, various pH and different light regimes. They observed that fungus produced a luxuriant mycelia growth and large number of sclerotia on PDM medium at pH 6 and in alternate light and dark conditions.

16.6.3   MANAGEMENT

Mridha and Alamgir (1987) reported that ammonia solution of 0.5% gave effective control when applied in soil. Brahmankar et al. (2011) reported that wilt or root rot management was done by integrated approach of disease management including soil solarization during summer followed by application of *Trichoderma* 10 kg/ha (set and soil application), neem cake 200 kg/ha, NPK @ 100:50:50 kg/ha and deep irrigation found significantly superior not only in controlling the wilt causing fungus and nematode but also had a favorable effect on growth parameters. Mopsin proved to be the superior in controlling the collar rot disease in field as it gave 100% disease control at 0.1% concentration (Singh, 2002). Rhizolex 50 WP (0.4%) used for soil drenching were effective against basal rot disease (Gangwar and Dasgupta, 1989). HaralPatil and Raut (2008) reported the fungicides, Metalaxyl + Mancozeb (0.1%), Difenoconazole (0.05%), Propiconazole (0.05%), Hexaconazole (0.05%) and Metiram (0.1%) to be effective in controlling *S. rolfsii*. In biological control, *Trichoderma lignorum*, *Gliocladium virens* and *T. viride* were most effective against *S. rolfsii*. However, in case of botanicals, the plant extracts of *Allium sativum*, *Azadirachta indica* and

Catheranthus roseus (all at 10% concentration) were promising against *S. rolfsii.*

16.7   BACTERIAL LEAF SPOT/STEM CANKER/BACTERIAL BLIGHT

A bacterial leaf spot of betelvine was first reported by Patel et al. (1951) from India and named the pathogen as *Xantomonas betlicola.* The pathogen *X. c.* pv. *betlicola* has been renamed as *Xanthomonas axonopodis* pv. *betlicola* by Vauterin et al. (1995). This was considered as minor disease, even though there were subsequent reports of its occurrence in 1971 from Jabalpur area of M.P. (Singh and Chand, 1971) and in 1978 from Kerala by Mathew et al. (1978). Nadugala and Amarasinghe (2009) reported up to 60% incidence of this disease from major betel cultivating areas and it can be increase to about 75% during the rainy seasons. The phenotypic and virulence variation were also found among isolates of *Xanthomonas axonopodis* pv. *betlicola.*

16.7.1   SYMPTOMOMATOLOGY

The infection begins as minute water soaked spots all over the leave blade delimited by veins. Several of these coalesce to form larger irregular brownish spots. The advance lesions are invariably accompanied by yellowish halos and result in ultimate defoliation (Figure 16.7). The symptoms are found to vary with varieties. In certain cases the infection is confined to the leave margin, defused and quickly spreading in an irregular fashion. Often grayish black lesions are seen on stem and petioles. When the disease becomes serious the leaves and internodes fall off and finally the vine dies (Mathew et al., 1978).

16.7.2   MORPHOLOGICAL AND CULTURAL CHARACTERISTICS OF ORGANISM

The color of colony is yellow and slimy, colony shape-circular and shiny, shape of organism-short rods; motility-motile; temperature range-26–30–°C; temperature optimum- 28°C, gram reaction-negative; growth on nutrient broth-turbid yellow growth.

FIGURE 16.7   Symptom of bacterial leaf spot.

16.7.3   PHYSIOLOGICAL AND BIOCHEMICAL PROPERTIES

Starch hydrolysis-strong and positive; catalase production-positive; Kovac's oxidase test-negative; H_2S production-positive; fermentation of sugars, Lactose-fermented with acid production and no gas formation; sucrose fermented with acid production and no gas formation (Patel et al., 1951; Singh and Chand, 1971; Mathew et al., 1978).

16.7.4   MANAGEMENT

The leaf spot of betelvine has also been reported to be caused by a fungus-bacterium complex (Bhale et al., 1985; Deka et al., 2005). In such leaf spot, *C. capsici* is always associated with the bacterium *X. a.* pv. *betlicola*. The diseases complex can be effectively controlled by the chemicals like 0.5% Bordeaux mixture or 0.1% copper oxychloride (Yadav et al., 1993). *In vitro* screening of four bioagents *viz., Trichoderma harzianum, Aspergillus terreus, Bacillus subtilis* and *Pseudomonas fluorescens* and extracts of four plant species *viz., Chromolaena odorata, Ageratum houstonianum, Polygoum hydropiper* and *Tagetes erecta* was done against *Colletotrichum capsici, Xanthomonas axonopodis* pv. *betlicola* and complex of *C. capsici* + *X.*

axonopodis pv. *betlicola. B. subtilis* significantly inhibited the growth of *C. capsici* (79.5%) and *X. axonopodis* pv. *betlicola* (68.1%) and their complex (76.4%). Among the phytoextracts, *P. hydropiper* was the best in restricting the growths of both pathogens (72.3 and 72.2%) and their complex (70.4%) respectively over control. Under field condition, sanitation + *B. subtilis* provided best control strategy of leaf spot complex showing 9.7% percent disease index (PDI), followed by 11.1% PDI in sanitation alone and sanitation + *P. hydropiper* (Deka et al., 2008). Tripathi et al. (1984) reported that during field evaluation, streptomycin was found to be most effective (80–90% disease control) followed by menadione (78–80%) and resorcinol (68–72%). Incidence of bacterial leaf spot decreased significantly as phosphorus level (25–125 kg/ha) increased (Wasnikar et al., 1993). Balasubrahmanyam et al. (1988) reported Copper oxychloride 50 WP (0.25%) with 500 ppm streptocycline was best against bacterial leaf spots. It is reported that most of the betel varieties that are cultivated at commercial level are susceptible to *X. campestris* pv. *betlicola* (Nema, 1988). Maiti (1994) reported betelvine cultivar Simurali Bangla was resistant to bacterial leaf spot.

16.8   ROOT KNOT

The root knot nematode caused by *Meloidogyne incognita* is the most serious pest of betelvine orchard alone has been reported to result in 26–38% yield losses. High population of root knot was noted from March to June in betel orchard (Shahina and Erum, 2005). *Sesbania grandiflora* and *S. sesban* act as reservoir for the nematodes (Rao et al., 1991). Amer-Zareen (1999) reported that infestation of root knot in betelvine are due to the use of untreated green manure, infested tools and some time infested cultivars of *Piper betle.*

16.8.1   SYMPTOMATOLOGY

The main symptoms of root knot are reduction in size of betel leaves, chlorosis and black spot near the mid leaves. In root knot disease (Figure 16.8) there is formation of tumor in the root tips of host plants known as root galls (Doosani et al., 1992). The nematodes have been found vital role in

FIGURE 16.8 Symptom of root knot associated with sclerotium wilt.

the development of wilt and root rot diseases caused by fungi and bacteria (Ray et al., 1993). It was evident that the nematode predisposes the vine to *Phytophthora paimivora, which* infects through wounds caused by *Meloidogyne incognita* (Marimuthu, 1991; Jonathan et al., 1996). Sitaramaiah and Devi (1994) observed positive correlation between the populations of *M. incognita* with wilt disease incidence. More damage was observed when *Xanthomonas campestris* pv. *betlicoia* was combined with *M. incognita* and *S. roifsii* (Acharya et al., 1987). The absorption of minerals is disturbed due to choking of xylem vessels and the attack of other style bearing nematodes, which produces the holes in root-lets, causes the absorption of the spores of soil born fungi with minerals due to the pressure and finally plants become week (Maiti and Sen, 1979).

16.8.2 MANAGEMENT OF ROOT KNOT DISEASES

Solarization by mulching the land with 100 gauge black and white polythene before planting for 15 days was found to reduce plant parasitic nematode population in India (Shivakumar and Marimuthu, 1987; Rao et al., 1996). Application of neem oil cake at 1 ton/ha and saw dust at 2 tons/ha can reduced the nematodes population and root galls and increased the number of leaves harvested significantly (Jagdale et al., 1985; Acharya and Padhi, 1988a). Bed amendment with chopped and shade-dried leaves of *Calotropis gigantea* at 2.5 tons also reduced the 60% nematode population (Shivakumar and Marimuthu, 1986; Murthy and Rao, 1992). When different levels of Potash were tried, 75 kg/ha was observed to be most effective in reducing the larval population and number of root galls. A combination of neem cake (0.5 t/ha) + NPK (150:100:50) + Carbofuran @ 0.75 kg a.i./ha also reduced nematode population and root gall index (Nakat and Madne, 1993). Mixed cropping of *Tagetes* sp. Sadaphuli and baken neem (*Melia azadarichta*) in betelvine gardens suppressed the nematode population in soil (Nakat and Madne, 1993). Neem cake was the most effective in reducing the root galls and producing maximum number of marketable leaves (Acharya and Padhi, 1988a; Murthy and Rao, 1992). Rao et al. (1996) observed that *Calotropis* leaves @ 80 kg/ha significantly reduced the population of plant parasitic nematodes in soil including root knot nematode and was superior to neem and castor leaves. Recently *Paecilomyces lilacinus* has been found effective biocontrol agent in pots as well as under field conditions (Bhatt and Vadhera, 2004). Application of *Trichoderma viride* multiplied on linseed oil cake was applied in the soil at an interval of 60, 90 and 120 days after planting found to be highly effective in reducing root-knot nematodes in betelvine (Bhatt et al., 2002). Unfortunately none of the cultivars was found resistant to root knot nematode. Out of the cultivars tested Sanchi and Birkoli were moderately resistant and rest of the cultivars showed susceptible reaction against *M. incognita* (Acharya and Padhi, 1988b). Nakat and Madne (1993) recorded less number of galls in Kuljedu, Maghai, Bhadana and Karapaku cultivars of betelvine. Aldicarb was observed to be most effective nematicide in the nematode management (Murthy and Rao, 1994; Bhatt and Vadhera, 2004). Rao et al. (1993) observed that carbofuran @ 1.5 kg a.i./ha was effective in reducing larval population of root knot nematode.

16.9 CONCLUSION

The shady and moist atmosphere are favor the development of many diseases incited by fungi, bacteria, viruses and nematodes that greatly affect the growth of plants and causes heavy losses to farmers. The major diseases like leaf and foot rot caused by *P. palmivora or P. parasitica*, anthracnose caused by *Colletotrichum capsici* (Syd.) Butler and Bisby and bacterial leaf spot caused by *Xanthomonas campestris* pv. *betlicola* (Patel, Kulkarni and Dhande) Dye, basal rot or collar rot caused by *Sclerotium rolfsii*, root knot caused by *Meloidogyne incognita* and others plant parasitic nematodes are the main yield limiting factors of the betelvine cultivation all over the India. Betel leaves are generally used for mastication along with areca nut as raw form. The chemical pesticides when applied in higher doses or frequently in orchards for management of diseases give residual effects. Therefore, in perspective to environmental and human health and proper management of betelvine's disease in future will be required judicious application of chemical and eco-friendly management strategies or integrated approaches as well as proper identification pathogen by new emerging tools.

KEYWORDS

- **bacterial**
- **betelvine**
- **blight**
- **canker**
- **rot**
- **spot**

REFERENCES

1. Acharya, A., Das, J. N. (1995). Control of anthracnose of betelvine by fungicidal chemicals. *Current Agricultural Research* 8(2), 58–60.
2. Acharya, A., Padhi, N. N. (1988a). Effect of neem cake and saw dust against nematode, *Meloidogyne incognita* on betelvine (*Piper betle,* L.). *Indian, J. Nematol.* 18, 105–106.
3. Acharya, A., Padhi, N. N. (1988b). Screening of betelvine varieties for resistance against root-knot nematode, *Meloidogyne incognita. Indian Journal of Nematology* 18, 109.

4. Acharya, A., Dash, S. C., Padhi, N. N. (1987). Pathogenic association of *Meloidogyne incognita* with *Sclerotium rolfsii* and *Xanthomonas betlicola* on betelvine. *Indian Journal of Nematology* 17(2), 196–198.

5. Akhter, M., Nawab, B., Perveen, S., Khan, H. A. (2011). Plant parasitic nematodes associated with diseased betel vine conservatories in Karachi, Thatta, Hub area and Ketty bunder. *Pak. J. Nematol.* 29(1), 53–61.

6. Amer-Zareen., Shahina, F., Shahzad, S. (1999). New records of nematodes associated with betelvine in Pakistan with observation s on *Aphelenchoides helicosoma*. *Pak. J. Nematol.* 17, 47–50.

7. Arulmozhiyan, R., Chitra, R., Prabhakar, K., Jalaluddin, S. M., Packiaraj, D. (2005). SGM. BV.2 – A new promising betel vine variety. *Madras Agric. J.* 92(2), 498–503.

8. Ashthana, R. P. (1947). The role of cutting in the dissemination of foot rot of *Piper betle*. *Indian Journal of Agriculture science* 18, 223–225.

9. Ayavoo, R., Samiyappan, R., Seshadri, K. (1981). Chemical controls of betelvine wilt disease. *Pesticides* 16, 14–15.

10. Balasubrahmanyam, V. R., Chaurasia, R. S., Tripathi, R. D., Johri, J. K. (1988). Evaluation of some fungicides and antibiotics against fungal and bacterial pathogens of betelvine (*Piper betle*, L.). *Tropical Pest Management* 34(3), 315–317.

11. Basak, A. B., Mridha, M. A. U., Jlali, M. A. (1992). Studies on the leaf spot disease of *Piper betel*, L. caused by *Colletotricum piperi* Petch. *Chittagony University Studies Part II: Science Volume* 16(2), 87–91.

12. Bhale, M. S., Chaurasia, R. K., Nayak, M. L. (1985). Association of *Colletotrichum capsici* with *Xanthomonas campestris* pv. *betlicola* incident of leaf spot of betelvine (*Piper betle*, L.). *Indian Phytopath.* 38, 565–566.

13. Bhale, M. S., Khare, M. N., Nayak, M. L., Chaurasia, R. K. (1987). Diseases of betelvine (*Piper betle*, L.) and their management. *Rev. Trop. Plant Pathol.* 4, 199–220.

14. Bhatt, J., Vadhera, I. (2004). Nematodes of betelvine and their management – A review. *Agric. Rev.*, 25(3), 231–234.

15. Bhatt, J., Sengupta, S. K., Chaurasia, R. K. (2002). Management of *Meloidogyne incognita* by *Trichoderma viridae* in betelvine. *Indian Phytopathology* 55, 348–350.

16. Brahmankar, S. B., Dange, N. R., Tathod, D. G. (2011). Remove from marked records integrated management of betel vine wilt in Vidarbha. *International Journal of Plant Protection* 4(1), 146–147.

17. Chattapdayay, S. P., Maiti, S. (1967). Diseases of Betelvine and species. ICAR New Delhi.

18. Chattopadhayay, S. B., Maiti, S. (1990). Diseases of betel vine and Spices. ICAR, New Delhi. pp. 60.

19. Chaurasia, S. C. (1976). Studies on foot rot and leaf rot diseases of Pan (*Piper betle*) with special reference to pathogenesis and control measure. PhD Thesis, Univ. of Sagar.

20. Chaurasia, S. C., and Vyas, K. M. (1977). *In vitro* effect of some volatile oil against *Phytophthora parasitica* var. *piperina*. *Madhya Bharati* 22, 63–65.

21. Chawdhury, S. (1944). Disease of Pan (*Piper betle*) in Sylhet, Assam. I. The problem an its economic importance. *Proc. Indian Acad. Sci.* 19, 147–151.

22. Chopra, R. N., Nayar, S. L., Chopra, I. C. (1956). Glossary of Indian Medicinal Plant. CSIR, New Delhi. Pp. 194.

23. Dasgupta, B., Maiti, S. (2008). Research on betel vine diseases under AINP on betel vine. Proc. National Seminar on "*Piperaceae-Harnessing Agro-technologies for Accelerated Production of Economically Important Piper Species*," 21–22 November, 2008, Indian Institute of Spices Research, Calicut – 673012, Kerala, India, pp. 270–79.

24. Dasgupta, B., Sen, C. (1999). Assessment of *Phytophthora* root rot of betelvine and its management using chemicals. *Journal of Mycology and Plant Pathology* 29, 91–95.

25. Dasgupta, B., Sen. C. (1985). Relationship of inoculums density of *Colletotricum capsici* and moisture on leaf spot development in betel vine. *Indian Phytopathology* 38(2), 364.

26. Dasgupta, B., Dutta, P., Das, S. (2011). Biological control of foot rot of betelvine (*Piper betle, L.*) caused by *Phytophthora parasitica* Dastur. *The Journal of Plant Protection Sciences* 3(1), 15–19.

27. Dasgupta, B., Dutta, P. K., Muthuswamy, S., Maiti, S. (2003). Biological control of foot rot of betelvine (*Piper betle* Linn.). *Journal of Biological Control* 17, 63–67.

28. Dasgupta, B., Mohanty, B., Dutta, P. K., Maiti, S. (2008). *Phytophthora* diseases of betelvine (*Piper betle, L.*): a menace to betelvine crop. *SAARC Journal of Agriculture* 6, 71–89.

29. Dasgupta, B., Roy, J. K., Sen, C. (2000). Two major fungal diseases of betelvine, in: M. K. Dasgupta (ed.), *Diseases of Plantation Crops, Spices, Betelvine and Mulberry.* pp. 133–137.

30. Dastur, J. F. (1927). A short note on the foot rot diseases of pan in central provinces. *Agric. J. India* 22, 105–108.

31. Dastur, J. R. (1935). Disease of pan (*Piper betle*) in the Central Provinces. *Proceedings of Indian Academy of Sciences* 1, 26–31.

32. Datta, P., Dasgupta, B., Majumder, D., Das, S. (2010). Disease prediction model of foot rot and leaf rot of betelvine (*Piper betle, L.*) caused by *Phytophthora parasitica* in West Bengal. *SAARC Journal of Agriculture* 8(2), 87–96.

33. Deka, S. N., Dutta, P. K., Bora, L. C., Saikia, L. (2005). Studies on leaf spot of betelvine caused by *Xanthomonas-Colletotrichum* in Assam. *Karnataka, J. Hort.* 2, 14–15.

34. Deka, U. K., Dutta, P. K., Gogoi, R., Borah, P. K. (2008). Management of leaf spot disease complex of betelvine by bioagents and plant extracts. *Indian Phytopath.* 61(3), 337–342.

35. Doosani, Z. A., Shahzad, S., Vahidy, A. A., Ghaffar, A. (1992). Diseases of betel vine in and around Karachi area and their control, in: Ghaffar, A., Saleem, S. (eds.), *Status of Plant Pathology in Pakistan.* Proc. National Symposium, 3–5th Dec, 1991. Dept. of Botany, University of Karachi, Karachi, pp. 87–92.

36. Gangwar, S. K., Dasgupta, B. (1989). Rhizolex 50 WP-a new fungicide to control the foot rot of betelvine (*Piper betle, L.*) caused by *Sclerotium rolfsii* Sacc. *Proceedings: Plant Sciences* 99(3), 265–269.

37. Goswami, B. K., Kader, K. A., Rahman, M. L., Islam, M. R., Malaker, P. K. (2002). Development of leaf spot of betelvine caused by *Colletotrichum capsici. Bangladesh, J. Plant Pathol.* 18(1&2), 39–42.

38. Guha, P. (2006). Betel leaf: The neglected green gold of India. *J. Hum. Ecol.* 19(2), 87–93.

39. Guha, P., Jain, R. K. (1997). *Status Report On Production, Processing and Marketing of Betel Leaf (Piper betle, L.)*. Agricultural and Food Engineering Department, IIT, Kharagpur, India (1997).

40. HaralPatil, S. K., Raut, S. P. (2008). Efficacy of fungicides, bio-agents and botanicals against collar rot of betelvine. *Journal of Plant Disease Sciences* 3(1), 93–96.

41. Huq, M. I. (2011). Studies on the epidemiology of leaf rot and leaf spot diseases of betel vine (*Piper Betle, L.*). *Bangladesh, J. Sci. Ind. Res.* 46(4), 519–522.

42. Jagdale, G. B., Pawar, A. B., Darekar, K. S. (1985). Effect of organic amendments on root-knot nematodes infesting betelvine. *International Nematology Network Newsletter* 2, 7–10.

43. Johri, J. K., Chaurasia, R. S., Balasubrahmanium, V. R. (1984). Status of betelvine pests and diseases in India, in: Khanduja, S. D., Balasubrahmanium, V. R. (eds.), *Proc. Group Discussion-Improvement of betelvine cultivation*. National Botanical research Institute, Lucknow. pp. 13–24.

44. Jonathan, E. I., Sivakumar, M., Padmanabhan, D. (1996). Interaction of *Meloidogyne incognita* and *Phytophthora palmivora* on betelvine. *Nematologia Mediterranea* 24, 341–343.

45. Khanra, S. (1997). Paan Vittik Silpakendra (In Bangali) "Betel leaf Based Indurtry" Nabanna Bharati, 30(2), 169.

46. Magdum, S. G., Shirke, M. S., Kamble, B. M., Salunkhe, S. M., Tambe, B. N. (2009). Integrated management of major diseases of betelvine. *Advances in Plant Sciences* 22(1), 35–36.

47. Maiti, S. (1994). Diseases of betel vine, in: Chadha, K. L., Rethinam, P. (eds.), *Advances in Horticulture Vol. 10, Plantation Crops and Spices Crops—Part 2*. Malhotra Publishing House, New Delhi.

48. Maiti, S., Sen, C. (1979). Leaf rot and foot rot of *Piper betel* caused by *Phytophthora palmivora*. *Indian Phytopath*. 30, 438–439.

49. Maiti, S., Sen, C. (1982). Incidence of major diseases of betel vine in relation to weather. *Indian Phytopathology* 35(1), 14–17.

50. Maiti, S., Shivshankara, K. S. (1998). Betelvine Research Highlights. IIHR, Bangalore, pp. 21.

51. Maiti, S., Khatua, D. C., Sen, C. (1978). Chemical control of two major diseases of betelvine. *Pesticides* 12, 45–47.

52. Maity, P. (1989). *Extension Bulletin: The Betel vine*. All India coordinated Research project of betel vine. Indian Institute for Horticultural Research. Hessarghatta, Bangalore, India.

53. Marimuthu, T. (1991). Fungal nematode wilt complex in betelvine (*Piper betel, L.*). *Pl. Disease Res*. 6, 85–88.

54. Mathew, J., Chirian, M., Abraham, K. (1978). Bacterial leaf spot of betel vine (*Piper betle* L) incited by *Xanthomonas betlicola* Patel et al., in Kerala. *Curr. Sci.* 16, 592–593.

55. Mehrotra, R. S., Aggarwal, A. (2003). Rots, Damping offs, Downy mildews and White Rusts, in: Melhotra, R. S., Aggarwal, A. (eds.), Plant Pathology, 2nd Edn. Tata McGraw- Hill Publishing Company Limited, New Delhi. pp. 313–368.

56. Mehrotra, R. S., Tiwari, D. P. (1967). Studies on the foot rot of *Piper betle*-I: Behavior of *Phytophthora parasitica* var. *piperina* in soil. *Indian Phytopath*. 20, 161–67.

57. Mehrotra, R. S., Tiwari, D. P. (1976). Organic amendments and control of foot rot of *Piper betle* caused by *Phytophthora parasitica* var. *piperina*. *Annals of Microbial Research* 27, 415–21.

58. Mithila, J., Shivashankara, K. S., Satyabrata, M. (2000). Growth and morphology of vegetative and reproductive branches in betel vine (*Piper betle,* L.). *Plant Crops* 28, 50–54.

59. Mitra, M. (1930). New disease reported during the year 1929. *Internal Bulletin of Plant Protection* 4, 103–104.

60. Mohanty, B., Dasgupta, B. (2008). Management of foot rot and leaf rot of betelvine (*Piper betle*) caused by *Phytophthora parasitica* by using safer fungicides. *Journal of Mycopathological Research* 46, 81–84.

61. Mridha, M. A. U., Alamgir, S. M. (1987). Studies on foot rot of betel vine (*Piper betle,* L.) caused by *Sclerotium rolfsii* in Bangladesh, in: *12th Annual Bangladesh Science Conference* (January, 1987), Dhaka (Bangladesh).

62. Murthy, M. M. K., Rao, K. T. (1992). Control of root knot nematode, *Meloidogyne incognita* on betelvine with certain oil cake amendments and non-volatile nematicides. *Indian, J. Pl. Protec.* 20(2), 171–173.

63. Murthy, M. M. K., Rao, K. T. (1994). Chemical control of root knot nematode, *Meloidogyne incognita* in betelvine with non-volatile nematicides. *Indian Journal of Entomology* 56(1), 44–50.

64. Nadugala, L. M. N. S., Amarasingh, B. H. R. R. (2009). Diversity among different isolates of *Xanthomonas campestris* pv. *betlicola* on the basis of phenotypic and virulence characteristics. *J. Natn. Sci. Foundation Sri Lanka* 37 (1), 77–80.

65. Naik, M. K., Hiremath, P. C. (1986). Unrecorded pathogen on betelvine causing anthracnose. *Current Sciences* 55(13), 625–626.

66. Nakat, R.V and Madne, N. P. (1993). *Proc. of National Symposium on Betelvine Production Technology,* held at JNKW, Jabalpur 22–23 Sept. 1993, pp. 66–67.

67. Narasimhan, M. J. (1933). *Annual report of the mycological section for the year 1931–32*. Rept. Mys. Agr. Dept. for the year ending 30th June, 1932, pp. 32–35.

68. Nema, A. G. (1988). Varietal resistance in betel vine against *Xanthomonas campestris* pv. *betlicola* and its manipulation with crop nutrients. *Indian Phytopathology* 41(3), 344–350.

69. Patel, M. K., Kulkarny, Y. S., Dhande, G. W. (1951). Bacterial leaf spot of castor. *Curr. Sci.* 20, 106.

70. Paul, W. R. C. (1939). *A leaf spot disease of betelvine*. Div. Plant Pathol. Adm. Reptr. Div. Agric., Ceylon, p. 41–45.

71. Ramamurthi, K., Usha Rani, O. (2012). Betel leaf: Nature's green medicine. *Fact for You*, September 2012, pp. 8–10.

72. Rangaswami, G., and Mahadevan, A. (2006). Diseases of cash crops, in: Rangaswami, G., Mahadevan, A. (eds.), *Diseases of crop plants in India*. Prentice-Hall of India Limited, New Delhi, 408–459.

73. Rao, B. N., Sultan, M. A., Reddy, K. N. (1993). Residues of carbofuran in betelvin. *Indian, J. Pl. Protec.* 21, 217–219.

74. Rao, D. V. S., Sitaramaiah, K., Maiti, S. (1996). Management of plant parasitic nematodes in betelvine garden through non-chemical methods. *Indian J, Nematol.* 26, 93–97.

75. Rao, M. S., Reddy, P. P., Khan, R. M., Rao, N. N. R. (1991). Role of standards in building of *Meloidogyne incognita* in betelvine. *Curr. Nematol.* 2, 143–144.

76. Ray, P., Das, S. C., Acharya, A. (1993). Pathogenic association of *Colletotrichum* sp. with *Meloidogyne incognita* in causation of lethal yellowing complex of betelvine. *Orissa, J. Agric. Res.* 6, 5–6.

77. Roy, T. C. (1948). Anthracnose disease of *Piper betle* caused by *Colletotricum dasturi. Roy. J. India Bot. Soc.* 27, 96–100.

78. Saksena, S. B. (1977). *Phytophthora parasitica*, the scourge of '*pan.*' *Indian Phytopathology* 30, 1–16.

79. Saksena, S. B., Melhotra, R. S. (1970). Studies on foot rot and leaf rot of *piper betle*. Host range and role of cutting in the spread of disease. *J. Indian Bot. Sci.* 49, 24–29.

80. Saleem, S. (2000). Anthracnose of betelvine in Pakistan. *Pakistan, J. Bot.* 32(1), 41–44.

81. Samanta, C. (1994). *Paan chasher samasyaboli -o- samadhan. India: Ekti samikkha* (In Bengali), A Report on Problem and Solutions of Betelvine Cultivation. A booklet published by Mr. H. R. Adhikarri, C-2/16 Karunamoyee, Salt Lake City, Kolkata- 64 (WB), India.

82. Satyabrata, M., Kakam, A. S., Sengupta, K., Punekar, L. K., Das, J. N. (1995). Effect of sources and leaves of nitrogen on growth and yield of betel vine (*Piper betle*, L.). *J. Plant Crops* 23, 126–125.

83. Sengupta, D. K., Dasgupta, B., Datta, P. (2011). Management of foot rot of betelvine (*Piper betle* L) caused by *Phytophthora parasitica* Dastur. *Journal of Crop and Weed* 7, 179–83.

84. Shahina, F., Erum, Y. I. (2005). A review on betelvine crop and nematodes associated with it. *Pak. J. Nematol.* 23, 111–114.

85. Shivakumar, M., Marimuthu, T. (1986). Efficacy of different organic amendments against phytonematodes associated with betevine. *Indian, J. Nematol.* 16, 278.

86. Shivakumar, M., Marimuthu, T. (1987). Preliminary studies on the effect of solarization on phytonematodes of betevine. *Indian, J. Nematol.* 17, 58–59.

87. Singh, A. (2002). Studies on the coller rot of betelvine (piper betle, L.). PhD Thesis, Lucknow, p. 187.

88. Singh, B. P., Chand, J. N. (1971). Studies on the diseases of Pan (Piper betle, L.) in Jabalpur (Madhya Pradesh) IV. A new fungal-bacterial complex. *Science and Culture* 37(7), 344.

89. Singh, B. P., Chand, J. N. (1972). Assessment of losses of betelvine (*Piper betle*) caused by *Sclerotium rolfsii* in Jabalpur (M. P.). *Science and Culture* 38(12), 526–527.

90. Singh, B. P., Chand, J. N. (1973). A note on the estimation of losses of betelvine *Piper betle, L.* due to *Phytophthora parasitica* Dastur in Jabalpur Madhya Pradesh. *Science and culture* 39(4), 201–202.

91. Singh, B. P., Joshi, L. K. (1971). Studies on the diseases of betel vine. *Indian, J. Mycol. Plant Pathol.* 1(2), 150–151.

92. Sitaramaiah. K., Devi, G. P. (1994). Nematodes, soil micro-organisms in relation to Phytophthora wilt diseases in betelvine. *Indian, J. Nematol.* 4, 16–25.

93. Stevenson, J. A. (1926). *Foreign Plant Diseases.* U.S.D.A, Washington.

94. Su, M. T. (1931). Report of the Mycologist, Mandalay for the year ended the 31st March, 1931, p. 9.

95. Thyagarajan, P., Venkata Rao, A., Varadarajan, S. (1972). Studies on betelvine wilt disease-influence of nitrogen and phosphorus in the control of betelvine wilt disease. *Madras Agricultural Journal.* 59(3), 187–189.

96. Tiwari, D. P., Mehrotra, R. S. (1974). Studies on foot rot and leaf rot of *Piper betle*-saprophytic colonization of dead organic substrate in soil by betel vine *Phytophthora. Journal of the Indian Botanical Society* 53(1/2), 119–123.

97. Tripathi, R. D., Johri, J. K., Balasubrahmanyam, V. R. (1984). Evaluation of chemicals inhibiting the bacterial leaf spot pathogen of betel vine. *Tropical Pest Management* 30(4), 440–443.

98. Turner, G. J. (1969). *Phytophthora palmivora* from *Piper betle,* L. in Sarawak. *Trans. British Mycol. Soc.* 52, 411–418.

99. Uppal, B.N and Kamat, M. N. (1938). Powdery mildew of betel vine. *Curr. Sci.* 12, 611.

100. Vauterin, L., Hoste, B., Kersters, K., Swings, J. (1995). Reclassification of *Xanthomonas. International Journal of Systematic Bacteriology* 45, 472–489.

101. Vijayakumar, J., Arumugam, S. (2012). Foot rot disease identification for vellaikodi variety of betelvine plants using digital image processing. *Ictact Journal on Image and Video Processing* 3(2), 495–501.

102. Wasnikar, A. R., Khatia, S. K., Nayak, M. L., Vishwakarma, S. K., Punekar, L. K. (1993). Effect of phosphorus on bacterial leaf spot disease incidence, and chemical composition and storage quality of *Piper betle* leaves. *Phytoparasitica* 21(1), 75–78.

103. Waterhouse, G. M. (1963). Key to the species of *Phytophthora* deBary. C. M. I. Paper. No.92, Kew, Surrey, England.

104. Yadav, B. P., Ojha, K. L., Prasad, Y. (1993). Cultivation of betelvine in Bihar, in: *National Symposium on Betel Vine Production Technology* (22–23 Sept, 1993), pp. 10–12.

PART IV

GENERAL ISSUES

CHAPTER 17

DISEASES OF MAKHANA AND THEIR MANAGEMENT

SANGITA SAHNI,[1] BISHUN D. PRASAD,[2] and SUNITA KUMARI[3]

[1]*Department of Plant Pathology, T.C.A., Dholi, Muzaffarpur, Bihar, India*

[2]*Department of Plant Breeding and Genetics, B.A.C., Sabour, Bihar, India*

[3]*Krishi Vigyan Kendra, Kishanganj, BAU, Sabour, Bihar, India*

CONTENTS

17.1 INTRODUCTION

Euryale ferox Salisb (also known as fox nut, foxnut, makhana, or gorgon plant) is one of the most valuable cash aquatic fruit grown in thousands of fresh water stagnant pools (both natural and man-made) of Northern and Eastern Asia, Europe, America, etc. and is very common in the freshwater habitats of Northern, Eastern and Western India. In India, it is grown in the various tracts of N.E. Assam, Tripura, West Bengal, Orissa, Bihar, and Meghalay, different parts in MP and UP and in Alwar (Rajasthan). North Bihar and lower Assam are probably the only states in India where the plant is cultivated for commercial purpose along with *Nelumbium, Trapa* species and fish (Ahmad and Singh, 1991, 1997; Datta Munshi, et al., 1991; Dehadrai, 1994; Jha, 2000). For geographical and climatic reasons Bihar is the heaven for Makhana Production. The cultivation of the crop is confined to Dharbhanga, Motihari, Madhubani, Samastipur, Sitamarhi, Saharsa, Supaul, Araria, Kishanganj, Katihar and Purnea districts of Bihar. Makhana is cultivated in more than 2000 ha in Darbhanga and Madhubani districts. Around 75% of the total Makhana production comes from Bihar Swetlands. Approximately 2000 tons of popped makhana worth Rs. 100 million are exported outside north Bihar (Thakur, 2005). Makhana is valued for its nutritional, medicinal and ritualistic significance also. In north Bihar, the seed is consumed in popped form, but in Manipur, other parts (leaves and stalks) are consumed as vegetables. Seeds are also used in many dishes of India. It is highly estimated for its spermatogenic and aphrodisiac properties and is used in rheumatic disorders. Makhana is superior to dry fruits such as almonds, walnut, coconut and cashew nut in term of sugar, protein, ascorbic acid and phenol content. It has 80% carbohydrate mainly in the form of starch and 10–12% protein (Jha et al., 1991a, b). It is rich in minerals and is almost fatless (0.1%). Like other crops, several diseases attack makhana plants, but unlike elsewhere in field, these cannot be sprayed with chemicals, for they would kill the fish. There is no systematic study has been done for correct identification and diagnosis of different diseases of makhana and their management. Therefore, there is urgent need of correct diagnosis and comprehensive understanding of the life cycles of the various diseases so that they can be managed at the most appropriate time, if they are causing economic loss. The most common diseases in Makhana are leaf blight, black and brown leaf spot, botrytis gray mold, root rot and development of tumors.

17.2 LEAF BLIGHT

Leaf blight caused by *Alternaria alternata*, is a most serious fungal disease of makhana (Haidar and Nath, 1987; Dwivedi et al., 1995). Symptoms of Alternaria leaf blight appear on the upper surface of leaves, which include leaf spots, and blighting. Leaf spots are light tan to brown and usually have a concentric ring or target pattern with a yellow halo. As leaf spotting increases, blighting and premature defoliation occur (Haider and Mahto, 2003).

17.2.1 MANAGEMENT

Remove infected leaves foliage, and destroy it safely, will prevent their spread. In pool, where there are no fish, foliar sprays with copper oxychloride or dithane M-45 @ 0.3% twice or thrice at fortnightly interval have been found very effective to check the disease.

17.3 LEAF SPOT

There are two species of leaf spot disease that damage makhana leaves. One causes random spotting and dark patches on the surface of the leaves, which eventually enlarge or merge together, and may be reddish to grayish brown; the other tends to start at the outer edges of the leaves, causing them to turn brown and crumble. Both are debilitating and disfiguring, but they are not very serious problems. Their incidence will vary considerably from year to year, depending upon the prevailing conditions (Figure 17.1).

FIGURE 17.1   Symptoms of leaf spot in makhana leaves.

17.3.1   MANAGEMENT

Remove spotted leaves on lightly diseased plants. Rake up and destroy infected fallen leaves. The removal of this leaf material will minimize the chances of the disease reoccurring the next season. If chemical control is needed, most fungal leaf spots can be controlled with sprays of fungicides containing chlorothalonil, thiophanate-methyl or mancozeb @ 0.2%. Apply when symptoms first appear and repeat every 10 to 14 days as needed. It is advisable not to undertake chemical treatment unless it is causing economical losses as applied chemicals will affects the water and fishes present in that pool.

17.4   ROOT ROT

The more recent strain of root rot disease is believed to be caused by a number of pathogens, which have not yet been fully identified. One has been known over the years as crown or root rot which is believed to be caused by *Phytophthora* species. Apparently, healthy plants suddenly showing wilting and yellowing. The roots appear dark brown or black, which become soft and rotten. The outer layer of cells easily strips off the roots leaving only the central strand of water conducting tissue.

17.4.1 MANAGEMENT

At present, there is no cure, so it is vital to remove and destroy any infected plants. Then the pool should be cleaned thoroughly and sterilized with a solution of sodium hypochlorite or swirling a muslin bag filled with copper sulfate crystals through the water before being flushed out with fresh water. Once clear water has been run back in, new plants can be safely introduced. All fish must be removed before any such treatment begins, and not returned until the pool has been emptied, swilled out and refilled with fresh water.

17.5 BOTRYTIS GRAY MOLD

Botrytis blight or "gray mold" is a widely distributed disease caused by the fungus *Botrytis* species. Botrytis, are ubiquitous and opportunistic; they attack only physiologically-weakened plants. Botrytis, gray mold is common on seeds and seedlings. *Botrytis* at first appears as a white growth on the plant but very soon darkens to a gray color. Smoky-gray "dusty" spores form and are spread by the wind or in water. Buds or small leaves with gray mold on them may influence growth and photosynthesis. The risk of disease and plant loss can be minimized by reducing pathogen populations and increasing plant vigor.

17.5.1 MANAGEMENT

Sanitation is the first important step in controlling this disease. Remove dead or dying tissue from the plants and from the soil surface. Sanitation alone is not sufficient to control this fungus.

Seeds should be soaked in Hydrogen peroxide and water at a ratio of 1:10. Fresh solution should be added every day until the seed has swollen and is ready to split it's skin and extend stems and roots. This is a time for extreme patience. Scarification or cutting the shell of the seed hurries things along. Just don't cut too deep. It can still take 20 days.

17.6   DEVELOPMENT OF TUMORS

The crop suffered a heavy damage to its leaves and other parts including the petiole, pedicel, etc., from a smut fungus known as *Doassansiopsis euryaleae* (Verma and Jha, 1999). The infection usually extended from leaf lamina to petiole and from pedicel to the basal part of flower causing great distortion in shape due to hypertrophy. At the beginning the hypertrophy was small but later attained a major dimension. Infection to the basal part of the flower including the ovary greatly reduced the number of viable seeds and caused an economic loss to the farmers (Verma et al., 2003). Sometimes it affects fruits and causes fruit galls. There is no control measure reported for this disease (Figure 17.2).

Other diseases like chlorosis, nutrient deficiencies and poor growth of plants is normally due to fertilization, CO_2 or lighting conditions. These can lead to other problems like bacterial rots, etc. because the plants are weak.

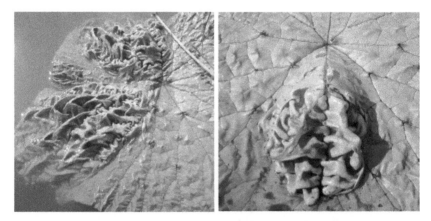

FIGURE 17.2   Different symptoms of development of tumor in makhana leaves.

KEYWORDS

- blight
- fruit
- mold
- rot
- spot
- tumor

REFERENCES

1. Ahmad, S. H., Singh, A. K. (1991). Fishery development in ox-bow lakes (Mans) of Bihar. *Fishing Chimes*, 6, 59–62.
2. Ahmad, S. H., Singh, A. K. (1997). Prospects of integration of Makhana (*Euryale ferox*) with fish culture in north Bihar. *Fishing Chimes*, 16, 45–50.
3. Datta Munshi, J. S., Datta Munshi, J., Choudhary, L. K., Thakur, P. K. (1991). Physiography of the Kosi river basin and formation of wetlands in north Bihar: A unique freshwater system. *J. Freshwater Biol.* 3, 105–122.
4. Dehadrai, P. V. (1994). Swamps of north Bihar. *Bull. Nat. Inst. Ecol.* 7, 17–21.
5. Dwivedi, A. K. Shekhar, R., Sharma, S. C. (1995). Ultrastructural studies of *Euryale ferox* leaf infected by *Alternaria alternata*. *Indian Phytopath.* 48, 61–65.
6. Haidar, M. G., Nath, R. P. (1987). Chemical control of *Alternaria* leaf blight of *Makhana (Euryale ferox). Nat. Acad. Sci. Lett.* 10, 301–302.
7. Haider, M. G., Mahto, A. (2003). Fungal leaf blight and nematode diseases of Gorgon nut (*Euryale ferox*) and their management. In: Mishra, R. K., Jha, Vidyanath and Dharai, P. V. (eds.) *Makhana*, ICAR, New Delhi, pp. 159–162.
8. Jha, V. (2000). Natural resource management in the flood zones of Bihar. In: H. K. Patra (ed.). *Environment and Disaster Management*, Utkal University, Bhubaneshwar, pp. 73–79.
9. Jha, V., Kargupta, A. N., Dutta, R. N., Jha, U. N., Mishra, R. K., Saraswati, K. C. (1991a). Utilization and conservation of *Euryale ferox* Salisb. In Mithila (North Bihar), India. *Aquatic Botany*, 9, 295–314.
10. Jha, V., Barat, G. K., Jha, U. N. (1991b). Nutritional evaluation of *Euryale ferox* Salisb. (Makhana). *J. Food Sci. Technol.* 28, 326–328.
11. Thakur, B. (2005). Urban and Regional Development in India: Essays in Honor of Prof. L. N. Ram, pp.526–527.
12. Verma, R. A. B., Jha, V. (1999). New *Doassansiopsis* associated with freshwater plant *Euryale ferox* Salisb. (*Makhana*) in north Bihar (India). *J. Freshwater Biol.* 11, 7–10.
13. Verma, R. A. B., Jha, V., Devi, S. (2003). Leaf and floral hypertrophy of Makhana caused by *Doassansiopsis euryaleae*. In: Mishra, R. K., Jha, Vidyanath and Dharai, P. V. (eds.) *Makhana*, ICAR, New Delhi, pp. 163–168.

CHAPTER 18

DISEASES OF JUTE AND SUNHEMP CROPS AND THEIR MANAGEMENT

SANTOSH KUMAR,[1] V. K. SINGH,[1] A. N. TRIPATHI,[2] and PRABHAT KUMAR[3]

[1]*Jute Research Station, Katihar, Bihar Agriculture University, Sabour, Bihar, India; E-mail: santosh35433@gmail.com*

[2]*Division of Crop Protection, Central Research Institute for Jute and Allied Fibers, Barrackpore, Kolkata – 700120, West Bengal, India*

[3]*Betelvine Research Centre, Islampur, Nalanda, Bihar – 801303, India*

CONTENTS

18.1 INTRODUCTION

Jute (*Corchorus capsularis* L. and *Corchorus olitorius* L.) and sunnhemp (Crotalaria juncea L.), are the most important bast fiber producing commercial/industrial cash crops in India. Jute ranks second next to cotton among all the natural fiber in case of production (Talukder et al., 1989). In India jute mostly cultivated in the Eastern-Indian States namely, Assam, Bihar, Eastern Utter Pradesh, Orissa and West Bengal. West Bengal alone contributes lion share (77%) of the Indian jute (Sinha et al., 2004). Jute fiber is extensively used all over the world for its diversified value added products like hessians, sacking, gunny bags, carpets, mat, rope, false ceiling boards and many geotextile products while the young green leaves of jute are edible and popular as vegetables.

Sunnhemp is also grown for fiber as well as green manure crops which able to fix nitrogen and could reduce the build-up of nematodes populations. The fiber obtained from the sunnhemp used for various purposes like making ropes, strings, twins, floor mat, fishing nets, handmade paper, etc. in cottage industry. Acreage, productivity and production of under jute and sunhemp started declining over the years due to high incidence of pests and disease complex under changing climatic scenario and low fluctuating economic return when compared with other competitive crops. Among biotic stresses diseases viz, stem root, root rot, soft rot, anthracnose, Hooghly wilt, mosaic and root knot on jute while *Fusarium* wilt, anthracnose and mosaic on sunnhemp which are infected crop from beginning with the seedling stage to harvesting stage.

Integrated disease management (IDM) strategy is a holistic approach/ strategy for developing location specific innovative and effective solution to manage diseases and present economic losses due to diseases it causes under changing climatic scenario (Table 18.1).

18.2 JUTE DISEASES

18.2.1 STEM ROT

Stem rot of jute caused by *Macrophomina phaseolina* (Tassi) Goid. is the most destructive wide host range soil and seed borne fungal disease. In

TABLE 18.1   Diseases and Their Causal Organisms

S.No.	Name of Disease	Causal Organism	Categories of Pathogen
Jute			
01	Stem Rot	*Macrophomina phaseolina*	Fungus
02	Black Band	*Botryodiplodia theobromae*	Fungus
03	Anthracnose	*Colletotrichum corchori*	Fungus
04	Soft Rot	*Sclerotium rolfsii*	Fungus
05	Powdery Mildew	*Oidium sp.*	Fungus
06	Tip blight	*Curvularia subulata*	Fungus
07	Hooghly wilt	*Ralstonia solanacearum*	Bacteria
08	Jute leaf Mosaic	*Jute leaf Mosaic Virus* (JLMV)	Virus
09	Yellow vein virus	*Corchorus yellow vein virus*	Virus
10	Root Knot Nematode	*Meloidogyne javanica/M. incognita*	Nematode
Sunhemp			
01	Anthracnose	*Colletotrichum crotolariae*	Fungus
02	Wilt	*Fusarium udum f. sp. Crotolaiae*	Fungus
03	Sunhemp mosaic	Sunhemp mosaic virus (SMV)	Virus
04	Sunhemp leaf curl	Indian tomato leaf curl virus	Virus

jute (both *Chorchorus olitorious* L. and *C. capsularis* L.) the pathogen incites disease complex *viz.* seed rot, collar rot, seedling blight, stem rot and root rot in various growth stages of the crop from seedling to till harvest (Roy et al., 2008). Incidence of stem rot (*M. phaseolina*) on jute crop results in fiber yield loss around 11–20% and affecting the fiber quality to its commercial value. It is more prevalent during in hot (34 ± 1°C) and humid weather condition during the cropping season (Mandal, 1990).

18.2.1.1 The Pathogen

Macrophomina phaseolina (Tassi) Goid is the pycnidial stage of the pathogen. The sclerotial stage is *Rhizoctonia bataticola* and perfect stage is *Orbilia obscura.* The conidia are hyaline, aseptate, thin-walled, and elliptical. Under favorable conditions, hyphae germinate from the sclerotia and infect the roots of the host plant by penetrating the plant cell wall through mechanical pressure and/or chemical softening (Ammon et al., 1974).

18.2.1.2   Pathogenesis

Cell wall degrading pectinolytic and cellulolytic enzymes were produced by *M. phaseolina* which play a significant role the pathogenesis of stem rot in jute. These enzymes were produced constitutively and inducible. They were extracted from infected and surrounding area of the diseased part. Distinct lesions were observed in 14 day old jute seedling kept in enzyme solution at 21°C for 48 hours (Chattopadhyay and Raj, 1978).

18.2.1.3   Symptomatology

The *M. phaseolina* hyphae initially invade the cortical tissue of jute plants, followed by sclerotia formation, causing stem rot. Gray-black mycelia and sclerotia are produced in the infected area of the plant. The disease is characterized by the initiation of the small lesions on the stem as blackish brown depression, which increase in size. On mature plants the leaves and stem infected with black colored lesions which girdle and break the stem and causes shredding of the bark fiber. In case of jute seed crop pycnidia and sclerotia are farmed on capsule and seed. Diseased capsules/pods discolored black and seeds become discolored and small. Pathogen causing wilt and root rot symptoms were found to be pathogenic to seedlings of jute causing pre-emergence killing and post-emergence damping off (Figures 18.1–18.3). In general disease incidence has been found to be more in *C. olitorius* cultivars then *C. capsularis* cultivars.

18.2.1.4   Disease Cycle and Epidemiology

Macrophomina phaseolina, (Tassi.) Goid. a global devastating necrotrophic fungal pathogen, infects more than 500 economically important monocotyledeons and dicotyledeons plant sp. (Wyllie, 1988). The pathogen is soil and seed borne in nature. It can survive for more than 4 years

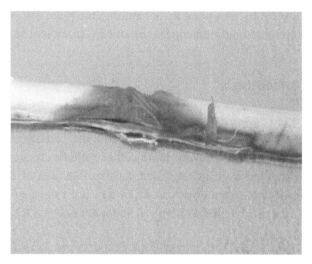

FIGURE 18.1   Stem rot symptom on jute plant.

FIGURE 18.2   Leaf rot symptom caused by *M. phaseolina*.

as sclerotia in the soil and plant debris (Short, 1980). The hyphae initially invade the cortical tissue of jute plants, followed by sclerotia formation, causing stem rot disease. Under favorable conditions, hyphae germinate from the sclerotia and infect the roots of the host plant (Ammon, 1974). It is transmitted from seed to seedling/ and seedling/plant to seeds

FIGURE 18.3   Root rot complex caused by *Rhizoctonia solani*, and *Fusarium oxysporum.*

(Akanda and Fakir, 1985). Islam (1987) has been reported that, disease development in jute plants due to *M. phaseolina*, increases gradually with the increase of seed borne infections. The pathogen favored low pH, High level of nitrogen, High rainfall, higher temperature (30–32°C), low soil moisture and high humidity favor infection (Ghosh and Bask, 1965).

18.2.1.5   Management

IDM is a sustainable approach for eco-friendly management of soil and seed borne; fungal diseases of jute under changing climatic scenario.

18.2.1.6   Cultural Practices

(1) Use of disease free certified seed;
(2) Deep plowing during the summer season;
(3) Field sanitation/clean cultivation;
(4) Liming of soil with 2–4 ton/ha applied in field 3–4 weeks before sowing;
(5) Judicious use of fertilizer @ 60:30:30 kg, NPK/ha. Stem rot increased with increasing level of N and it was maximum at 80 kg/ha. Thakur Ji, (1974) reported that stem rot decreased with application of micronutrients (Zn, Fe, Bo) along with NPK;

(6) Field sanitation viz. rouging of diseased plants, timely weeding and thinning of the field;

(7) Prefer disease/pest resistant/tolerant varieties for cultivation viz. JRO-524 (Navin), JRO- 8432, JRO −128, JRO-204, JRC-212 (Basudev), S-19 (Subala and IRA- 2003.

18.2.1.7 Chemical Control

(1) Seed treatment with fungicide Carbendazim @ 2.0 g/kg or captan @ 5.0 g/kg. The use of vitavax-200 as good seed treating fungicide has been reported by Akanda and Fakir, (1985).

(2) Prophylectic or curative spray of fungicide viz. Blitox or Fytolan @ 5.0 g/L, Bavistin (Carbendazim) @ 2.5 g/L, Dithane M 45 (Mancozeb) @ 5.0 g/L, 3.

18.2.1.8 Biological

Seed treatment with *Trichoderma viride* @ 10 g/kg or *Pseudomonas fluorescens* @ 10g/kg of seed or *T. viride* @ 1 kg/ha mix with FYM @ 100 kg applied in the field 3–4 weeks before sowing. Strains of *T. viride*, *Aspergillus niger* (Strain AN-27) and some species of fluorescent *Pseudomonas* have been established as very effective biocontrol agents for stem and root rot in jute (Anonymous, 1990 & 2006; Srivastava and Singh, 2009 and Roy et al., 2008). The fungitoxic effects of garlic have been reported by many scientists (Dubey and Dwivedi, 1991; Fakir and Khan, 1992 and Hossain et al., 1993). Soil application of *Trichoderma viride* thrice, for example, 7, 15, 30 DAS was found best in controlling seedling blight, collor rot, stem rot and root rot diseases giving minimum percent disease incidence (1.45, 3.07, 4.70 and 4.92, respectively) as compared to control (16.12, 9.47, 16.67 and 16.34, respectively) (Srivastava et al., 2010).

18.2.2 ANTHRACNOSE

Anthracnose of jute caused by *C. corchori,* which is considered as one of the destructive disease of fiber and seed crop of jute (*C. capsularis* than

C. olitorius). The disease is of regular occurrence in the *capsularis* belt of India, viz., Assam, North Bangal, Bihar and Uttar Pradesh. It is also prevalent in Bangladesh (Ghosh, 1957 and Mandal, 1990).

18.2.2.1   Symptomatology

Diseased plants produces black, brown depressed dot like spots on stem, which crack the bast fiber bundle. Infected pods/capsules are discolored with black necrotic lesions and the seeds become discolored and diseased. Infected seed are lighter, shrunken and poor in germination. Anthracnose affected plants yield poor quality fiber, mostly knotty in nature with adherent barks, which resist retting. The disease is seed-borne and thus seedling blight and pre-emergent death show gaps in the field (Ahmed, 1966). In mature stage, the plants do not die, but the disease badly affects the fiber quality. As a result, the market values of this fiber become 30–50% less than that from healthy plants (Khan and Strange, 1975). Such poor quality fiber is classed in the market as 'Crossbottom' (Ghosh, 1957). It transmitted from seed (Akanda and Fakir, 1985). Islam (1987) has been reported that, disease development in jute plants due to *C. corchori* increases gradually with the increase of seed borne infections. Continuous rain, high relative humidity and temperature of around 35°C are congenial for the faster development of this disease.

18.2.2.2   Management

Various cultural practices such as field sanitation use of disease free seed, crop rotation and field drainage help in reduction of disease incidence. Seed treatment with Bavistin @ 2g/kg of seed or Captan @ 5g/kg eliminate primary source of inoculum/infection. Seed lots having 15% or more infection should not be used as seed. Foliar spray of Bevestin (0.5%) or Dithane M 45 @ 5 g/L or copper oxichloride (0.75%) is suggested for the management of anthracnose.

18.2.3   *HOOGHLY WILT*

Ghosh (1961) coined the name "Hooghly wilt" and it is caused by plant pathogenic bacteria *Ralstonia solanacearum* (Mandal, 1986; Mandal and Ghosh, 2002). The disease is most prevalent in the areas where jute is followed by potato or other solanaceaous crops. In India, the disease was observed in the district of Hooghly, parts of Howrah, North 24 Parganas, Burdman and Nadia districts of West Bangal. *Olitorius* jute cv. more prone for bacterial wilt than capsularis jute (Mandal and Khatua, 1986). The benchmark survey estimated 30–34% loss of jute crop each year between 1950 and 1954 (Annonymous, 1949–56). During late eighties and early nineties, 5–37% disease was recorded in Kamarkundu area of Hooghly district and 2–20% in some area of Nadia and North 24 Parganas districts (Mandal and Mishra, 2001).

18.2.3.1 Symptomatology

In jute, Hooghly/bacterial wilt starts with drooping of the leaves at the base and proceeds upwards. The stem on pressing produces slimy, turbid fluid (bacterial ooze) and causing wilt (Figure 18.4).

18.2.3.2 Management

Crop rotation with Potato and other solanaceaous crops are to be avoided this disease. Break cropping system "jute after potato" followed crop

FIGURE 18.4 Bacterial wilt in jute.

rotation with non-host crop viz. paddy and wheat for two years. The diseased plants or plant parts should be rogue out and destroy them, burning the solanaceaous plants and rejecting rotten potato tubers (Mandal, 1986) are important cultural practices to control the disease. Applied organic manure @ 5 ton/ha. By adopting cultural practices particularly the appropriate crop rotation in Hooghly district the disease is came down to 1–2% compared to above 40% in the late eighties (Mandal and Ghosh, 2002). Seed treatment with bevistin (carbendazim) @ 2g/kg of seed and spraying the same fungicide @ 2 g/L of water helps to reduce root rot incidence, which favors the entry of the bacteria.

18.2.4   JUTE LEAF YELLOW MOSAIC (JYLM)

This vector born virus disease was firstly reported by Finlow (1917). Ghosh et al. (2008) reported that the causal agent is a virus belonging to member of begomovirus under family Geminiviridae. The leaf mosaic of jute has widespread occurrence in the major jute growing countries of the world, namely Bangladesh, Burma, India (Ghosh and Basak, 1951). In India, it is a major problem in West Bengal, Orissa, Assam, Bihar and Eastern Uttar Pradesh. It has been reported to be the most important biotic stress of jute cultivation (Harender et al., 1993). It is observed from different surveys that infection reduces plant height to the extent of 20% and thus adversely affects the yield of the fiber (Ghosh et al., 2008). Leaf mosaic infected plants have lower percentage of cellulose, lignin, and pectin, thus the fiber strength becomes weak (Biswas et al., 1989). Biswas et al. (1989) reported that the infected plants raised from infected seeds yielded 16.8–65.9% less fiber. The incidence of the disease has been found to be around 50% on some of the leading *C. capsularis* cultivars.

18.2.4.1   Symptomatology

Infected plant show yellowing of leaves and stunting of the plant height. Variegated appearance due to yellow and light yellow patches on the leave surface. The disease is characterized by symptoms such as small yellow flakes on the lamina during the initial infection stage which gradually increases in size to form green and chlorotic intermingled patches producing a yellow and light yellow patches and mosaic appearance. Leaves may be reduced in size and may be curled. The symptom bearing true leaves crinkled, leathery and sometimes, at the top of the plant, some-what needle like (Figure 18.5). The floral organs are more or less deformed. Internodes and branches become proliferated. These symptoms develop quickly and are more pronounced on younger leaves. Lower leaves are subjected to "mosaic burn" especially during periods of hot and dry weather. Severe infestation of whitefly may result in defoliation of jute and it causes reduction of yield through secretion of wax and honeydew, which significantly reduces the photosynthetic area of the plant (Alam, 1998).

18.2.4.2 Transmission

The disease has been reported to be transmitted through grafts, seed and pollen (Ghosh and Basak, 1951). White fly (*Bemisia tabaci*) transmission of the disease has been reported by many workers (Verma et al., 1966; Ahmed, 1978; Ahmed et al., 1980). The virus-vector relationship is of circulative and non-propagative (no evidence of multiplication of virus in

FIGURE 18.5   Typical leaf mosaic symptom on jute plant.

the insect vector). The whitefly has a minimum acquisition-feeding period (AFP) and minimum inoculation-feeding period (IFP) of 30 Minutes for successful transmission of JLYMV. It was found that 15 viruliferous whiteflies could transmit JLYMV to a range of hundred percent transmission (Dastigeer et al., 2012).

18.2.4.3 Management

Sanitation is one of the common control methods for JLYMV, which includes removing the infected plants and weeds from field. Mishra (1986) reported that cultural control such as water management, soil pH, fertilizer use, weeding, thinning, rouging and removal of infected stubble reduced the mosaic disease incidence. Application of neem cake @ 250 kg/ha before sowing was found effective in controlling whitefly. Neem oil (1%), fish oil resin soap (2.5%) and neem seed kernel extract (NSKE) 5% also gave effective control of whitefly (Veenila et al., 2007). Yellow sticky traps can be used to detect and monitor whitefly activity. Around 3–5 traps should be placed in a block of 2–3 ha. level with the tops of the plants. Baskey (1983) observed that transparent and light blue plastics mulches decreased the number of mosaic-infected plants by 70 and 77%, respectively. Need based, judicious and safe application of pesticides are the most vital tripartite segments of chemical control measures under the ambit of IDM. Seed treatment with Imidacloprid 70% WS @ 1g/kg of jute seeds and use of Imidacloprid @ 0.25 ml/L as spray for insect vector control. Triazophos 40 EC @ 600 g a.i./ha, Ethion 50 EC @ 1000 g a.i./ha and acetamaprid 20 SP @ 30–40 g/ha ware effective against whiteflies (Veenila et al., 2007).

18.2.5 YELLOW VEIN DISEASE OF JUTE

This is newly occurring minor viral disease of jute is noticed in *C. capsularis* cultivars from all jute growing areas of India. Beside India, the disease was observed from many other countries such as Vietnam, Yucatan Peninsula and Mexico. A bipartite begamovirus from Vietnam was identified to be associated with the disease (Cuong Ha et al., 2006). Analysis of the DNA A and DNA B genomic component of this virus showed that it was more similar to New World *Begomoviruses* than to viruses from the Old World and named as Corchorus yellow vein virus (CoYVV).

18.2.5.1 Symptomatology

The first visual symptom of the disease is the clearing bright yellow network of veins, which usually starts at various points near the margins of top leaves (Figure 18.6). In severe cases chlorosis extends to interveinal areas resulting in complete yellowing of leaves. The leaves become reduced in size.

18.2.5.2 Transmission

CoYVV is transmitted by White fly (*Bemisia tabaci*) in circulative, non-propagative manner. The vector acquires the virus after feeding on an infected plant for at least 15 to 30 minutes. There is latent period of several hours (more than 20 hours) after which the virus can be inoculated into a healthy plant. Whiteflies remain viruliferous for about 2 weeks.

FIGURE 18.6   Yellow vein symptom on jute leave.

18.2.5.3   Management

Destruction of alternative hosts and weed plants, uprooting and burying of infected plants and control of whiteflies through insecticide to reduce damage caused by virus. Seed treatment with imidacloprid 70% WS @ 1g/kg of jute seeds and use of neem cake @ 250 kg/ha before sowing of the crop was much effective against whiteflies. Application of insecticides like Metasystox (0.02%, Rogor (0.05%) and Imidacloprid @ 0.25 ml/L was found effective in controlling whitefly.

18.2.6   *ROOT KNOT NEMATODE*

Root-knot of jute caused by *Meloidogyne* spp. is one of the most important diseases of jute affected at various stages of growth.

18.2.6.1   Symptomatology

Root knot nematode infested plants produce knot like globular swelling in roots (Figure 18.7). As a result of this, translocation of water and

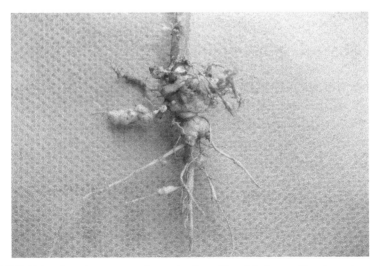

FIGURE 18.7   Root knot nematode causing knot like globular swelling in roots.

nutrients is blocked. Infected plants show yellowing and stunting of the plant. Hot and humid climate is a suitable for the predominantly occurring *Meloidogyne javanica* and *Meloidogyne incognita.* Chattopadhary and Sengupta (1955) reported that stunting, wilting, defoliation and death to the attack of root knot nematode, *Meloidogyne javanica.* Ahmed and Timm (1961) found that *M. incognita* and *M javanica* were causal organisms of root knot of jute.

18.2.6.2   Management

A number of nematicides were tried for reducing the infestation of root knot nematodes in jute of which Thiometon, Nematox and Nemagon are important. Various organic amendments, namely, cakes of karanj, mahua, groundnut castor cowdung manure, etc., were tried for checking the nematode infestation. Sound cultural practices viz. removal of plant debris from the field, weeding, thinning, deep plowing during summer, crop rotation with paddy and wheat for two years reduced nematode population in jute field. Screening of resistant lines against root knot nematodes results to develop few tolerant lines in both species of jute (Laha et al., 1995a, 1995b).

18.3   SUNNHEMP DISEASES

Sunnhemp is not affected by too many diseases and pests. Sunnhemp is attacked by diseases viz. anthracnose, caused by *Colletotrichum curvatum* Briant and Martyn (Mitra, 1934; Whiteside, 1955), wilt caused by *Fusarium udam* f. sp. *crotalariae* (Mitra, 1934; Kundu, 1964).

18.3.1   ANTHRACNOSE

Anthracnose is the most serious diseases of sunnhemp grown in India caused by *Colletotrichum curvatum.* It occurs in all sunnhemp-growing areas of India.

18.3.1.1   Symptomatology

First infection of the disease is appeared on cotyledons and fallowed by stem and growing point. Soft discolored areas on the cotyledons are the characteristic symptoms of this disease. Later brownish spot are formed on all part of host except underground parts. The affected seedling drops from the point below cotyledon. The infected seedlings usually die. When older plants are infected the disease restricted on leaf and stem and the badly infected leaves fall off. The spot on older leaves appear on one side of the leaf but gradually enlarge and extent to the opposite side. These spot are grayish brown to dark brown, round or irregular. Several spots coalesce and cover the entire leaves or large portion of leaf.

18.3.1.2   Disease Cycle and Epidemiology

Cloudy weather and continues rain favor the disease to spread quickly in thickly populated crop. The infection was severe during the seedling stage. Rain splash helps the spore of the fungus with in the field. The pathogen required high rainfall, higher temperature (30–32°C) and high relative humidity favor infection.

18.3.1.3   Management

Seed treatment with carbendazim @ 2g/kg seed to check the seed borne infection. Secondary infection is checked by foliar spray with carbendazim @ 1.5%. Dey et al. (1990) reported that sources of resistance to anthracnose have been identified, indicating the potential to reduce disease losses through the development of anthracnose resistant varieties.

18.3.2   *FUSARIUM WILT*

Wilt in sunhemp is caused by *Fusarium udum* f. sp. *crotolar*iae is an important disease of sunhemp. Incidence of the disease is about of 10–12% but it may be 60–70% under favorable condition. Average loss due to this disease is 11–15% (Bandopabhyay et al., 1982).

18.3.2.1 Symptomatology

Characteristic symptoms of the disease is the plant gradually withers, droops down and ultimately dies with in a days. It attacks the young plants causing wilt and necrosis and in the older plants causing yellowing of the leaves which leads to eventual necrosis. It was stated that wilt is caused by continuous planting of sunnhemp on the same land and recommended crop rotation (Medina, 1959).

18.3.2.2 Disease Cycle and Epidemiology

The fungus survives in the soil as well as in crop residues as facultative parasite for long period of time. It cause infection in the plant through thinner roots, rootlets and even through the cracking in the basal portion of the stem with germtubes arising from micro or macroconidia or chlamydospores reaching the vascular tissues to establish and multiply rapidly, causing wilting of parts or all the plant. When the crop is harvested, the plants are cut at the stem base, leaving the entire root system and stubble to the soil. As a saprophyte, the fungus continues to multiply in soil and remains there until the next crop is grown. If the crop is sown every year in the same field the fungus builds up, increasing the disease incidence. The fungus was also noticed on the pod and in many cases in the seed of diseased pod. The infected discolored seed also initiate the infection in the field. The incidence increase with decreasing temperature, prolonged drought, low soil moisture and increasing crop age.

18.3.2.3 Management

Crop rotation, clean cultivation and seed treatment with fungicides and prophylactic spray of fungicides are the most effective and feasible remedy. Application of neem cake (25q/ha) and zinc sulfate (25q/ha) with *Rhizobium japonicum* culture inoculated seed was found better for reducing wilt up to 40% and increased 12% flower, pod and fiber production in sunnhemp (Bandopadhyay, 2002). Seed treatment with carbendazim

@ 3 g/kg of seed and spraying with carbendazim @ 2 g/L of water found most effective (Bandopadhyay, 2002). Seed treatment with carbendazim @ 2g/kg seed or *Trichderma viride* @ 5g/kg seed reduce the disease at seedling stage. Showing of early variety like K12 yellow and SH4 is tolerant against this disease.

18.3.3   SUNNHEMP MOSAIC

Sunnhemp mosaic disease is caused by Sunnhemp mosaic virus which is transmitted by silver leaf whitefly (*Bemisia tabaci*), causes yellowing and crinkling of the leaves and weak stems that lodge easily, resulting in low fiber yields. The virus particle is rod shaped particle of 300 nm long and 17 nm wide. Sunnhemp mosaic virus (SMV) has few antigenic determinants in common with strain of tobacco mosaic virus or with other well studied of *Tobamovirus*.

18.3.3.1   Symptomatology

Characteristic symptoms of the disease are mottling, severe mosaic, puckering, and malformation with enations on the undersides of the leaves. As the disease progressed patches of light and dark green areas become more prominent. Infected plants are shorter in height (Capoor, 1962). Diseased leaves were smaller than the normal. In severely affected leaves mesophyll is incompletely differentiated, chloroplasts remain indistinct and a few phloem cells are hypertrophied, the affected plants become dwarf and produce a few seeds.

18.3.3.2   Management

As the virus is transmitted and having wide host range, the disease may initiate from wild or weed hosts. Sanitation is one of the common control methods for SMV, which includes removing the infected plants and weed host from field. April sowing of tolerant variety like K-12 yellow and SH-4 largely escapes the disease.

Other diseases of sunhemp of minor importance are rust (*Uromyces decoratus*), *Sclerotinia* rot, powdery mildew (*Oiduim* sp.), wilt (*Ceratocystis fimbriata*), and twig blight (*Choanephora cucurbitarum*).

18.3.3.3   Conclusions and Future Perspective

Developments of integrated disease management (IDM) modules for the management of jute and sunhemp diseases have been utter importance for the production of these fiber crops under changing climatic scenario. Earlier research was mainly focused on resistant sources and chemical control of few diseases. Now the major emphasis is on identifying, evaluating and integrating location specific components of IDM. IDM packages of food legumes have been successfully refined and validated in partnership with stakeholders and end users. The chemical pesticides when applied in higher doses or frequently in jut field for management of diseases give residual effects which have adverse effect on soil microbiota and below underground diversity Therefore, in perspective to changing climatic scenario it is very important to develop greener and compatible management strategies for jute and sunnhemp diseases in future will be required.

KEYWORDS

- fiber
- jute
- mildew
- rot
- sunhemp
- virus
- wilt

REFERENCES

1. Ahmed, M. (1978). A white fly vectored yellow mosaic of jute. *FAO Plant Prot. Bull.*, 26, 169–171.

2. Ahmed, Q. A. (1966). Problems in jute plant pathology. *Jute Fab. Pak.* 5, 211–213.
3. Ahmed, Q. A., Timm. (1961). Studies on wilting of jute. The influence of root knot and lance nematodes for growth and wilting of jute. *Pak. Jour. Biol. & Agric. Sci.* 3, 19–21.
4. Ahmed, Q. A., Biswas, A. C., Farukuzzaman, A. K. M., Kabir, M. Q., Ahmed, N. (1980). Leaf mosaic disease of jute. *Jute Fabrics Bangladesh* 6, 9–13.
5. Akanda, M. A. M., Fakir, G. A. (1985). Prevalence of major seed-borne pathogens of jute. *Bangladesh, J. Plant Pathol.* 1, 76–76.
6. Alam, M. (1998). Effectiveness of three insecticides for the control of the spiraling white fly, *Aleurodicus dispersus*, (Homoptera: Aleyrodidae) of guava. *Bangladesh, J. Entomol.* 8, 53–58.
7. Ammon, V. Wyllie, T. D., Brown, M. F. (1974). An ultrastructural investigation of pathological alterations induced by, M. phaseolina (Tassi) Goid. in seedlings of soybean, *Glycine max* (L.) Merril. *Physiol. Plant Pathol.* 4, 1–4.
8. Annonymous, (1949–56). Annual Report, Jute Agriculture Research Institute (JARI), Barrackpore.
9. Anonymous, (1990–2006). *Annual Report.* Central Research Institute for Jute and Allied Fibres, Barrackpore.
10. Anonymous, (1999). Fifty years or research on jute and allied fibers agriculture (1948–1997). Central Research Institute for Jute and Allied fibers (CRIJAF), Barrackpore.
11. Bandopadhaya, A. K. (2002). A current approach to the management of root diseases in bast fiber plants with conservation of natural and microbial agents. *J. Mycopathol. Res.* 40 (1) 57–62.
12. Bandopadhyay, A. K., Prakash, G., Som, D. (1982) Screening of germplasm of sunhemp against Fusarium wilt disease. *Bangladesh j. Bot.* 11 (1) 14–16.
13. Baskey, Z. 1983. The effect of reflective mulches on virus infection in seed cucumber. Zoldsegtermesztesi Kutato Intezet Bulletinje, 16, 23–31.
14. Biswas, A. C., Asaduzzaman, M., Sultana, K., Taher, M. A. (1989). Effect of leaf mosaic disease on loss of yield and quality of jute fiber. *Bangladesh, J. Jute Fiber Res.* 14, 43–46.
15. Capoor (1962). Crotalaria juncea, Southern sunhemp mosaic virus. *Phytopathol.* 52, 393.
16. Chattopadhay, S. B., Raj, S. K. (1978). Role of hydrolytic enzyme in seedling blight of jute incited by *Macrophomina phaseolina. Madras Agric. J.* 65 (5), 320–324.
17. Chattopadhay, S. B., Sengupta, S. K. (1955). Root knot disease of jute in West Bengal. *Curr. Sci.* 24, 276–277.
18. Chaudhury, J., Singh, D. P., Hazra, S. K. (1978). Sunnhemp. Central Research Institute for jute and allied fibers (ICAR). Accessed June 10, 2007.
19. Cuong Ha., Coombs, S., Revill, P., Harding, R., Vu, M., Dale, J. (2006). Corchorus yellow vein virus, a New World Gemini virus from the Old World. *J. Gen. Virol.* 87, 997–1003.
20. Dastogeer, K. M. G., Ashrafuzaman, M., Ali, M. A. (2012). The Virus-Vector Relationship of the Jute Leaf Mosaic Virus (JLMV) and its Vector, *Bemisia tabaci* Gen. (Hemiptera: Aleyrodidae). *Asian Journal of Agricultural Sciences* 4 (3), 188–192, 2012.
21. Dey, D. K., Banerjee, K., Singh, R. D. N., Kaiser, S. A. K. M. (1990). Sources of resistance to anthracnose disease of sunnhemp. *Environ. Ecol.* 8, 1217–1219.

22. Dubey, R. C., Dwivedi, R. S. (1991). Fungitoxic properties of some plant extracts against vegetative growth and sclerotial variability of *Macrophomina phaseolina*. *Indian Phytopathol.*, 44, 411–413. Ghosh, T and Bask, M. N. (1965). Possibility of controlling stem rot of jute. *Indian Journal of Agriculture Sciences* 35, 90–100.

23. Fakir, G. A., Khan, A. A. (1992). Control of some selected seed borne fungal pathogens of jute by seed treatment with garlic extract. *Proc. BAU Res. Prog.*, 6A: 176–180.

24. Finlow, R. S. (1917). Annual reports of the department of agriculture. Bengal chlorosis. *Bengal Agric. J.* 4, 118–118.

25. Ghosh, R., Paul, S., Das, S. Palit, P., Acharyya, S., Das, A., Mir, J. I. Ghosh, S. K., Roy, A. (2008). Molecular evidence for existence of a New World begomovirus associated with yellow mosaic disease of Corchorus capsularis in India. *Aust. Plant Dis. Note* 3, 59–62.

26. Ghosh, T. (1957). Anthracnose of jute. *Indian phyropathology.* 10, 63–70.

27. Ghosh, T. (1961). Stidies on disease of jute caused by *Macrophomina phaseolina* (Maubi) Ashby. PhD Thesis, Calcutta University.

28. Ghosh, T., Basak, M. (1951). Chlorosis of jute. *Sci. Cult.* 17(6), 262–264.

29. Ghosh, T., Bask, M. N. (1965). Possibility of controlling stem rot of jute. Indian journal agriculture Sciences 35, 90–100.

30. Harender, R., Bhardwaj, M. L., Sharma, I. M., Sharma, N. K. (1993). Performance of commercial okra (*Hibiscus exculentus*) varieties inrelation to disease and insect pests. *Indian, J. Agricul. Sci.* 63 (11), 747–748.

31. Hossain, I., Ashrafuzzaman, H., Khan, M. H. H. (1993). Biocontrol of Rhizoctonia solani and Bipolaris sorokiniana *Proc. BAU Res. Prog.*, 7A: 264–269.

32. Islam. F. (1987). Study of transmission of seed-borne fungal pathogens from seed to plant to seed (*Corchorus capsularis* L.). M.Sc. Thesis, Department Plant Pathology, Bangladesh Agriculture University Mymensingh, Bangladesh.

33. Khan, S. R., Strange, R. N. (1975). Evidence of the role of a fungal stimulant as a determinant of differential susceptibility of jute cultivars to *Colletotrichum corchori*. *Physiol. Plant Pathol.* 5, 157–164.

34. Kundu, B. C. (1964). Sunn-hemp in India. *Proc. Soil Crop Soc. Florida.* 24, 396–404.

35. Laha, S. K., Pradhan, S. K., Sasmal, B. C., Dasgupta, M. K. (1995b). *Nematode Medit.* 23 (1), 51–52.

36. Laha, S. K., Mandal, P. K., Dasgupta, M. K. (1995a). Studies on the varietal susceptibility of jute against root knot nematodes under field conditions. *Inndian Biol.* 26 (2), 27–29.

37. Mandal, R. K. (1986). Wilt disease of jute and its management (in Bangali) Sabuj sona, March issue.

38. Mandal, R. K. (1990). Jute diseases and their control. In "Proceeding of National workshop cum training on jute, mesta, sunhemp and ramie." CRIJAF, Barrackpore.

39. Mandal, R. K. (1990). Jute diseases and their control. In: Proceedings of National Workshop cum Training on Jute, Mesta, Sunn hemp and Ramie. Central Research Institute for Jute and Allied Fibres, Barrackpore, West Bengal, India.

40. Mandal, R. K., Ghosh, T. (2002). Hooghly Wilt of jute in retrospect (1949–2000). *J. Mycopathol. Res.* 40 (1), 67–69.

41. Mandal, R. K., Khatua, D. C. (1986). Incidence of two important diseases of jute in high rainfall areas. *Jute Dev. J. Agril.* June issue.
42. Mandal, R. K., Mishra, C. D. (2001). Role of different organism in inducing Hooghly wilt symptom in jute. *Environ. Ecol.* 19 (4), 969–972.
43. Mishra, C. B. P. (1986). Strategies for control of jute and allied fiber diseases. *Pesticides* 20, 19–21.
44. Mitra, M. (1934). Wilt disease of *Crotalaria juncea* Linn. (sunn-hemp). *Indian, J. Agr. Sci.* 4, 701–714.
45. Roy, A., De, R. K., Ghosh, S. K. (2008). Diseases of bast fiber crops and their management in jute and allied fibers, pp. 327. In: Karmakar, P. G., Hazara, S. K., Subramanian, T. R., Mandal, R. K., Sinha, M. K., Sen, H. S (Eds.). Updates Production Technology, Central Research Institute for Jute and Allied Fibres, Barrackpore, West Bengal, India.
46. Short, G. E., Wyllie, T. D., Bristow, P. R. (1980). Survival of *M. phaseolina* in soil and residue of soybeans *Phytopathology* 70, 13–17.
47. Sinha, M. K., Sengupta, D., Sen, H. S., Ghosh, T. (2004). Jute and jute-like fibers: *current situation. Sci. Cult.* 70 (1–2), 32–37.
48. Som, D. (1977). Recent concept on jute diseases and control measures. *Jute Bull.* April-March issue. pp. 1–4.
49. Srivastava, R. K., Singh, R. K. (2009). Evaluation of consortium of bio-agents for the management of *Macrophomina phaseolina* disease complex in the jute (*Corchorus olitorius*). 5th International Conference on Plant Pathology in the Globalized Era, Nov. 10–13, 2009, Organized by Indian Phytopathological Society, Indian Agricultural Research Institute, New Delhi, India. Abstract 656 (S-15), 351.
50. Srivastava, R. K., Singh, R. K., Kumar, N., Singh, S. (2010) Management of *Macrophomina* disease complex in jute (*Corchorus olitorius*) by *Trichoderma viride. Journal of Biological Control* 24 (1), 77–79.
51. Talukder, D., Khan, A. R., Hasan, M. (1989). Growth of *Diacrisia oblique* [Lepidoptera: Arctiidae] with low doses of *Bacillus thuringiensis* Var. Kurstaki. *Entomophaga* 34 (4), 587–589.
52. Thakur Ji, (1974). Influence of NPK on stem rot of Capsularis jute. *Indian, J. Mycol. Plant pathol.* 4, 117–120.
53. Veenila, S., Birendra, V. K., Sabesh, M., Bambawale, O. M. (2007). Know Your Cotton Insect Pest Whiteflies. *Crop Protection folder series*: 4 of 11, pp. 1–2.
54. Verma, P. M., Rao, G. G., Capoor, S. P. (1966). Yellow mosaic of Corchorus trilocalaris. *Sci. Cult.*, 32, 466–466.
55. Whiteside, J. O. (1955). Stem break (*Colletotrichum curvatum*) of sunnhemp in southern Rhodesia. *Rhodesian Agr. J.* 52, 417–425.
56. Wyllie, T. D. (1988). Charcoal rot of soybean-current status In Soybean Diseases of the North Central Region. Edited by Wyllie TD, Scott DH. APS, St. Paul; 106–113.

CHAPTER 19

MESTA PATHOSYSTEMS: AN OVERVIEW

A. N. TRIPATHI,[1] R. K. DE,[1] P. N. MEENA,[1] V. RAMESH BABU,[1]
Y. R. MEENA,[3] S. KUMAR,[2] and V. K. SINGH[3]

[1]Division of Crop Protection, ICAR-Central Research Institute
for Jute and Allied Fibers, Barrackpore, Kolkata–700120, West
Bengal, India

[2]Department of Plant Pathology, Bihar Agricultural University,
Sabour–813210, Bhagalpur, Bihar, India;
E-mail: antripathi_patho@rediffmail.com

[3]Jute Research Station (Katihar), Bihar Agriculture University, Sabour,
Bihar, India

CONTENTS

19.1   INTRODUCTION

Mesta (*Hibiscus cannabinus* [kenaf] and *H. sabdariffa* [roselle]) belongs to the family Malvaceae. Mesta is a third most important bast fiber producing commercial crops grown in India. Mostly two species of *Hibiscus* namely *sabdariffa* (Roselle) and *cannabinus* (Kenaf) are cultivated for fiber yield. Mesta is cultivated in many countries and geographically distributed in Africa, Australia, Brazil, Caribbean Islands, Egypt, Hawaii, Saudi Arabia, Sudan, Gambia, India, Indonesia, Latin America, Malaysia, Mali, Myanmar, Namibia, Nigeria, Panama, Philippines, Senegal, Congo, France, Thailand, and United States of America. In India, it is cultivated mainly in the states of Assam, Andhra Pradesh, Bihar, Madhya Pradesh, Maharashtra, Orissa, Punjab, Tamil Nadu, Uttar Pradesh and West Bengal (Mahadevan et al., 2009). In India its cultivation occupies 1.5 lakh hectares acreage and production 8.5 lakh bales (1 bale = 180 kg) with productivity of 11 q/ha. The highest productivity of mesta was recorded in West

Bengal 17 q/ha. Acreage, productivity and production of Mesta started declining over the years due to high incidence of pests and disease and non-availability of effective integrated disease management strategies.

19.2 ECONOMIC IMPORTANCE

Mesta fiber is also used for making ropes, twines, fishing nets and also in the paper pulp from stalks especially for fine paper, structural boards. It is also used for production of value added products of food and beverage like tea, syrup, jams and jellies, vegetable sauce, seed oil in soap, paints, cosmetics industries and cut flowers for the export.

19.3 PATHOSYSTEM OF MESTA

Common pathosystems of mesta are reported as foot and stem rot, white stem rot, leaf blight and leaf spot, *Fusarial* wilt, bacterial wilt, mosaic and leaf curl and Phyllody (Biswas et al., 2011; Eslaminejad et al., 2012; Biswas et al., 2013) (Table 19.1; Figures 19.1–19.8).

19.3.1 FOOT AND STEM ROT

Foot and stem rot of mesta caused by *Phytophthora parasitica var. subdariffae* is the most important disease of mesta (Ghosh, 1983). In general

TABLE 19.1 Different Mesta-Pathosystem

S. No.	Diseases	Pathogen
1	Foot and Stem rot	*Phytophthora parasitica* var. *subdariffae*
2	Leaf blight and leaf spot	*Phoma sabdariffae*
3	White stem rot	*Sclerotinia sclerotiorum*
4	Collar rot and Wilt	*Sclerotium rolfsii*
5	Root knot	*M. incognita/M. javanica*
6	Bacterial wilt	*Ralstonia solanacearum*
7	Yellow mosaic	Mesta yellow vein mosaic virus (MYVM V)
8	Phyllody	Phytoplasmas

FIGURE 19.1 White Stem Rot and Sclerotia on stem (Symptoms of mesta diseases).

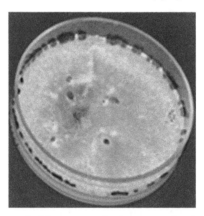

FIGURE 19.2 Colony of *Sclerotinia sclerotiorum* and sclerotia on culture plate (Symptoms of mesta diseases).

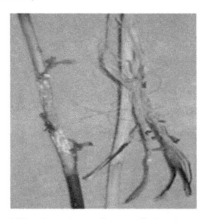

FIGURE 19.3 Exposed fiber (symptoms of mesta diseases).

FIGURE 19.4   *Phoma* Leaf Spot (symptoms of mesta diseases).

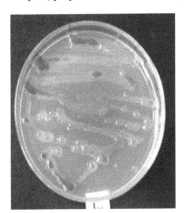

FIGURE 19.5   Virulent colonies of *Ralstonia solanacearum* [fungal diseases (colonies of *Ralstonia solanacearum*) bacterial wilt pathogen].

FIGURE 19.6   Yellow Vein Mosaic (virus diseases).

FIGURE 19.7   Leaf Curl (virus diseases).

FIGURE 19.8   Phyllody.

fiber yield losses due to this disease is reported upto 25%. In severe cases more than 40% crop losses was documented (De and Mondal, 2007). Disease development is favored by high humidity (70–93%) and temperature (24–33°C).

19.3.1.1   Symptom

Disease symptoms were appeared in the form of water soaked lesions at the lower portion of the plant stem. These light brown colored lesions are spread in the infected portion of the stem. A typical black rot spreading was occurred on the basal stem upto 1m above the ground level. Severely affected plants showings stem breaking and look ash gray in color. Partial rotting of basal stem is also observed in some of the plants.

19.3.1.2   Management

Deep summer plowing. Maintenance of proper field drainage to prevent water logged condition. Seed treatment with Mancozeb 75% WP or Metalaxyl 35% WS @ 3 g/kg or *Pseudomonas fluorescens* @ 20g/kg seed. Foliar spraying with ridomylmz 72% WP @ 0.2%.

19.3.2   *LEAF BLIGHT AND LEAF SPOT*

Phoma sabdariffae causes leaf spot on mesta. Disease Symptom can be observed on early and late stage of the crop. Infection started at the leaf apex in seedlings as a small spot which gradually spread towards the petiole along the midrib. In some cases the infection initiates from the base of the lamina and moves along the sides of the midrib towards the apex and become rotted.

19.3.2.1   Management

Spraying of mancozeb 75% or copper oxychloride 50% WP @ 0.3%.

19.3.3   *WHITE STEM ROT*

White stem rot of mesta was recorded first time in seed crops with incidence upto 50% at CRIJAF Research Farm, Barrackpore, West Bengal.

The disease appeared in severe form during the winter season particularly in the month of December to January. White stem rot of mesta caused by *Sclerotinia sclerotiorum*. The pathogen infects number of economically important plants including mesta. Pathogen produces typical large irregular black colored rat dropping like sclerotia (0.5–1.5 cm), on diseased stems and bolls/capsules and on culture media (Tripathi et al., 2013). The pathogen is survive in the form of sclerotia in the soil/infected plant debris and with falling of temperature in the month of December-January, it germinates and developed apothecia in which ascospores are formed. Upon discharge the ascospore causes new infection in the crop.

19.3.3.1 Symptom

The disease symptoms were appeared as water soaked lesion on stem which later turn into brown colored patches. Initially the portion of the stem above or below the patches looks healthy. Finally the infection girdles the stem completely, which extend as much as a foot or more. The rot causes the tissues to become soft and easily peel off into shreds. The portions above the affected part may ultimately wilted die backed and break away. The surface of the affected parts is covered with white stands of fungus mycelium, which form mycelial mat/cushion in the axils of the branches. Black colored sclerotia also imbedded on this mycelial mat. The pith region filled with scleroria. The sclerotia were also noticed in bolls, which contaminate seed lots.

19.3.3.2 Management

Foliar spraying with Azoxystrobin 23% SC (0.08%).

19.3.4 COLLAR ROT AND WILT

Sclerotium rolfsii causes collar rot of mesta near the soil line. Pathogen produced cottony mycelial growth with black colored mustered like sclerotia on stem. Vascular wilt of Roselle caused by *F. oxysporum* was reported in Malaysia by Ooi and Salleh (1999) and Ooi et al. (1999).

In general important fungal spp. were infect mesta crop and causing root rot and wilt are; *Fusarium decemcellulare, F. sarcochroum, F. solani, F. vasinfectum, Phoma sabdariffae, P. exigua, Phymatotrichum omnivorum, Phythophtora parasitica, P. terretris, Pythium perniciosum* and *Rhizoctonia solani* (Eslaminejad and Zakaria, 2011).

19.3.4.1 Management

Spraying with copper oxychloride 50% WP (0.5%) and azoxystrobin 23% SC (0.1%).

19.3.5 ROOT KNOT

Mesta has been seriously attacked by root-knot nematodes such as *Meloidogyne arenaria, M. incognita* and *M. javanica* which causing qualitative and quantitative losses.

19.3.5.1 Management

Granular nematicide viz. aldicarb 10G @ 1 kg a.i./ha, carbofuran, @ 1kg a.i./ha and soil amendment with neem cakes powder @ 10,000 kg/ha. have been reduced nematode population and remarkable increase crop performance.

19.3.6 BACTERIAL WILT

A bacterial disease has been reported on Roselle plants caused by *Ralstonia solanacearum*. Bacterial wilt pathogen *Ralstonia solanacearum* is a gram negative; rod shaped, mesophilic and non-fluorescent bacteria. Yabuuchi et al. (1995) proposed new genus *Ralstonia* for *Pseudomonas/ Burkholderia. R. solanacearum* comprising 5 races and 6 biovars (Hayward, 1994; 1991). Virulent isolates of *R. solanacearum* forms white and irregular fluidal/mucoid colonies with pinkish center and produce abundant extracellular polysaccharide (EPS) on Kelmans triphenyl tetrazolium chloride (TZC) medium (Kelman, 1995; Tripathi and Sood, 2005). A virulent type produces round, nonflidal/nonmucoid dark red colonies and are deficient in production of EPS. The colony morphology showed wide phenotypic plasticity and its associated collective alterations are called "phenotype conversions (PC)." *R. solanacearum* is a soil-borne pathogen. Inoculum survival of *R. solanacearum* in soil depending on soil type and physico-chemical properties, soil pH reaction (pH 4.3–6.8), cropping sequences practiced and the native weed flora present. From infected plants, secondary spread over short, medium and long distances takes place by root contacts, flood/irrigation water, farm implements, storage and packaging material. High soil moisture content favoring survival and inoculum production; high soil organic matter content leading to decline in pathogen population; high temperatures decreasing pathogen population; presence of alternate hosts favoring survival as the pathogen.

19.3.6.1 Symptom

Symptoms of the disease were drooping of leaves, wilting of plants and brownish discoloration of the vascular tissue. The disease may damage the crop due to premature/mature wilting of crop.

19.3.6.2 Management

Wide host range, exceptional survival ability, high level of genetic diversity, cultural reversion and genome/plasimid plasticity, complex etiology and epidemiology of bacterial wilt pathogen *R. solanacearum* make it a very difficult to manage. Information on the ecology of the disease and variation in the pathogen is important in establishing rational disease management strategies, including the development of resistant varieties, shifting of cropping pattern and appropriate cultural management. The better understanding of ecology, biology and epidemiology of bacterial wilt pathogen is play very important role for more rigorous studies on the disease and formulation of location specific integrated disease management strategies. The various core component of IDM approach described as below:

Avoidance of the disease is possible through the use of crop rotation with non-host plants. Manipulation of soil pH through soil amendment and soil disinfection by using chemicals like Bordeaux mixture and copper sulfate have been controlled bacterial wilt in some extent (Kelman, 1953). Certain bacteria like *Pseudomonas fluorescens, Bacillus polymyxa, Bacillus* spp. and actinomycetes delayed wilt development and reduced incidence of bacterial wilt. Use of Chemicals for management of bacterial wilt has been attempted by many workers but it was found non effective and non-feasible. Use of resistant varieties is on of the best and effective way for managing the disease. Thus, breeding for disease resistance is the most appropriate solution. There should be much more emphasis on local breeding programs to identify material suited for particular ecosystems.

19.3.7 MESTA YELLOW VEIN MOSAIC (MYVM)

Mesta yellow vein mosaic disease (MYVM) was first time documented in Bahraich district of Uttar Pradesh and has gradually spread fast in several other areas of the country, incurring heavy losses. Now it is considered as an important limiting factor for mesta cultivation. The causal agent of the disease has been identified as monopartite begomovirus associated with satellite DNA b (Chatterjee et al., 2005; Paul et al., 2006). The disease is transmitted by cleft grafting and whitefly (*Bemisia tabaci*). The full-length sequence of DNA-A and DNA b of the virus infecting mesta was cloned and sequenced. DNA-A had 2728 nucleotides in length having 83.5% identity with Cotton leaf curl Bangalore virus (CLCuBV) and 83.3% identity with CLCuRV. There were six-conserved ORFs identified in both orientations of the sequence. Sequence analyzes proved that this new DNA-A was typical of Old World begomoviruses. It was therefore identified as a separate species of Begomovirus and was named as Mesta yellow vein mosaic virus (Chatterjee and Ghosh, 2007).

19.3.7.1 Symptom

Infected plants showed stunted growth and abundant pinhead spots formed on the leaf lamina, including veins. These spots gradually enlarge and coalesce to form chlorotic to yellow flecks. Affected leaves become leathery, smooth and reduced in size. Diseased plants defoliated, mature late and bear very few flowers and capsules.

19.3.7.2 Management

Sowing of mesta in first fortnight of June and spraying with thiamethaxam (0.1 g/L) or imidacloprid (0.25 ml/L) at 50 days after sowing to check the white fly which is vector for the disease.

19.3.8 MESTA PHYLLODY

The phytoplasmas are group of plant pathogenic cell wall-less, phloem-inhabiting prokaryotes in the class mollicutes (Hoat et al., 2012). They have been found in more than 1,000 plant species worldwide including several economically important crops, vegetables, fruit and fiber crops, ornamental plants, weeds, and timber and shade trees and cause devastating damage to plants by loss in biomass and quality of plant products (Camerota et al., 2012; Li et al., 2012; Mitrović et al., 2012; Win and Jung, 2012; Biswas et al., 2013).

19.3.8.1 Symptoms

Infected plants were stunted. Leaves deformed due to hypotrophic growth and these were turned into clusters. Leave margins were showed reddening. Stems were malformed and proliferated in to many broom like small and thick secondary branches.

19.4 CONCLUSIONS AND FUTURE PERSPECTIVE

India has comprised a varied pathometerological parameters and diverse agroclimatic conditions from tropical, subtropical to temperate. Mesta is a third most important bast fiber crop in India. Disease free seed play important role for pathogen free conservation and exchange of germplasm for various crop improvement program. Molecular and Serodiagnostic techniques are potential tool for development of reliable, sensitive and quick diagnostic for detection of the plant pathogens. Information on the ecology of the disease and variation in the pathogen is important in establishing national disease management strategies. Accurate and quick diagnosis of viral pathogens is imperative for the success of mapping of epidemics, breeding for resistance, and developing quarantine and other control measures (Nene, 2006; Rishi, 2006). Research needs on the ecology and epidemiology of the disease to be focused. Better understandings of molecular genetics elucidate the genetic basis of pathogenicity and variability in pathogens. Such studies may play a much more important role in the future for development of location specific sustainable diseases management modules.

KEYWORDS

- **fiber**
- **management**
- **phyllody**
- **rot**
- **spot**
- **wilt**

REFERENCES

1. Biswas, C., Dey, P., Satpathy, S. Kumar, M., Satya, P., Mahapatra, B. S. (2013). Phytoparasitica Molecular identification of a Candidatus phytoplasma (Group-16SrV-D) coding partial uvrB gene and degV gene on a new host – mesta (*Hibiscus sabdariffa*) – with phyllody and reddening of leaves in India. *Phytoparasitica*: DOI 10.1007/s12600-013-0314-0.
2. Biswas, C., Sarkar, S. K., De, R. K. (2011). Diseases of jute and mesta: Present status and management options. In, S. K. Biswas, S. R. Singh (Eds.), Sustainable disease management of agricultural crops (pp. 62–79). New Delhi, India: Daya Publishing House.

3. Camerota, C., Raddadi, N., Pizzinat, A., Gonella, E., Crotti, E., Tedeschi, R., et al. (2012). Incidence of 'Candidatus Liberibacter europaeus' and phytoplasmas in Cacopsylla species (Hemiptera: Psyllidae) and their host/shelter plants. *Phytoparasitica*, 40, 213–221.

4. Chatterjee, A., Ghosh, S. K. (2007). A new monopartite begomovirus isolated from Hibiscus cannabinus, L. in India. *Arch Virol* 152, 2113–2118.

5. Chatterjee, A., Roy, A., Padmalatha, K. V., Malathi, V. G., Ghosh, S. K. (2005) Occurrence of a Begomovirus with yellow vein mosaic disease of mesta (Hibiscus cannabinus and Hibiscus sabdariffa). *Australas Plant Pathol*, 34, 609–610.

6. De, R. K., Mondal, R. K. (2007). Effect of seed treatment with fungicides on foot and stem rot disease caused by *Phytophthora parasitica* var. *sabdariffae* in *Hibiscus sabdariffa*. *J. Interacademicia*. 11(2), 161–165.

7. Eslaminejad, T., Zakaria, M. (2011) Morphological characteristics and pathogenicity of fungi associated with Roselle (*Hibiscus Sabdariffa*) diseases in Penang, Malaysia. *Microbial Pathogenesis*. 51(5), 325–37, Epub 2011/08/16.

8. Ghosh, R., Paul, S., Das, S., Palit, P., Acharyya, S., Das, A., Mir, J. I., Ghosh, S. K., Roy, A. (2008). Molecular evidence for existence of a New World begomovirus associated with yellow mosaic disease of Corchorus capsularis in India. *Aust. Plant Dis. Note* 3, 59–62.

9. Ghosh, T. (1983). Handbook on jute. FAO. Plant Production and Protection paper. 51, 219.

10. Ghosh, T., Basak, M. (1951). Chlorosis of jute. *Sci. Cult.* 17(6), 262–264.

11. Harender, R., Bhardwaj, M. L., Sharma, I. M., Sharma, N. K. (1993). Performance of commercial okra (*Hibiscus exculentus*) varieties inrelation to disease and insect pests. *Indian, J. Agricul. Sci.* 63 (11), 747–748.

12. Hayward, A. C. (1991). Biology and epidemiology of bacterial wilt caused by Pseudomonas solanacearum. *Annu. Rev. Phytopathol.* 29, 65–87.

13. Hayward, A. C., Hartman, G. L. (1994). Bacterial Wilt: The Disease and its Causative Agent, Pseudomonas solanacearum, CAB International Wallingford, United Kingdom, 259 p.

14. Hoat, T. X., Bon, N. G., Quan, M. V., Hien, V. D., Thanh, N. D., & Dickinson, M. (2012). Detection and molecular characterization of sugarcane grassy shoot phytoplasma in Vietnam.. *Phytoparasitica*, 40, 351–359.

15. Kelman, A. (1953). The bacterial wilt caused by Pseudomonas solanacearum. *Tech Bull No.* 99, North Carolina Agril. Exp. Sta., North Carolina, USA, 194p.

16. Kelman, A. (1954). The relationship of pathogenicity of Pseudomonas solanacearum to colony appearance in a tetrazolium medium. *Phytopathology* 44, 693–695

17. Li, Z.-N., Zhang, L., Bai, Y. B., Liu, P., Wu, Y. F. (2012). Detection and identification of the elm yellows group phytoplasma associated with Puna chicory flat stem in China. *Canadian Journal of Plant Pathology*, 34, 34–41.

18. Mahadevan, N., Shivali and Kamboj, P. (2009). *Hibiscus sabdariffa*, L.—an overview. Natural Product Radiance, 8, 77–83.

19. Mitrović, M., Jović, J., Cvrković, T., Krstić, O., Trkulja, N., Toševski, I. (2012). Characterization of a 16SrII phytoplasma strain associated with bushy stunt of hawkweed oxtongue (Picris hieracioides) in south-eastern Serbia and the role of the

leafhopper Neoaliturus fenestratus (Deltocephalinae) as a natural vector. *European Journal of Plant Pathology*, 134, 647–660.

20. Nene, Y. L. (2006). Prologue: status of plant pathology in India before 1905. In: Chahal SS, Khetarpal, R. K., Thind, T. S. (eds.) *One hundred years of plant pathology in India: an overview.* ISMPP, Scientific Publishers, Jodhpur, pp. 1–18.

21. Ooi, K. H., Salleh, B. (1999). Vegetative compatibility groups of Fusarium oxysporum, the causal organism of vascular wilt on Roselle in Malaysia. *Biotropia.* 12, 31–41.

22. Ooi, K. H., Salleh, B., Hafiza, M. H., Zainal, A. A. A. (1999). Interaction of *Fusarium oxysporum* with *Meloidogyne incognita* on Rosella. *Journal Indonesia Plant Protection.* 5, 83–90.

23. Paul, S., Ghosh, R., Roy, A., Mir, J. I., Ghosh, S. K. (2006). Occurrence of a DNA b-containing begomovirus associated with leaf curl disease of kenaf (Hibiscus cannabinus, L.) in India. *Aust Plant Dis Notes* 1, 29–30.

24. Rishi, N. (2006). Significant achievements and current status: virology. In: Chahal, S. S., Khetarpal, R. K., Thind, T. S. (eds.) *One hundred years of plant pathology in India: an overview.* ISMPP, Scientific Publishers, Jodhpur, pp. 143–206.

25. Roy, A., De, R. K., Ghosh, S. K. (2008). Diseases of bast fiber crops and their management in jute and allied fibers, pp. 327. In: Karmakar, P. G., Hazara, S. K., Subramanian, T. R., Mandal, R. K., Sinha, M. K., Sen, H. S (Eds.). Updates Production Technology, Central Research Institute for Jute and Allied Fibres, Barrckpore, West Bengal, India.

26. Tripathi, A. N., Sarkar, S. K., Sharma, H. K., Karmakar, P. G. (2013). Stem rot of roselle: A major limitation for seed production. *JafNews* 11 (1), 14.

27. Tripathi, A. N., Sood, A. K. (2005). Variability studies on *Ralstonia solanacearum* causing bacterial wilt in Himachal Pradesh. *Indian Phytopathology* (58) 3, 350.

28. Win, N. K. K., Jung, H. Y. (2012). The distribution of phytoplasmas in Myanmar. *Journal of Phytopathology*, 160, 139–145.

29. Yabuuchi, E., Kosako, Y., Yano, I., Hotta, H., Nishiuchi, Y. (1995). Transfer of two Burkholderia and an Alcaligenes species to Ralstonia gen. nov: proposal of Ralstonia pickettii (Ralston, Palleroni and Doudoroff, 1973) comb. nov., Ralstonia solanacearum (Smith, 1896) comb. nov. and Ralstonia eutropha (Davies. 1969) comb. *nov. Microbiol. Immunol.* 39, 897–904.

CHAPTER 20

CHARACTERIZATION OF SPECIES AND PARASITISM IN THE GENUS *ALTERNARIA*

UDIT NARAIN,[1] SANDHYA KANT,[2] and GIREECH CHAND[3]

[1]*Department of Plant Pathology, C. S. Azad University of Agriculture and Technology, Kanpur–208002, India*

[2]*Former Research Scholar, Department of Botany, D.A.V. College, Kanpur–208001, India*

[3]*Department of Plant Pathology, Bihar Agricultural University, Sabour, Bhagalpur–813210, India*

CONTENTS

20.1   INTRODUCTION

Genus *Alternaria* Nees ex Fr. is represented by the multitude of species ranging from saprophytes to strong parasites having polyphagous nature. Due to the wide prevalence in nature, several workers have studied the morphological and cultural characters of this group of fungi since the erection of the genus by Nees in 1817 with a single species, *A. tenuis*.

20.2   TAXONOMY

Elliott (1917), after a century of the first description of the genus *Alternaria*, emphasized the form of conidia, mainly, obclavate, pointed, often with a long beak, as a generic characteristic and stated that the chain formation under unfavorable conditions might be suppressed. Bolle (1924) isolated a fungus from crucifers and other plants and identified it as *A. tenuis* showing very short beaks and recognized Elliott's (1917) concept of the genus.

Wiltshire (1933 and 1938) being pioneer in basic studies on this group of Hyphomycetes published the results of his examination of the available type specimens and descriptive literature, which were fundamental to the then current concept of *Alternaria* and *Macrosporium*. His major conclusions were that *Macrosporium* should be suppressed as a *Nomen ambiguum* in favor of *Alternaria* typified by *A. tenuis* Nees. 'This suggestion was accepted by later workers (Groves and Skolko, 1944; Neergaard, 1945).

Neergaard (1945) made a monographic study on the taxonomy, parasitism and economic significance of the genus *Alternaria* and the systematic study of this difficult group of fungi made by him is one of the best works so far produced.

Joly (1959) described the morphological variation of species in the genus *Alternaria*. Further, Joly (1964) also made the monographic study of this genus and proposed a simple key for determination of the most common species dividing them in three sections (Joly, 1967). Simmons (1969) described typification of the genus *alternaria*.

Ellis (1971 and 1976) has given the characters of the most common species of *Alternaria* in his monographs of Dematiaceous Hyphomycetes. The characters and classification of *Alternaria* spp. from India have also been described by Subramanian (1971) in his monograph on Hyphomycetes.

The genus *Alternaria* is characterized by the formation of conidia either singly (solitary) or in short or longer chains (catenate) and provided with both cross as well as longitudinal or oblique septa (muriform) and having longer or short beaks.

It belongs to the Family Dematiaceae of Class Hyphomyces of Subdivision Deuteromycotina. The teleomorph (sexual stage) is known in a very limited species and is placed in the genus *Pleospora* of Class Loculoascomycetes of Subdivision Ascomycotinia, in which in bitunicate asci, sleeper shaped ascospores are produced, which are muriform too.

20.3 CHARACTERIZATION

The genus *Alternaria* is a large group of fungi with great diversity and differences in the mode of formation of spores, their shape, size, septation, ornamentations and beak formation. Beak itself plays an important role in species differentiation. Beaks are provided with swellings at their tip or apex and formed abruptly or gradually from the spore body (Neergaard, 1945; Ellis, 1971, 1976). Gradual transition of beak from the spore body is another characteristic feature for species differentiation and beak the length also as not more than 1/3 or 1/2 of spore body, to be equal or many times in the length of spores.

20.3.1 CHAIN FORMATION

In majority of the species, the conidia are produced in short or longer chains (catenate) and Neergaard (1945) has categorized them into three sections one as *Brevicatenatae* (having short chains of conidia) and the other as *Longicatenatae* (with longer chain of conidia) and third one as *Noncatenatae* with no chain formation (solitary).

There are two species, *A alternata* and *A. brassicicola* in which long chains of conidia (up to 20 or even more) are produced but in the latter, the conidia are unbeaked.

20.3.2 UNBEAKED CONIDIA

There are certain species *viz. A. helianthi, A. radicina, A. chrysanthemi, A. papaveris, A. brassicicola and A. pluriseptata* in which beak is almost absent or there is no existence of beak. In *A. brassicicola*, the apical cell is being more or less rectangular or resembling a truncate cone. In *A. helianthi* and *A. chrysanthemi* the cells are more or less rounded at the end while in *A. radicina*, the cells are rounded or conical and in *A. papaveris* and *A. pluriseptata*, the apical cell of conidia is somewhat rounded or rectangular, resembling as a very shot beak. In *A. brassicicola* long chains of conidia are produced and in these species, the conidia are mostly solitary.

20.3.3 CONIDIAL BEAK

(a) **Beak short:** So far as the length of beak is concerned, in *A. sonchi* (parasitic on *Sonchus*), the beak is very short and fat (Narain et al., 1988). In *A. longipes* and *A. alternata* the beak is about 1/2 or 1/3 of spore body and is conical or cylindrical but in former species, it has terminal swellings and conidia sometimes solitary but usually in short chains where as the latter is with long and often branched chain. In case of *A. raphani* and *A. tenuissima*, the beaks are also short in comparison to their spore bodies but in latter the beaks are frequently with swollen apex.

(b) **Beak long:** In many of species of the genus *Alternaria*, like *A. dauci, A. cucumerina, A. ricini* and *A. zinniae*, long conidial beak is produced and the conidia are solitary or seldom a secondary conidium is formed. The beak is flexuous, rapidly narrowing or abruptly formed from the spore body in *A. porri, A. cucumerina, A. ricini* but in *A. zinniae*, the beak is often swollen at apex and in *A. crassa*, the beak is tapering gradually from the spore body.

20.3.4   BIFURCATION OF BEAK

It has been quite interesting to observe that in some species, the conidial beak has the branching habit and it may be bifurcated as in *A. solani, A. sesami, A. dauci, A. carthami* and *A. fallax.*

20.3.5   GRADUAL TAPERING OF BEAK

The beaks may be thick and unbranched (simple) in *A. brassicae, A. cinerariae* and gradually tapering from the spore body in *A. tenuissima, A. dianthicola* and *A. petroselini.*

20.3.6   TERMINAL SWELLING IN BEAKS

In majority of the species of genus, short or longer beaks are produced in conidia which may be thick, cylindrical, conical, flat or truncate or pointed but in some of the species the beaks are provided with terminal swellings (*A. longipes, A. tenuissima, A. dianthi, A. dianthicola, A. zinniae, A. triticola*).

20.3.7   CONIDIA OF LARGER SIZE

There are two species in which the conidia are solitary or catenate and many conidia of which are very long *Cercospora* like. Conidia of *A. longissima* are quite variable and may be longer upto 500 mm (Deighton and McGarvie, 1968 and Bilgrami, 1972) while in *A. saparva*, the conidia are 150–300 mm in length and conidophores are aggregated in synnemata (Ellis, 1976).

20.3.8   SHAPE OF CONIDIA

The shape of conidia is also quite variable in many of species. Spores may be obclavate or ellipsoidal (*A. raphani*), straight or slightly curved

(*A. dianthi, A. ricini*), obclavate rostrate (*A. crassa, A. dianthicala, A. cucumerina*), oblong or ellipsoidal (*A. solani*), cylindrical (*A. helianthi*) and even polymorphic or variable is shape in *A. alternata, A. radicina* and *A. cheiranthi* (Neergaard, 1945; Ellis, 1971, 1976, Groves and Skolko, 1944).

20.3.9 SEPTATION IN CONIDIA

The number of cross and longitudinal septa depend on the length and breadth (width) of spore, longer the conidia, greater the number of cross septa and the wider spore, with comparatively greater number longitudinal/oblique septa. There is one species (*A. cheiranthi*) parasitic on wallflower (*Cheiranthus cheiri*), in which the conidia are with numerous transverse and longitudinal and oblique septa and sometimes cannot be counted easily (Narain and Singh, 1981).

The characteristic feature of the genus is the formation of muriform conidia provided with both cross and longitudinal or oblique septa. The number of the septa is quite variable according to species. So far as their number are concerned, the cross or horizontal septa are normally many in a conidium but the formation of longitudinal septa may be occasional in species like, *A. helianthi, A. chrysanthemi* and *A. flagelloideum, A. dennisii*. Because of the reason of over looked longitudinal septa, *A helianthi* and *A. flagelloideum* were originally described as the species of *Helminthosporium, H. helianthi* (Hansford, 1943) and *H. flagelloideum* (Atkinson, 1897), respectively.

20.3.10 FORMATION OF SCLEROTIA

Exceptionally, in *A. padwickii*, the asexual fruiting bodies, sclerotia are produced, which are spherical or subspherical, black with reticulate walls and 50–200 mm in diameter (Ellis, 1971).

20.3.11 FORMATION OF SYNNEMATA

Interestingly in one species of the genus, *Alternaria saparva*, the synnemata are produced in which conidiophores almost always are aggregated and up to 1 mm high and 60–200 mm wide with individual threads, 5–7 mm thick (Ellis, 1976).

20.3.12 FORMATION OF CHLAMOYDOSPORES

In few species of the genus, chlamydospores are frequently produced in culture. *A. chlamydospora* is characterized by the production of abundunt multicellular chlamydospores in culture (Mouchacca, 1973; Narain et al., 1991). Chlamydospores of *A. phragmospora* are different from those of the former species as some of them closely resemble the conidia, characteristic of the genus *Monodictys* (Emden, 1970). In *A. raphani*, they are also formed abundantly in culture either singly or in chains or sometimes in groups also and conidiophores often develop from them (Narain and Saksena, 1975). The conidial beaks of *A. carthami* have been reported to be changed into a number of chlamydospores (Narain, 1982) and in *A. neergaardii*, their frequent formation has been observed in beaks (Mehrotra and Narain, 1969).

20.4 PARASITISM

The species of the genus *Alternaria* frequently parasitic on a number cultivated plants and weeds, common components of the flora of seeds, constitute an important group of fungal pathogens. *Alternaria* species are such a versatile group of fungi, which parasite a number of plant species from seedling to maturity stage causing different types of diseases in field crops, fruits, vegetables, spices and ornamental and medicinal plant (Rotem, 1994; Narain and Srivastava, 2004b).

The diseases caused by *Alternaria* species have been considered to be of minor importance in the past, but now-a-days, due to newer agriculture technology and extensive cultivation of newer varieties of crops, a number of its species are coming up into prominence and they are causing enormous losses due to leaf spots and blights, lesions on stem, branches and petiole, blossom blight, rotting and decay of fruits in field and storage and in their transit.

Some of the species of the genus parasitize the plants of specific host families and certain are confined to the particular genus of the family.

Species of *Alternaria* associated with specific families		Species of *Alternaria* associated with specific host genus	
Host family	*Alternaria* species	Host genus	*Alternaria* species
Solanaceae	*Alternaria solani*	*Chrysanthemum*	*Alternaria chrysanthemi*
Cruciferae (Brassicaceae)	*Alternaria brassicae, A. brassicicola*	*Cheiranthus*	*Alternaria cheiranthi*
Malvaceae	*Alternaria macrospora*	*Datura*	*Alternaria crassa*
Compositae (Asteraceae)	*Alternaria zinniae*	*Dianthus*	*Alternaria dianthicola*
Cucurbitaceae	*Alternaria cucumerina*	*Nicotiana*	*Alternaria longipes*
Liliaceae	*Alternaria porri*	*Sonchus*	*Alternaria sonchi*
Umbelliferae	*Alternaria dauci, A. radicina*	*Sesamum*	*Alternaria sesami*
Papaveraceae	*Alternaria papaveris*	*Helianthus*	*Alternaria helianthi*
Linaceae	*Alternaria linicola*	*Ricinus*	*Alternaria ricini*
Rutaceae	*Alternaria citri*	*Carthamus*	*Alternaria carthami*

A. alternata and *A. tenuissima* are of common occurrence and cosmopolitan in their distribution *A. alternata* is a common saprophyte to parasitic in nature, found in many kinds of plants and other substrata including food stuffs, soil and textiles, etc. *A. tenuissima* is also extremely common and recorded on a very wide range of plants as a primary parasite or secondary invader (Ellis, 1971).

The family Asteraceae (Compostae) includes many ornamental and wild plants and some are used for oil extraction. Eight species of the genus *Alternaria* have been reported on the plants of this family (Narain et al., 1988). *A. zinniae* is the important one parasitizing a large number of host species. *A. carthami* (Ellis, 1971) and *A. neergaardii* (Metrotra and Narain, 1969) are confined to their single host where as *A. chrysanthemi* is known to attack species of the host genus *Chrysanthemum* and *A. sonchi* on *Sonchus*, *A. helianthi* is known to be associated with species of host genus *Helianthus* (Narain and Saksena, 1973; Anil Kumar et al., 1974). On sunflower itself, four species *viz. A. alternata*, *A. helianthi*, *A. tenuissima* and *A. zinniae* are of common occurrence (Narain and Srivastava, 1996).

Alternaria species associated with oilseed crops

Eleven species of *Alternaria* have been reported with the following eight oleiferous crops:

Oilseed crops	*Alternaria* species	Oilseed crops	*Alternaria* species
Sunflower	A.alternata	Groundnut	*A.alternata*
	A.helianthi		*A.tenuissima*
	A.tenuissima	Castor	*A.ricini*
	A. zinniae	Sesame (Til)	*A.sesami*
Mustard/Rapeseed	A.brassicae	Linseed	A.linicola, A. lini
	A.brassicicola	Soybean	*A.alternata*
	A.raphani	Safflower	*A.carthami*

Alternaria species associated with pulse crops

Out of 10 Leguminous plants, *A. alternata* is found to parasitize nine hosts to cause leaf spots and blight (Mehrotra and Narain, 1960). *A. cyamopsidis* is confined only to cluster bean and pigeonpea and chickpea are parasitized by *Alternaria alternata* and *A. tenuissima* (Narain, 1995; Narain et al., 1990, 1994).

Pulse crops	*Alternaria* species	Pulse crops	*Alternaria* species
Broad (faba) bean	*Alternaria alternata*	Chickpea	*Alternaria alternata*
			Alternaria tenuissima
Clusterbean	*Alternaria cyamopsidis*	Pea	*Alternaria tenuissima*
Frenchbean	*Alternaria alternata*		*Alternaria alternata*
Ricebean	*Alternaria alternata*	Pigeonpea	*Alternaria alternata*
Soybean	*Alternaria alternata*		*Alternaria tenuissima*
Wingedbean	*Alternaria alternata*	Lentil	*Alternaria alternata*

The plants of family Solanaceae are parasitized by four species of the genus *Alternaria* viz., *A. alternata*, *A. crassa*, *A. longipes* and *A. solani* in which the latter one is the most widely distributed to have a wide-host range including many important vegetables, ornamental and medicinal plants (Deshwal, 2004). *A. solani* is wide spread to cause leaf-spots, blight and fruit rots in many vegetable crops (Chupp and Sherf, 1960 and Singh, 1999), *Alternaria alternata* is serious pathogen and causes great loss in chili (Narain and Bhale, 2000).

Alternaria species associated with vegetable crops

A. Solanaceous vegetables		B. Bulbous vegetables	
Chilli	*Alternaria solani, A.alternata*	Onion	*Alternaria porri*
Potato	*Alternaria solani, A.alternata*	Garlic	*Alternaria porri*
Tomato	*Alternaria solani, A.alternata*	Leek	*Alternaria porri*
Brinjal	*Alternaria solani, A.alternata*	Shallot	*Alternaria porri*

A. porri (Angell, 1929) is known to cause an important disease "purple blotch" and leaf spot in onion, garlic, leek and shallot, and confined only to the hosts of Liliaceae (Patil and Patil, 1991). *A. dauci* causes leaf spot and blight of carrot and coriander (Narain and Srivastava, 2004a) and has been recorded on other Umbelliferous host plants. Another species, *A. radicina* causes black rot of carrot (Lauritzen, 1926; Yoshii, 1929) and has been reported on other hosts of this family like celery, dill and parsnip (Ellis, 1971).

Key to *Alternaria* spp. parasitic on Solanaceous hosts

- **Conidia formed in chains:**
    - (a) Conidial chain longer, conidia usually polymorphic, often with short, conical or cylindrical beaks.....................*A. alternata*
    - (b) Conidial chain shorter, conidia obclavate, rostrate, beaks often slightly swollen. .... *A. longipes*

- **Conidia solitary or seldom formation of secondary conidium:**
    - (a) Beak flexuous and sometimes branched, culture on PDA chromogenic...*A. solani*
    - (b) Beak long, unbranched, tapering gradually, culture on PDA non-chromogenic...*A. crassa*

C. Brassicaceous Vegetables

Cabbage	*Alternaria brassicae*	Turnip	*A. brassicae*
	A. brassicicola		*A. raphani*

Cauliflower	*Alternaria brassicae*	Knolkhol	*Alternaria brassicae*
	A. raphani		*A. brassicicola*
	A. brassicicola	Broccoli	*A. alternata*
Radish	*Alternaria brassicae*		*A. brassicae*
	A. alternata		*A. brassicicola*
	A. raphani		*A. raphani*

The family Brassicaceae (Cruciferae) is fairly important from economic point of view. A large number of vegetable crops as well as garden and wild flowers are included in this family and seeds of several plant species yield vegetable oil of multipurpose use. Five species of *Alternaria* are known to be associated with the plants of this family (Gupta and Basuchaudhary, 1992; Verma and Saharan, 1993; Narain, 1986). *A. alternata, A. brassicae, A. brassicicola* and *A. raphani* have been reported on rapeseed and mustard and broccoli (Chand, 2007; Chand and Narain, 2005; Chand et al., 2005; 2012; Prasad and Narain, 2007) and with cabbage, cauliflower and knolkhol, *A. brassicae* and *brassicicola; A. raphani* with radish (Atkinson, 1950, Changsri and Weber, 1963; Narain, 1986; Sangwan et al., 2002), turnip (Narain and Saksena, 1975) and *A. brassicae* with taramira (Verma and Saharan, 1993). In ornamental plants, *A. cheiranithi* from wallflower (Narain and Singh, 1981) and from candytuft, *A. brassicae* and *A. raphani* have been reported (Narain et al., 1982). *A. alternata* is also known to parasitize radish (Suhag et al., 1985).

Key to *Alternaria* spp. parasitic on Brassicaceous hosts

   I. Conidia solitary or occasionally in chains upto four, obclavate, rostrate, tapering gradually into thick cylindrical beak.....*A. brassicae*

  II. Conicia in long chains consisting of twenty or even more in a chain.

     (a) Conidia usually cylindrical, basal cell rounded and apical cell more or less rectangular and beak usually almost non-existent..*A. brassicicola*

     (b) Conidia usually polymorphic, often with short conical or cylindrical short beaks.................................*A. alternata*

  III. Conidia 3–4 in chain, straight or curved, obclavate, generally with short beak, chlamydospores formed abundantly in culture.......

....*A. raphani*

D. Cucurbitaceous Vegetables

The members of family Cucurbitaceae are the major source of vegetables in India and almost all the plants of the family are attacked by *A. cumerina* (Narain et al., 2002). to cause leaf spot/blight (Ahmad and Narain, 2000 and Narain et al., 2003). Apart from the association of *A. cucumerina*, two more species, *A. alternata* and *A. tenuissima* have also been found to cause leaf spot and blight (Sharma and Bhargava, 1977; Narain and Prasad, 1981b; Narain and Srivastava, 2000; Narain et al., 2003). Incidence of fruit rot in some of the cucurbits have also been observed in the field and storage and during their transit (Sharma and Bhargava, 1977).

Alternaria species parasitic on Cucurbitaceous vegetables	
Alternaria cucumarina	*Alternaria alternata*
Citrullus vulgaris (Water melon)	*Citrullus vulgaris* (Water melon)
C. vulgaris var *fistulosus* (Round gourd)	*Coccinia indica* (Scarlet gourd)
Coccinia indica (Scarlet gourd)	*Cucurbita maxima* (Red gourd)
Cucumis melo (Musk melon)	*Cucumis sativus* (Cucumber)
C. melo var. *momordica* (Phoot)	*Lagenaria vulgaris* (Bottble gourd)
C. melo var. *utilissimus* (Kakri)	*Luffa acutangula* (Ridge gourd)
Cucumis sativus (Cucumber)	*Momordica charantia* (Bitter gourd)
Cucubita maxima (Redgourd)	*Trichosanthes dioica* (Pointed gourd)
Cucurbita pepo (Vegetable marrow)	
Lagenaria vulgaris (Bottle gourd)	*Alternaria tenuissima*
Luffa acutangula (Sponge gourd)	*Citrullus melo* var. *momordica* (Phoot)
Momordica charantia (Bitter gourd)	*C. melo* var. *utilissimus* (Kakri)
Momordica dioica (Kareli)	*Momordica charantia* (Bitter gourd)
Trichosanthes dioica (Pointed gourd)	*Trichosanthes anguinea* (Snake gourd)

Key to *Alternaria* spp. occurring on Cucurbitaceous hosts

- **Conidia formed in long and often branched chains consisting upto 20 or even more:**
  Conidia usually polymorphic, often with short conical or cylindrical beaks...*A. alternata*

- **Conidia formed in short chains (3–6):**
  Conidia conical to oval with many cross and longitudinal septa, beaks sometimes swollen terminally and may be of upto the length of spore body..*A. tenuissima*

- **Conidia formed singly or rarely up to two:**
  Conidia obclavate, rostrate, comparatively larger with many cross and longi-septa and with long filiform beaks......................*A. cucumerina*

The family Umbelliferae comprises of a number of plants like carrot, coriander, fennel, Indian dill, celery, cumin, etc. which are used as salad, vegetables and spices and they are also affected from a number of *Alternaria* species from seedling to maturity stage of plant growth. Two species of *Alternaria* are known to be associated with carrot *viz.*, *A. dauci* which causes the foliar disease (Roy, 1969; David, 1988; Pryor and Standberg, 2001) and *A. radicina* causing black rot which was first described and reported in 1922 by Meir et al. (1922) and later from other parts of the word (Ellis and Holliday, 1972; Pryor and Gilbertson, 2002). Wearing (1980) also reported *A. radicina* on celery.

There is a great diversity and parasitism in *Alternaria* species infecting the crop of spices (Narain and Srivastava, 2004a).

An elaborate account of *Alternaria* spp. associated with some Umbelliferous crops has been given by Kumar et al. (2010, 2013).

Alternaria alternata and *A. tenuissima* can also infect the inflorescences, seeds and developing seedling in saunf and coriander (Mehrotra and Narain, 1969 and Kumar et al., 2013).

Alternaria species associated with spices

Spices	Alternaria species
Onion (*Allium cepa*)	*Alternaria porri*
Garlic (*Allium sativum*)	*Alternaria porri*
Cumin (*Cuminum cyminum*)	*Allernaria burnsii*
Coriander (*Coriandrum sativum*)	*Alternaria dauci*
Turmeric (*Curcuma longa*)	*Alternaria alternata*
Methi (*Trigonella foenum graecum*)	*Alternaria alternata*
Fennel (*Foemiculm vulgare*)	*Alternaria dauci*
Ginger (*Ziner officinale*)	*Alternaria alternata*

Key to *Alternaria* spp. parasitic on Umbelliferous vegetables and spices
Conidia in chains (catenate):

  (a) Chain very long consisting of up to 20 or even more, conidia poly-
      morphic with short conical or cylindrical beaks. ....*A. alternata*
  (b) Chain short with 3–6 spores, beak up to the length of spore body with
      terminal swellings...*A. tenuissima*

Conidia solitary (acatenate):

  (a) Conidia without beak (unbeaked)...........................*A. radicina*
  (b) Conidia with long filiform, often branched beaks...........*A. dauci*

E. *Alternaria* species associated with ornamental/medicinal plants

Hosts	*Alternaria* species	Hosts	*Alternaria* species
Antirrhinum	*A. fallax var. linariae*	Dahlia	*A. zininiae*
Linum	*A. linicola*	Marigold	*A. zininiae*
Hollyhock	*A. macrospora*	Calendula	*A. zininiae*
Candytuft	*A. brassicae,*	Pothas	*A. tenuissima*
	A. raphani	Gerbera	*A. zininiae*
Poppy	*A. papaveris*	Mirabilis jalpa	*A. tenuissima*
Petunia	*A. solani*	Dianthus	*A. dianthicola*
Alocasia indica	*A. tenuissima*		*A. dianthi*
Allium porrum	*A. porri*	Celosia	*A. tenuissima*
Physsalis minima	*A. solani*	Datura	*A. crassa*
Wallflower	*A. cheiranthi*	Sunflower	*A. alternata*
Chrysanthemum	*A. chrysanthemi*		*A. tenuissima*
Cineria	*A. cineriae*		*A. helianthi*
Rose	*A. alternata*		*A. zinniae*
Solanum khasianum	*A. solani*	Jasminum	*A. alternata*
Ipomera arborescence	*A. tenuissima*	Tuberose	*A. polyanthi*
Withania sominifera	*A. solani*	Stock	*A. raphani*
Achnia malvariscus	*A. tenuissima*	Tobacco	*A. longipes*

Several *Alternaria* species also parasitize the ornamental and medicinal plants like, *A. linicola* on linum (Narain and Koul, 1982), *A. brassicae*
and *A. raphani* on candytuft (Narain et al., 1982), poppy (Narain, 1991),

A. cheiranthi on wallflower (Narain and Singh, 1981), *A. alternata,*
A. tenuissima, A. helianthi, A. zinniae on sunflower (Narain and Saksena,
1973; Narain and Chauhan, 1981; Narain and Srivastava, 1996; Narain
et al., 1988); *Alternaria tenuissima* on Celosia (Singh and Narain, 1980),
A. chryanthemi on *Chrysanthemum indicum, A. sonchi* on *Sonchus olera-
ceus, A. zinniae* on *Ageratum conyzoides, Helianthus cucumerifolius,
Centauria cyanus, Tagetes erecta, Zinia elegans* (Narain et al., 1988). *A.
tenuissima* on *Mirabilis jalpa* (Narain and Prasad, 1981a), *A. porri* on
Allium porrum, A. solani on *Withamia somnifera, Physalis minima; A.
tenuissima* on *Achania malvariscus, Alocasia indica, Ipomoea arbores-
cens, Lawsonia inermis, Pothas aureus* (Narain, 1983).

F. *Alternaria* species associated with fruit crops

A number of *Alternaria* spp. have been reported to be associated with fruit
crops (Singh, 2000 and Gupta and Sharma, 2000). Alternaria core rot and
black rot of citrus (Keily, 1964; Agarwal and Hasija, 1967); leaf spot of
apple and pear (Koul and Narain, 1981) and their post-harvest decay (Kaul
and Munjal, 1981) and leaf spot of mulberry (Narain and Sinha, 1994) are
common diseases.

Fruit crops	*Alternaria* species	Fruit crops	*Alternaria* species
Citrus	*Alternaria citri*	Apple	*Alternaria alternata*
Mango	*Alternaria tenuissima*	Peach	*Alternaria alternata*
Cherry	*Alternaria alternata*	Papaya	*Alternaria alternata*
Loquat	*Alternaria eriobotryae*	Litchi	*Alternaria alternata*
Pear	*Alternaria kickuchiana*	Mulberry	*Alternaria alternata*

Sometimes it becomes much more difficult to diagnose a disease and
to identify the associated species correctly when there is the involve-
ment of one or two or more species with the causation of disease in a
particular crop *viz., A. alternata, A. brassicicola, A. brassicae, A raphani*
with rapeseed and mustard (Narain, 1986) and broccoli (Chand, 2005);
A. alternata, A. helianthi, A. tenuissima, A. zinniae with sunflower (Narain
and Srivastava, 1996); *A.alternata, A. tenuissima* and *A. cucumerina* with
cucurbits (Narain et al., 2002).

After comparative study of symptomatology and etiology of disease and observations of morphological characters of *Alternaria* spp. in

nature (host) and in culture (PDA), a very simple and suitable keys have been framed for ready identification of *Alternaria* spp. associated with Crucifers (Narain, 1986), Brassicaceous vegetables (Khalid, 2003 and Khalid et al., 2004), Cucurbitaceous vegetables (Narian et al., 2002), hosts of Compositae (Narain et al., 1988) and Solanaceous vegetables (Deshwal, 2004) and crops used for spices (Narain and Srivastava, 2004).

KEYWORDS

- **Alternaria**
- **conidia**
- **host**
- **parasitism**
- **species**
- **taxonomy**

REFERENCES

1. Agarwal, G. P., Hasija, S. K. (1967). Alternaria rot of citrus fruits. *Indian Phytopath.*, 20, 259.
2. Ahmad, Shahid, Narain, U. (2000). A new host record of *Alternaria cucumerina* on bottlegourd. *Indian Phytopath.*, 53, (2) 234.
3. Angell, H. R. (1929). Purple blotch of onion (*Macrosporum porri* Ell.). *J. Agric. Res.*, 38, 467–487.
4. Anil Kumar, T. B., Urs, S. D., Seshadri, V. S., Hegde, R. K. (1974). Alternaria leaf spot of sunflower. *Curr. Sci.*, 43, 93–94.
5. Atkinson, G. T. (1897). Some fungi from Alabama. *Bull. Cornell University (Sci)*, 3(1), 1–50.
6. Atkinson, R. G. (1050). Studies on parasitism and variation of *Alternaria raphani*. *Can. J. Res. Sect. C.*, 28, 288–317.
7. Bilgrami, R. S. (1972). First record of *Alternaria longissima. Curr. Sci.*, 41, 722.
8. Bolle, P. C. (1924). Plant diseases caused by the blackening fungi (Phacodictyae). *Meded. Phytopath. Lab. Willi. Commelin Scholten Baarn*, 7, 77.
9. Chand, Gireesh (2005). Studies on Alternaria leaf spot of broccoli. PhD Thesis, C.S.A.U.A. & T., Kanpur.
10. Chand, Gireesh (2007) Symptomatology, etiology and eco-friendly management of Alternaria leaf spots and blight of broccoli. In: *Ecofriendly management of plant*

diseases, (Eds. Ahmad, Shahid & Narain, U.). Daya Publishing House, Delhi, pp. 461–472.

11. Chand, Gireesh, Narain, U. (2005). Characterization of *Alternaria* spp. and their symptom differentiation in broccoli. Paper presented in 6ᵗʰ Nat. Symp. on "Sustainable plant protection strategies: Health & environment concern" held at, K. K. V., Ratnagiri (M. S.) during Oct. 15–17, 2005.

12. Chand, Gireesh, Narain, U., Kumar, M. (2005). Taxonomy and parasitism of *Alternaria* spp. occurring on broccoli in India. Paper presented in Nat. Seminar on "New-horizon in life science" held at Govt. P. G. College, Narsingpur (M. P.) during Sept. 2–3, 2005.

13. Chand, Gireesh, Yadav, S. P., Yadav, G. C., Kumar, Sanjeev (2012). Eco-friendly and innovative approaches in management of Alternaria blight of broccoli. In: *Eco-friendly innovative approaches in plant disease management* (Eds. V. K. Singh, Y. Singh and, A. Singh). International Book Distributors, Dehradun, pp.419–430.

14. Changsri, W., Weber, G. P. (1963). Three *Alternaria* species pathogenic on certain cultivated crucifers. *Phytopathology*, 53, 643–648.

15. Chupp, C., Sherf, A. F. (1960). *Vegetable diseases and their control.* The Ronald Press Company, New York, 693 pp.

16. David, J. C. (1988). *Alternaria dauci.* No. 951. CMI description of pathogenic fungi and bacteria. Commonwealth Mycological Institute, Kew, England.

17. Deighton, F. C., MacGarvie, Q. D. (1968). *Alternaria longissima* sp. nov. Mycological paper No. 113, CMI, Kew, Surrey, England, 1–15pp.

18. Deshwal, Kuldip (2004). Taxonomy and parastism of *Alternaria* species associated with Solanaceous hosts. M.Sc. (Ag.) Thesis, C. S. A. U. A. & T. Kanpur.

19. Elliott, J. A. (1917). Taxonomic characters of the genera *Alternaria* and *Macrosporium. Am. J. Bot.*, 4, 439–476.

20. Ellis, M. B. (1968a). *Alternaria brassicae.* Commonwealth Mycological Institute, No.162, CMI, Kew, England.

21. Ellis, M. B. (1968b). *Alternaria brassicicola.* Commonwealth Mycological Institute, Description of Pathogenic fungi and bacteria. No.163, CMI, Kew, England.

22. Ellis, M. B. (1971). *Dematiaceous Hyphomycetes.* Commonwealth Mycological Institute, Kew, Surrey, England, 464–497.

23. Ellis, M. B. (1976). *More Dematiaceous Hyphomycetes.* Commonwealth Mycological Institute, Kew, England, 411–427 pp.

24. Ellis, M. B., Holliday, P. (1972). *Alternaria radicina.* No.346 CMI description of pathogenic fungi and bacteria. Commonwealth Mycological Institute, Kew, England.

25. Emden, J. H. van (1970). *Alternaria phragmospora* nov. spec. *Acta Bot.*, 19 (3), 393–400.

26. Groves, J. W., Skolko, A. J. (1944). Notes on seed borne fungi II. *Alternaria. Can. J. Res. Sect. C.*, 12, 217–234.

27. Gupta, D. K., Basuchaudhary, K. C. (1992). Occurrence and prevalence of *Alternaria* species in Crucifers grown in Sikkim. *Indian, J. Hill Farming*, 5, 129–131.

28. Gupta, V. K., Sharma, S. K. (2000). *Diseases of fruit crops.* Kalyani Publishers, New Delhi, 344 pp.

29. Hansford, C. G. (1943). *Proc. Lin. Soc. Lond.*, 155, 49.

30. Joly, P. (1959). Morphological variations and the idea of species in the genus *Alternaria. Bull. Soc. Mycol.*, 75, 149–158.

31. Joly, P. (1964). *Le Genre Alternaria*. Editions Paul Lechevalier, Paris, 150 pp.
32. Joly, P. (1967). Key for determination of most common species of the genus *Alternaria* (Nees) Wiltsh. Emend Joly. *Plant Dis. Reptr.*, 51, 296–298.
33. Kaul, J. L., Munjal, R. L. (1981). post-harvest fungal diseases of apple in Himachal Pradesh. *Indian Phytopath.*, 34, 80.
34. Keily. T. B. (1964). Brown rot of Emperior mandarins. *Agric. Gaz. N. S. Wales*, 75, 854.
35. Khalid, Abdul (2003). Taxonomy and parasitism of *Alternaria* species associated with Brassicaceous vegetables. MSc (Ag.) Thesis, C. S. A. U. A. & T., Kanpur.
36. Khalid, Abdul, Akram, Mohammad, Narain, U., Srivastava, M. (2004). Characterization of *Alternaria* spp. associated with Brassicaceous vegetables. *Farm Sci. J.*,13, 195–196.
37. Koul, A. K., Narain, U (1981). Alternaria leaf spot of pear in India. *Indian Phytopath.*, 34, 257–258.
38. Kumar, M., Narain, U., Chand, Gireesh (2010). Detection and diagnosis of *Alternaria* spp. associated with Umbelliferous spices and vegetables. Paper presented in 8th Nat. Symp. on "Problems and perspectives in eco-friendly innovatives to plant protection" held at, C. S. A. Univ. of Agric. & Tech., Kanpur during Jan. 24–25, 2010.
39. Kumar, M., Narain, U., Chand, Gireesh (2013). Study of *Alternaria* spp. associated with some Umbelliferous crops. Lambert, Germany.
40. Lauritzen, J. I. (1926). Relation of black rot to the storage of carrots. *J. Agric. Res.*, 33, 1025–1041.
41. Mehrotra, B. S., Narain, U. (1969). Studies on the genus *Alternaria*, I. Some new records and a new species. *Indian Phytopath. Soc. Bull.*, 5, 1–7.
42. Meier, F. C., Drechler, C., Eddy, E. D. (1922). Black rot of carrots caused by *Alternaria radicina* n. sp. *Phytopathology*, 12, 157–168.
43. Mouchacca, Jeam (1973). Deux *Alternaria* des sols arides d'egypte: *A chalmydosporum* sp. nov. et, *A. phygmospora* van Emden. *Mycopath. et Mycologia applicata*, 50 (3), 217–225.
44. Narain, U. (1982). Chlamydospore formation in conidial beak of *Alternaria carthami*. *Indian Phytopath.*, 35, 172–173.
45. Narain, U. (1983). Some new records of *Alternaria tenuissima* from India. *Indian, J. Mycol. Pl. Pathol.*, 13, 94.
46. Narain, U. (1986). Studies on *Alternaria* spp. associated with leaf spots of Crucifers in India. *Adv. Biol. Res.*, 4 (1 & 2), 187–191.
47. Narain, U. (1991). A new record of *Alternaria papaveris* from India. *Indian Phytopath.*, 44, 147.
48. Narain, U. (1995). Detection, diagnosis and control of *Alternaria* spp. causing leaf spots and blight of beans and leafy vegetables. Paper presented in Global Conference on "Advances in Research on Plant Diseases and their Management" held at, R. C. A., Udaipur on Feb. 12–17, 1995. Abs. P. 136.
49. Narain, U., Banerjee, A. K., Singh, Mohit, Swarup, J. (1988). *Alternaria* species associated with Compositae plants. *Farm Sci. J.*, 3, 164–169.
50. Narain, U., Bhale, Usha (2000) Alternaria leaf spot and fruit rot of chili. In: *Advances in Plant Disease Management*. Eds. U. Narain, K. Kumar and, M. Srivastava, Advance Publishing Concept, New Delhi, 163–173 pp.

51. Narain, U., Chauhan, L. S. (1981). Leaf spot of sunflower caused by *Alternaria tenuissima*. F. A. O. *Plant Prot. Bull.*, 29, 29.
52. Narain, U., Koul, A. K. (1982). Alternaria leaf spot of *Linum grandiflorum* from India. *Indian, J. Mycol. Pl. Pathol.*, 12, 343–344.
53. Narain, U., Prasad, R. (1981a). Some new host records of *Alternaria* from India. *Plant Science*, 13, 83–84.
54. Narain, U., Prasad, R. (1981b). Alternaria leaf spot of Kundru from India. *Plant Science*, 13, 96.
55. Narain, U., Saksena, H. K. (1973). Occurrence of Alternaria spot sunflower in India. *Indian, J. Mycol. Pl. Pathol.*, 3, 115–116.
56. Narain, U., Saksena, H. K. (1975). A new leaf spot of turnip. *Indian Phytopath.*, 28, 98–100.
57. Narain, U., Singh, J. (1981). Alternaria leaf spot of Cheiranthus from India. *Nat. Acad. Sci. Lett.*, 4, 3–4.
58. Narain, U., Singh, J., Koul, A. K. (1982). Leaf spot of candytuff caused by *Alternaria raphani*. *Nat. Acad. Sci. Lett.*, 5, 13.
59. Narain, U., Sinha, A. K. (1994). Leaf spot of mulberry from India. *Van Anusandhan*, 9–10, 76–77.
60. Narain, U., Sinha, A. K., Yadav, L. B. (1991). New record of *Alternaria chlamydospora* from India. *Nat. Acad. Sci. Lett.*, 14, 401–402.
61. Narain, U., Srivastava, Mukesh (1996). Taxonomy and parasitism of *Alternaria* spp. occurring on sunflower in India. *Ann. Pl. Protec. Sci.*, 4, (2) 95–98.
62. Narain, U., Srivastava, Mukesh (2000). Detection and diagnosis of *Alternaira* spp. associated with fruits of Cucurbitaceous vegetables. Proc. Indian Phytopathological Society–Golden Jubilee Volume II: 718–719.
63. Narain, U., Srivastava, Mukesh (2004a). Disease diagnosis and species diversity in the genus *Alternaria* infecting spices in Uttar Pradesh. Paper presented in National Seminar on Opportunities to potentials of spices for crop diversification. held during Jan. 19–21, 2004 at JNKVV, Jabalpur; pp. 105–106.
64. Narain, U., Srivastava, M. (2004b). Diversity of parasitism and species in the genus *Alternaria*. Paper presented in Nat. Symp. on "Detection and management of plant diseases using conventional and modern tools" held at CISH, Lucknow on Dec. 31, 2004.
65. Narain, U., Srivastava, Mukesh, Rani, Pinky (2002). Taxonomy and parasitism of *Alternaria* spp. associated with Cucurbitaceous vegetables. In: *Frontiers of Fungal Diversity in India*. Eds. Rao, G. P., Manoharachari, C., Bhat, D. J., Rajak, R. C., Lakhanpal, T. N. International Book Distributing Co. Lucknow, 351–366.
66. Narain, U., Srivastava, Mukesh, Rani, Pinky (2003). A new record of *Alternaria cucumerinai* on Cucurbitaceous hosts. *Farm Sci. J.*, 12 (1), 80–81.
67. Narain, U., Swarup, J., Dwivedi, R. P. (1994). Diagnosis of foliar diseases of pulses caused by *Alternaria* spp. Paper presented in International Symp., on pulses research, held at Delhi during April 2–6, 1994.
68. Narain, U., Yadav, L. B., Sinha, A. K. (1990). A new blight of chickpea. *Nat. Acad. Sci. Lett.*, 13, 401–402.
69. Neergaard, P. (1945). *Danish species of Alternaria and Stemphylium:* taxonomy, parasitism and economical significance. Oxford Univ. Press, London, 560.

70. Nees, VonEsenbeck, C. G. (1917). *System der Plize Und Schwamme, Wurzburg,* pp. 234.

71. Patil, A. O., Patil, B. C. (1991). Survey of Alternaria leaf blight and other diseases of onion. *Maharashtra, J. Hort.,* 5, 71–72.

72. Prasad, R., Narain, U. (2007). Integrated management of Alternaria blight of rapeseed and mustard: An overview. In: *Ecofriendly management of plant diseases.* (Eds. Ahamad, Shahid, Narain, U.). Daya Publishing House, Delhi, pp. 201–214.

73. Pryor, B. M., Gilbertson, R. L. (2002). Relationship and taxonomic status of *Alternaria radicina, A. carotiicultae* and, *A. petroselini* based upon morphological, biochemical and molecular characteristics. *Mycologia,* 94, 49–61.

74. Pryor, B. M., Strandberg, J. O. (2001). Alternaria leaf blight of carrot. In: *Compendium of Umbelliferous crop diseases.* (Eds. R. M. Davis and, R. N. Raid). American Phytopath. Society, St. Paul, M. N.

75. Rotem, J. (1994). The genus *Alternaria*: Biology, epidemiology and pathogenicity. American Phytopath. Society, St. Paul, M. N.

76. Roy, A. K. (1969). Studies on leaf blight of carrot caused by *Alternaria dauci. Indian Phytopath.,* 22, 105–109.

77. Sangwan, M. S., Mehta, N., Gandhi, S. K. (2002). Some pathological studies on *Alternaria raphani* causing leaf and pod blight of radish. *J. Mycol. Pl. Pathol.,* 32, 125–126.

78. Sharma, N., Bhargava, K. S. (1977). Fruit rot of bitter gourd. *Indian Phytopath.* 30, 557–558.

79. Simmons, E. G. (1967). Typification of *Alternaria, Stemphylium* and *Ulocladium. Mycologia,* 59, 67–92.

80. Singh, J., Narain, U. (1980). Three new diseases of ornamental plants from India. *Nat. Acad. Sci., Letters,* 3 (9), 261–262.

81. Singh, R. S. (1999). *Diseases of vegetable crops.* Oxford & IBH Publishing Co. Pvt. Ltd., New Delhi, 406.

82. Singh. R. S. (2000). *Disease of fruit crops.* Oxford & IBH Publishing Co. Pvt. Ltd., New Delhi, 310.

83. Subramanian, C. V. (1971). *Hyphomycetes.* I. C. A. R. Publication, New Delhi.

84. Suhag, L. S., Singh, R., Malik, Y. S. (1985). Epidemiology of pod and leaf blight of radish caused by *Alternaria alternata. Indian Phytopath.,* 38, 148–149.

85. Verma, P. R., Saharan, G. S. (1993). *Monograph on Alternaria diseases of crucifers.* Research Branch, Agriculture and Agri-Food Canada, Saskatoon Res. Centre, 162pp.

86. Wearing, A. H. (1980). *Alternaria radicina* on celery in South Australia. *Aust. Plant Pathol.,* 9, 116.

87. Wiltshire, S. P. (1933). The foundation species of *Alternaria* and *Macrosporium. Trans. Br. mycol. Soc.,* 18, 135–160.

88. Wiltshire, S. P. (1938). The original and modern conceptions of *Stemphylium. Trans. Br. mycol. Soc.* 23, 211–239.

89. Yoshii, H. (1929). Black spot disease of carrot (*Alternaria radicina*)., *J. Plant Prot. Japan,* 16, 17.

INDEX

W

Warm temperature, 18
Water logging, 148
Watermelon, 226
Web blight, 73
Wet rot, 171
White rot, 198
Whitefly, 60, 62, 356–359, 363, 379
  *Bemisiatabaci*Glov., 63, 356, 358
Wilt, 28–31, 37, 41–44, 47, 52, 88–91,
  109, 110, 114, 132, 140, 173, 176–
  178, 182, 197, 208, 227, 251, 286,
  291, 307, 320, 321, 325, 347–349,
  353, 354, 360–364, 371, 373, 376–381
Windborne, 227
World Health Organization, 297

X

Xanthomonas axonopodis pv. *betlicola*,
  322, 323

Xanthomonas axonopodis pv. *glycines*,
  119, 120
Xanthomonas phaseoli, 74
Xanthophylls, 299
Xiphenema spp., 286, 288
Xylem vessels, 30, 132, 325

Y

Yellow vein disease of jute, 357
Yellow vein symptom on jute leave, 358
Young peduncles, 281

Z

Zea mays L., 5
Zinc, 154, 155, 160–162
  deficiency, 154
Zoospores, 34, 114, 117, 210, 211, 227,
  228, 236, 249, 252, 280, 314